In Situ Bioremediation of Perchlorate in Groundwater

SERDP and ESTCP Remediation Technology Monograph Series
Series Editor: C. Herb Ward, Rice University

In Situ Bioremediation of Perchlorate in Groundwater

Edited by

Hans F. Stroo
HydroGeoLogic, Inc., Ashland, OR

C. Herb Ward
Rice University, Houston, TX

Authors

Carol E. Aziz
Robert C. Borden
John D. Coates
Evan E. Cox
Douglas C. Downey
Patrick J. Evans
Paul B. Hatzinger
Bruce M. Henry
W. Andrew Jackson
Thomas A. Krug

M. Tony Lieberman
Raymond C. Loehr
Robert D. Norris
Valentine A. Nzengung
Michael W. Perlmutter
Charles E. Schaefer
Hans F. Stroo
C. Herb Ward
Carrolette J. Winstead
Christopher Wolfe

Editors
H. F. Stroo
HydroGeoLogic, Inc.
300 Skycrest Drive
Ashland, OR 97520

C. H. Ward
Rice University
6100 Main Street
Houston, TX 77005

ISBN: 978-1-4419-2746-0   e-ISBN: 978-0-387-84921-8

© 2009 Springer Science+Business Media, LLC
Softcover reprint of the hardcover 1st edition 2009
All rights reserved. This work may not be translated or copied in whole or in part without the written permission of the publisher (Springer Science+Business Media, LLC, 233 Spring Street, New York, NY 10013, USA), except for brief excerpts in connection with reviews or scholarly analysis. Use in connection with any form of information storage and retrieval, electronic adaptation, computer software, or by similar or dissimilar methodology now known or hereafter developed is forbidden.
The use in this publication of trade names, trademarks, service marks, and similar terms, even if they are not identified as such, is not to be taken as an expression of opinion as to whether or not they are subject to proprietary rights.
While the advice and information in this book are believed to be true and accurate at the date of going to press, neither the authors nor the editors nor the publisher can accept any legal responsibility for any errors or omissions that may be made. The publisher makes no warranty, express or implied, with respect to the material contained herein.

Cover photograph courtesy of Paul C. Johnson, Department of Civil and Environmental Engineering, Arizona State University.

Printed on acid-free paper

9 8 7 6 5 4 3 2 1

springer.com

SERDP/ESTCP Remediation Technology Monograph Series
Series Editor: C. Herb Ward, Rice University

SERDP and ESTCP have joined to sponsor the development of a series of monographs on remediation technology written by leading experts in each subject area. This volume provides a review of the state-of-the-art on *in situ* bioremediation of perchlorate. Additional volumes planned for publication in the near future include:

- *In Situ* Remediation of Dissolved Chlorinated Solvents in Groundwater
- Delivery and Mixing in the Subsurface: Processes and Design Principles for *In Situ* Remediation
- Bioaugmentation for Groundwater Remediation
- Chlorinated Solvent Source Zone Remediation
- Monitored Natural Recovery at Contaminated Sediment Sites
- Remediation of Munition Constituents in Soil and Groundwater

U.S. Department of Defense Strategic Environmental Research & Development Program (SERDP)
901 North Stuart Street, Suite 303
Arlington, VA 22203

U.S. Department of Defense Environmental Security Technology Certification Program (ESTCP)
901 North Stuart Street, Suite 303
Arlington, VA 22203

PREFACE

In the late 1970s and early 1980s, our nation began to grapple with the legacy of past disposal practices for toxic chemicals. With the passage in 1980 of the Comprehensive Environmental Response, Compensation, and Liability Act (CERCLA), commonly known as Superfund, it became the law of the land to remediate these sites. The U.S. Department of Defense (DoD), the nation's largest industrial organization, also recognized that it too had a legacy of contaminated sites. Historic operations at Army, Navy, Air Force, and Marine Corps facilities, ranges, manufacturing sites, shipyards, and depots had resulted in widespread contamination of soil, groundwater, and sediment. While Superfund began in 1980 to focus on remediation of heavily contaminated sites largely abandoned or neglected by the private sector, the DoD had already initiated its Installation Restoration Program in the mid 1970s. In 1984, the DoD began the Defense Environmental Restoration Program (DERP) for contaminated site assessment and remediation. Two years later, the U.S. Congress codified the DERP and directed the Secretary of Defense to carry out a concurrent program of research, development, and demonstration of innovative remediation technologies.

As chronicled in the 1994 National Research Council report, "Ranking Hazardous-Waste Sites for Remedial Action", our early estimates on the cost and suitability of existing technologies for cleaning up contaminated sites were wildly optimistic. Original estimates, in 1980, projected an average Superfund cleanup cost of a mere $3.6 million per site and assumed only around 400 sites would require remediation. The DoD's early estimates of the cost to clean up its contaminated sites were also optimistic. In 1985, the DoD estimated the cleanup of its contaminated sites would cost from $5 billion to $10 billion, assuming 400 to 800 potential sites. A decade later, after an investment of over $12 billion on environmental restoration, the cost to complete estimates had grown to over $20 billion and the number of sites had increased to over 20,000. By 2007, after spending over $20 billion in the previous decade, the estimated cost to complete the DoD's known liability for traditional cleanup (not including the munitions response program for unexploded ordnance) was still over $13 billion. Why did we underestimate the costs of cleaning up contaminated sites? All of these estimates were made with the tacit assumption that existing, off-the-shelf remedial technology was adequate to accomplish the task; that we had the scientific and engineering knowledge and tools to remediate these sites; and that we knew the full scope of chemicals of concern.

However, it was soon and painfully realized that the technology needed to address the more recalcitrant environmental contamination problems, such as fuels and chlorinated solvents in groundwater, and dense nonaqueous phase liquids (DNAPLs) in the subsurface, was seriously lacking. In 1994, in the "Alternatives for Ground Water Cleanup" document, the National Research Council clearly showed that as a nation we had been conducting a failed 15-year experiment to clean up our nation's groundwater and that the default technology, pump-and-treat, was often ineffective at remediating contaminated aquifers. The answer for the DoD was clear. The DoD needed better technologies to clean up its contaminated sites and better technologies could only arise through a better scientific and engineering understanding of the subsurface and the associated chemical, physical, and biological processes. Two DoD organizations were given responsibility for initiation of new research, development, and demonstrations to obtain the technologies needed for cost-effective remediation of facilities across the DoD: the Strategic Environmental Research and Development Program (SERDP) and the Environmental Security Technology Certification Program (ESTCP).

SERDP was established by the Defense Authorization Act of 1991, as a partnership of the DoD, the U.S. Department of Energy, and the U.S. Environmental Protection Agency with the mission "to address environmental matters of concern to the Department of Defense and the Department of Energy through support of basic and applied research and development of technologies that can enhance the capabilities of the departments to meet their environmental obligations". SERDP was created with a vision of bringing the capabilities and assets of the nation to bear on the environmental challenges faced by the DoD. As such, SERDP is the DoD's corporate environmental research and development program. To address the highest priority issues confronting the Army, Navy, Air Force, and Marine Corps, SERDP focuses on cross-service requirements and pursues high-risk and high-payoff solutions to the Department's most intractable environmental problems. SERDP's charter permits investment across the broad spectrum of research and development, from basic research through applied research and exploratory development. SERDP invests with a philosophy that all research, whether basic or applied, when focused on the critical technical issues, can impact environmental operations in the near term.

A DoD partner organization, ESTCP was established in 1995 as the DoD's environmental technology demonstration and validation program. ESTCP's goal is to identify, demonstrate, and transfer technologies that address the Department's highest priority environmental requirements. The program promotes innovative, cost-effective environmental technologies through demonstrations at DoD facilities and sites. These technologies provide a large return on investment through improved efficiency, reduced liability, and direct cost savings. The current cost and impact on DoD operations of environmental compliance is significant. Innovative technologies are reducing both the cost of environmental remediation and compliance, and the impact of the DoD's operations on the environment, while enhancing military readiness. ESTCP's strategy is to select laboratory-proven technologies with potential broad DoD application and use DoD facilities as test beds. By supporting rigorous test and evaluation of innovative environmental technologies, ESTCP provides validated cost and performance information. Through these tests, new technologies gain end-user and regulatory acceptance.

In the 10 to 15 years since SERDP and ESTCP were formed, much progress has been made in the development of innovative and more cost-effective environmental remediation technology. Since then, once recalcitrant environmental contamination problems for which little or no effective technology was available are now tractable. However, we understand that newly developed technologies will not be broadly used in government or industry unless the consulting engineering community has the knowledge and experience needed to design, cost, market, and apply them.

To help accomplish the needed technology transfer, SERDP and ESTCP have joined to sponsor the development of a series of monographs on remediation technology written by leading experts in each subject area. Each volume will be designed to provide the background in process design and engineering needed by professionals who have advanced training and five or more years of experience. The first volume on *In Situ* Bioremediation of Perchlorate in Groundwater will be followed by others on such topics as the remediation of both soluble phase and DNAPL chlorinated solvents, bioaugmentation to enhance bioremediation processes, delivery and mixing strategies and technologies to enhance subsurface remediation, and remediation of contaminated sediments. Additional volumes will be written as new remediation technologies are developed and proven to be effective.

This volume provides a review of the past decade of intensive research, development, and demonstrations on the *in situ* bioremediation of perchlorate. The intended audiences include the decision makers and practicing engineers and hydrogeologists who will select, design, and operate these remedial systems, as well as researchers seeking to improve the current state-of-the-art. Our hope is that this volume will serve as a useful resource to assist remediation professionals in applying and developing the technology as effectively as possible.

A brief technology overview is provided in Chapter 1. Chapter 2 summarizes the development of *in situ* bioremediation of perchlorate to illustrate how we arrived at our state-of-understanding today. Chapter 3, on the principles of *in situ* bioremediation of perchlorate, presents the current state-of-the-science, covering the microbial processes, abiotic processes, and the engineering and implementation issues underlying the technologies described.

Chapter 4 deals with the important characterization issues relevant to perchlorate contamination, including a discussion on perchlorate sources (i.e., both the anthropogenic and natural sources of perchlorate) and the methods available to distinguish between differing sources (particularly the use of compound specific isotopic analysis).

Chapter 5 initiates a more detailed discussion of the different methods for implementing *in situ* bioremediation, beginning with a summary of the primary methods available, and the factors affecting the selection of *in situ* bioremediation at a specific site. Chapter 5 also discusses the specific remedial approaches available, with discussion on their design and monitoring, and the advantages and disadvantages of each approach under different site-specific conditions.

Chapters 6, 7, 8 and 9 detail the different options for implementing *in situ* bioremediation. These chapters describe the design and operation of the particular option, the current stage of development, and case histories that illustrate the issues involved and provide examples of the performance that is achievable. Chapter 6 discusses active bioremediation, in which substrates are continuously circulated through the target treatment zone. Chapter 7 discusses what is often described as semi-passive bioremediation, in which the substrate is added at intervals and mixing is intermittent. The final Chapters (8 and 9) discuss two different approaches to passive bioremediation, in which there is no active mixing. In the first case, edible oil is injected into the subsurface, and in the second, a biowall is created by installing a trench across the contaminant plume filled with a degradable material such as mulch.

Chapter 10 provides cost information for each technology, using analyses of several template sites to aid the reader in estimating the economics of applying these technologies at other sites. Cost information includes capital costs, as well as costs for laboratory testing, pilot-scale demonstration, design, system operation, monitoring and maintenance during operations, and demolition and restoration after remediation. In addition, analogous cost data are presented for pump-and-treat systems for each template site to illustrate the potential cost savings associated with the use of alternative approaches.

The final chapter on emerging technology (Chapter 11), describes three innovative bioremediation technologies still in the developmental stages. These technologies (monitored natural attenuation, phytoremediation, and vadose zone bioremediation) are described, and field demonstrations are used to illustrate the current stage of maturity and the potential applicability of these approaches for specific situations.

Each chapter in this volume has been thoroughly reviewed for technical content by one or more experts in each subject area covered. The editors and chapter authors have produced a well-written and up-to-date treatise that we hope will prove to be a useful reference for those making decisions on remediation of perchlorate, remediation practitioners, and for those involved in development of advanced technology for the *in situ* remediation of perchlorate.

SERDP and ESTCP are committed to the development of new and innovative technologies to reduce the cost of remediation of soil, groundwater, and sediment contamination as a result of past operational and industrial practices. We are also firmly committed to the widest dissemination of these technologies to ensure that our investments continue to yield savings for not only the DoD but also the nation. In sponsoring this monograph series, we hope to provide the broader remediation community with the most current knowledge and tools available in order to bring these technologies to bear on the remediation of perchlorate.

Bradley P. Smith, Executive Director, SERDP
Jeffrey A. Marqusee, PhD, Director, ESTCP
Andrea Leeson, PhD, Environmental Restoration Program Manager, SERDP and ESTCP

ABOUT THE EDITORS

Hans F. Stroo

Dr. Stroo is a Principal Technical Advisor with HydroGeoLogic, Inc. He provides technical support on large remediation projects for private- and public-sector clients and has served as a technical advisor to the SERDP and ESTCP programs for over 10 years.

Dr. Stroo received his BS degrees in Biology and Soil Science from Oregon State University, an MS in Soil Science from West Virginia University, and a PhD in Soil Science from Cornell University.

He was formerly a Principal with Remediation Technologies, Inc. (RETEC). He has over 20 years of experience in the assessment and remediation of contaminated soil and groundwater, particularly in the development and use of *in situ* bioremediation.

Dr. Stroo has served on several Expert Review Panels for SERDP, other government agencies and private companies. Most recently, he served as Co-Chair of the SERDP workshops on Remediation of Chlorinated Solvents in Groundwater and Remediation of DNAPL Source Zones.

C. Herb Ward

Dr. Ward holds the Foyt Family Chair of Engineering in the George R. Brown School of Engineering at Rice University. He is also Professor of Civil and Environmental Engineering and Ecology and Evolutionary Biology.

Dr. Ward has undergraduate (BS) and graduate (MS, PhD, MPH) degrees from New Mexico State University, Cornell University, and the University of Texas School of Public Health, respectively. He is a registered professional engineer in Texas and a Board Certified Environmental Engineer by the American Academy of Environmental Engineers.

He has been a faculty member at Rice University for 42 years during which he has served as Chair of the Department of Environmental Science and Engineering and the Department of Civil and Environmental Engineering. He has also served as Director of the USEPA-sponsored National Center for Ground Water Research and the DoD-sponsored Advanced Applied (Environmental) Technology Development Facility (AATDF).

Dr. Ward has been a member of the USEPA Science Advisory Board and served as Chair of the SERDP Scientific Advisory Board. He is the founding and current Editor-in-Chief of the international scientific journal *Environmental Toxicology and Chemistry*.

The editors of this volume gratefully acknowledge the excellent organizational and editorial assistance of Catherine M. Vogel, PE, Noblis, Atlanta, Georgia.

ABOUT THE AUTHORS

Carol E. Aziz

Dr. Aziz is a Senior Environmental Engineer with Geosyntec Consultants. She specializes in the remediation of soil and groundwater contaminated with recalcitrant compounds. Dr. Aziz has led the research, development, design and/or implementation of several *in situ* bioremediation technologies including mulch biowalls, hydrogen biosparging, emulsified vegetable oil biobarriers, and bioaugmentation. She is also the lead author of the natural attenuation model, BIOCHLOR. Over the past seven years, she has served as Co-PI or Project Manager for a number of SERDP/ESTCP projects. Most recently, Dr. Aziz served as Co-PI on a SERDP project investigating anthropogenic sources of perchlorate.

Dr. Aziz received her BASc and MASc degrees in Chemical Engineering and Applied Chemistry from the University of Toronto and a PhD in Civil Engineering (Environmental) from the University of Texas at Austin. She is a registered professional engineer in Texas and Ontario.

Robert C. Borden

Dr. Borden is a Professor of Civil Engineering at North Carolina State University and Principal Engineer at Solutions-IES, Inc. His research and consulting work is focused on the natural and enhanced remediation of hazardous materials in the subsurface.

Dr. Borden received his BS and ME degrees in Civil and Environmental Engineering from the University of Virginia and a PhD in Environmental Engineering from Rice University.

He has been a faculty member at North Carolina State University since 1986, teaching and conducting research in surface water hydrology, groundwater hydrology, subsurface contaminant transport, and *in situ* remediation. His research includes laboratory studies, fieldwork and mathematical model development. Recently, much of his research has focused on remediation of chlorinated solvents, perchlorate, chromium and acid mine drainage using emulsified oils. At Solutions-IES, Dr. Borden supports many of the firm's projects including traditional remediation approaches, *in situ* bioremediation, *in situ* chemical oxidation, monitored natural attenuation, and expert witness testimony.

John D. Coates

Dr. Coates joined the faculty of the Plant and Microbial Biology Department at the University of California, Berkeley in 2002 where he is currently a Professor of microbiology. In addition, he holds a joint appointment in the Geological Scientist Faculty of the Earth Sciences Division, Ernest Orlando Lawrence Berkeley National Laboratory. He obtained an Honors BSc in Biotechnology in 1986 from Dublin City University, Ireland and his PhD in 1991 from University College in Galway, Ireland. His major area of interest is geomicrobiology applied to environmental problems. Specific interests include diverse forms of anaerobic microbial metabolism such as microbial perchlorate reduction, microbial iron oxidation and reduction, and microbial humic substances redox cycles. Other interests include bioremediation of toxic metals, radionuclides, and organics. He was the recipient of the 1998 *Oak Ridge Ralph E. Powe Young Faculty Enhancement Award*, joint recipient of the 2001 *DoD SERDP Project of the Year Award*, and the 2002 *Southern Illinois University College of Science Researcher of the Year Award*. He has given more than 90 invited presentations at national and international meetings and has authored and co-authored more than 80 peer-reviewed publications and book chapters. He has eight patent submissions based on technologies developed in his laboratory. He sits on the editorial boards of the journals *Applied and Environmental Microbiology* and *Applied Microbiology and Biotechnology*. He also sits on the microbiology editorial board of Faculty of 1000 databases.

Evan E. Cox

Mr. Cox is a Principal and Senior Remediation Microbiologist with demonstrated experience in the development, feasibility evaluation and application of innovative *in situ* remediation technologies, including monitored natural attenuation (MNA), enhanced *in situ* bioremediation (EISB), *in situ* chemical oxidation, and metal-catalyzed reduction of chlorinated and recalcitrant chemicals in subsurface environments. Mr. Cox received his BSc and MSc degrees in Microbiology from the University of Waterloo. Over the past 18 years, Mr. Cox has pioneered the use of MNA and EISB technologies, including bioaugmentation for remediation of chlorinated solvents in porous media and fractured bedrock, co-authoring several guidance documents and presenting educational courses on these subjects. Mr. Cox has pioneered the use of EISB to treat rocket fuel components such as perchlorate in soil and groundwater. He has served as the program manager or technical lead for more than 50 perchlorate projects at more than 20 perchlorate-impacted sites nationwide. Mr. Cox has published more than 30 articles regarding the degradation of hazardous contaminants in subsurface environments.

Douglas C. Downey

Mr. Downey is a Principal Technologist at CH2M HILL, responsible for providing technical direction for site remediation projects in the United States and 30 nations where CH2M HILL operates. He is a graduate of the U.S. Air Force Academy and received his MS Degree in Civil and Environmental Engineering from Cornell University.

Mr. Downey has over 30 years of engineering experience with an emphasis on developing and testing innovative remediation methods such as bioventing, monitored natural attenuation, biowalls and *in situ* bioreactors. He is currently developing sustainable remediation systems that operate on solar and wind power. Mr. Downey is a registered professional engineer in Colorado and has designed and completed over 300 remediation projects. Mr. Downey is working on several projects to establish clean drinking water supplies in developing nations.

Patrick J. Evans

Dr. Evans is a Vice President with CDM in Bellevue, Washington where he has been located for 14 years. He serves as a technical advisor on projects in remediation, drinking water treatment, and industrial wastewater treatment. Dr. Evans received his BS and PhD degrees in Chemical Engineering from The University of Michigan and his MS degree in Chemical Engineering from Rutgers, The State University of New Jersey. He also completed three years of postdoctoral research in environmental microbiology at the New York University Medical Center.

He has 20 years of experience in process engineering, chemistry, and microbiology. This experience has included research and development, bench-scale and pilot-scale evaluation of innovative technologies, design and construction of treatment facilities, and troubleshooting and optimization of operating processes. He has invented and holds patents for analytical methods and treatment processes.

About the Authors

Paul B. Hatzinger

Dr. Hatzinger is a Senior Research Scientist in the Biotechnology Development and Applications Group of Shaw Environmental, Inc. in Lawrenceville, New Jersey. He holds a BS in Biology and Environmental Science from St. Lawrence University, and both a MS and PhD in Environmental Toxicology from Cornell University.

Dr. Hatzinger was previously employed as a staff scientist at Envirogen, Inc. and has more than 15 years of experience in biodegradation, bioremediation, groundwater geochemistry and microbiology.

Dr. Hatzinger's research is focused on the development of *in situ* and *ex situ* bioremediation technologies for emerging contaminants. He has been studying perchlorate biodegradation since 1999, and has served as the Principal Investigator on several field projects evaluating *in situ* remedial approaches for the oxidant. In addition, Dr. Hatzinger has worked closely with the engineering group at Shaw that has designed and constructed five full-scale bioreactor systems for treatment of perchlorate in groundwater.

Bruce M. Henry

Mr. Henry is a Project Manager and Principal Geologist with Parsons Infrastructure & Technology Group, Inc. in Denver, Colorado. He has 15 years of experience in the investigation and remediation of contaminated soil and groundwater, and was formerly employed in the field of oil and gas exploration. Mr. Henry received a BA degree in Geology from the University of Colorado and a MS degree in Geology from Colorado State University. He is a registered professional geologist in Wyoming.

Mr. Henry provides project management and technical direction for the *in situ* remediation of fuel hydrocarbons and chlorinated solvents in groundwater. In particular, he has worked with the U.S. Air Force to develop technical protocols for enhanced *in situ* biroemediation of chlorinated solvents and is the primary author of the Air Force *Technical Protocol for Enhanced Anaerobic Bioremediation Using Permeable Mulch Biowalls and Bioreactors.* Mr. Henry also provides technical support on remediation projects for several private- and public-sector clients.

W. Andrew Jackson

Dr. Jackson is a Whitacre Faculty Fellow in the College of Engineering at Texas Tech University and an Associate Professor of Civil and Environmental Engineering. Dr. Jackson received his BS degree in Biology from Rhodes College and a MS and PhD in Engineering Science/Environmental Engineering from Louisiana State University. He is a registered professional engineer in Louisiana.

He has been a faculty member at Texas Tech University for 10 years. He is very active in the general area of contaminant fate (perchlorate, RDX, chlorinated solvents) in natural environments and the use of engineered wetlands as passive sustainable treatment systems. His work includes the design of biological reactors for source separated waste streams and their consequences on water quality in closed loop recycled systems.

Dr. Jackson has served on several Expert and Review Panels for NSF, USEPA, other government agencies and private companies. He is an associate editor for the journal *Water Air and Soil Pollution* and is a member of the steering committee for the International Conference on Environmental Systems.

Thomas A. Krug

Mr. Krug, P.Eng., is an Associate and senior environmental engineer with Geosyntec Consultants. He has an MS Degree in Chemical Engineering and over 20 years of professional experience dealing with innovative technologies to solve challenging environmental problems. He has extensive experience in the development and evaluation of innovative technologies for remediation of soil and groundwater at contaminated properties for *Fortune 500* companies, agencies of the DoD, and NASA.

The focus of his professional practice has been in taking new technologies for groundwater remediation from the early development stage to successful field-scale application and adapting new and conventional technologies to solve client's real world problems. He has been a pioneer in the development, demonstration, and application of permeable reactive barriers for the treatment of chlorinated solvents. In 2007, he was inducted into the Space Foundation Technology Hall of Fame for his contributions to the development of emulsified nano-scale zero-valent iron technology for treatment of chlorinated solvent dense nonaqueous phase liquid source zones.

M. Tony Lieberman

Mr. Lieberman is the Bioremediation Program Manager with Solutions-IES, Inc. He works extensively in the field of bioremediation and is also Project Manager on many soil and groundwater assessment and remediation projects, utilizing both biological and non-biological treatment approaches for various groundwater contaminants.

Mr. Lieberman received his BS degree in Biology from St. Lawrence University and his MS degree in Microbiology from Pennsylvania State University. In addition, he spent 11 years as a Research Support Specialist in the Laboratory of Soil Microbiology in the Department of Agronomy at Cornell University.

He was formerly a Principal with ESE Bioremediation, Inc., and has over 23 years of consulting experience, particularly in the laboratory and field application of bioremediation. He currently serves as Principal Investigator and Co-PI on ESTCP-funded projects separately evaluating the use of emulsified oil and the potential of MNA for remediating perchlorate in groundwater. Mr. Lieberman was a member of the ITRC Perchlorate Team and a contributor to its Tech-Reg documents, has co-authored chapters on *in situ* bioremediation and has prepared numerous presentations and publications on perchlorate remediation.

Raymond C. Loehr

Dr. Loehr is the H. M. Alharthy Centennial Chair and Professor Emeritus in the College of Engineering at The University of Texas at Austin. His teaching, research and professional interests have emphasized hazardous and industrial waste management of contaminated liquids, slurries, soils and sediments. He has over 300 technical publications and has authored or edited 14 books.

Dr. Loehr has BS and MS degrees from Case Institute of Technology and a PhD from the University of Wisconsin. He is a registered professional engineer in several states and is a Board Certified Environmental Engineer.

In addition to his academic activities, he has been a consultant to industries, engineering and law firms and government agencies. He has been the Chair of the USEPA Science Advisory Board, has been a member of the SERDP Scientific Advisory Board and has served on numerous National Research Council committees.

He is a member of the National Academy of Engineering and has received awards from organizations such as the American Society of Civil Engineers, the Water Environment

Federation, the Society of Environmental Toxicology and Chemistry and the American Academy of Environmental Engineers.

Robert D. Norris

Dr. Norris is a graduate of Beloit College (BS, 1966) and the University of Notre Dame (PhD in Chemistry, 1971). He has conducted research in the biomedical field, on development of flame retardants, and on new commercial uses of hydrogen peroxide in several fields including environmental applications.

He has 23 years of experience in development and implementation of remediation technologies. His experience includes development of soil gas survey techniques for site investigations. He co-managed the first full-scale demonstration of the use of hydrogen peroxide for *in situ* bioremediation for the American Petroleum Institute. He managed both laboratory evaluations and field demonstrations of *in situ* remediation of petroleum hydrocarbons. He has experience in various remediation methods including bioremediation of chlorinated solvents, air sparging, biosparging, *in situ* and *ex situ* soil vapor extraction, bioventing, *in situ* chemical oxidation, land farming of contaminated soils, aboveground soil cells, *in situ* chemical treatment, and monitored natural attenuation. He serves as corporate *in situ* technology director and provides senior technical advice and review for most of Brown and Caldwell's remediation projects.

He has co-authored nine books on remediation, served on committees for the National Academy of Sciences, SERDP, and the Electric Power Research Institute, published over 100 articles on remediation and related topics, and served on the Regenesis Science Advisory Board. He was co-editor of an MNA column in the journal *Remediation* and has served on the editorial boards for *Remediation* and the *Bioremediation Journal*.

Valentine A. Nzengung

Dr. Valentine Nzengung is Professor of Environmental Geochemistry in the Department of Geology at the University of Georgia. He is also the President and founder of PLANTECO Environmental Consultants, LLC.

Dr. Nzengung received his undergraduate (BS) degree in Geology from Georgia State University and graduate (MS and PhD) degrees in Environmental Engineering and Environmental Geochemistry, respectively from the Georgia Institute of Technology.

He has been a faculty member at the University of Georgia for 13 years during which he has developed and demonstrated at the field scale multiple innovative technologies for filtration and degradation of organic and inorganic contaminants in soil and water. He has taught environmental science courses at the undergraduate and graduate levels.

Dr. Nzengung has served on several Expert Peer Review Panels for the USEPA and other Federal agencies. Most recently, he has been involved in the preparation of technology guidance documents for the USEPA's Remediation Technologies Development Forum (Phytoremediation of Chlorinated Solvents) and the ITRC Perchlorate and Acid Mine Waste Teams.

Michael W. Perlmutter

Mr. Perlmutter is a remediation engineer and project manager with CH2M HILL in Atlanta, Georgia. He has 10 years of experience in the assessment and remediation of contaminated soil and groundwater and currently serves as lead remediation engineer for multiple federal and commercial clients. Mr. Perlmutter received his BS and MS in Civil Engineering from the Georgia Institute of Technology and University of Texas at Austin, respectively. He is a registered professional engineer in Tennessee, Georgia, Alabama, and New Mexico.

Charles E. Schaefer

Dr. Schaefer is a Senior Technology Applications Engineer for Shaw Environmental, Inc, where he has worked for over seven years on a variety of commercial, government, and research projects. He received his undergraduate and doctorate degrees in Chemical and Biochemical Engineering from Rutgers University, and served as a post-doctoral Research Associate in both the Civil and Environmental Engineering and Petroleum Engineering departments at Stanford University. Dr. Schaefer has over 15 years of experience working on the fate and transport of organic contaminants in soil and groundwater.

Carrolette J. Winstead

Ms. Winstead has a BS degree in Chemical Engineering from Purdue University. She worked in the environmental consulting industry for over 11 years and is now a Unit Manager in the Water Quality Division at the Arizona Department of Environmental Quality. Ms. Winstead's experience includes extensive work on remediation of chlorinated hydrocarbons and fuels using various remedial technologies. Her input into this monograph was costing analysis of remedial alternatives using RACER© software.

Christopher Wolfe

Mr. Wolfe is a Project Manager and Lead Engineer specializing in site remediation at HydroGeoLogic, Inc. He has over 13 years of experience in remediation of contaminated soil and groundwater, industrial wastewater treatment, and pollution prevention. In addition to completing projects for a variety of industry clients, Mr. Wolfe has worked for public clients including the DoD, USEPA, and the Delaware Department of Natural Resources and Environmental Control. His work has taken him to project sites throughout the United States as well as in Hungary and Israel.

Mr. Wolfe has undergraduate degrees in Industrial Engineering and International Relations from Lehigh University and an MS Degree in Environmental Engineering from Virginia Tech. He is a registered professional engineer in Delaware.

EXTERNAL REVIEWERS

Farrukh Ahmad
Visiting Scholar
Technology and Development Program
Massachusetts Institute of Technology
MIT 1-175, 77 Massachusetts Avenue
Cambridge, MA 02139
Email: ahmadf@mit.edu

Erica Becvar
Air Force Center for Engineering and the Environment
Technical Division
3300 Sidney Brooks
Brooks City-Base, TX 78235
Email: erica.becvar@brooks.af.mil

Elizabeth Edwards
Department of Chemical Engineering and Applied Chemistry
University of Toronto
200 College Street
Toronto, Ontario, Canada, M5S 3E5
Email: elizabeth.edwards@utoronto.ca

Mark Goltz
Air Force Institute of Technology
Department of Systems & Engineering Management
2950 Hobson Way, Bldg. 640
Wright-Patterson Air Force Base, OH 45433
Email: Mark.Goltz@afit.edu

Gregory Harvey
Air Force Research Laboratory
1801 Tenth Street
Bldg. 8, Suite 200, Area B
Wright-Patterson Air Force Base, OH 45433
Email: gregory.harvey@wpafb.af.mil

Michael C. Kavanaugh
Malcolm Pirnie, Inc.
2000 Powell Street, Suite 1180
Emeryville, CA 94608
Email: mkavanaugh@pirnie.com

David W. Major
GeoSyntec Consultants, Inc.
130 Research Lane, Suite 2
Guelph, Ontario, Canada N1G 5G3
Email: DMajor@GeoSyntec.com

Kevin Mayer
U.S. Environmental Protection Agency, Region 9
Mail Stop SFD-7-275
Hawthorne Street
San Francisco, CA 94105
Email: Mayer.Kevin@epa.gov

Elaine A. Merrill
Henry M. Jackson Foundation for the Advancement of Military Medicine
Air Force Research Laboratory/RHPB
Bldg. 837
2729 R Street
Wright-Patterson Air Force Base, OH 45433
Email: elaine.merrill@wpafb.af.mil

Scott L. Neville
Aerojet General Corporation
P.O. Box 13222/Dept 0330
Sacramento, CA 95813
Email: scott.neville@aerojet.com

Robert J. Steffan
Shaw Environmental, Inc.
17 Princess Road
Lawrenceville, NJ 08648
Email: Rob.Steffan@shawgrp.com

Harry Van Den Berg
ENSR Corporation
1220 Avenida Acaso
Camarillo, CA 93012
Email: hvandenberg@ensr.aecom.com

ACRONYMS AND ABBREVIATIONS

°C	degrees Celsius	DCE	dichloroethene
μg/kg	micrograms per kilogram	DDT	dichlorodiphenyltrichloroethane
μg/L	micrograms per liter		
μm	micrometer	DERP	Defense Environmental Restoration Program
μmoles/L	micromoles per liter		
μS/cm	microsiemens per centimeter	DHG	dissolved hydrocarbon gas
		DNAPL	dense nonaqueous phase liquid
1,1,1-TCA	1,1,1-trichloroethane		
1,1-DCA	1,1-dichlororethane	DO	dissolved oxygen
16S rRNA	ribosomal RNA	DOC	dissolved organic carbon
AWMA	Air & Waste Management Association	DoD	U.S. Department of Defense
		DPRB	dissimilatory perchlorate-reducing bacteria
AFB	Air Force Base		
AFCEE	Air Force Center for Environmental Excellence (renamed the Air Force Center for Engineering and the Environment)	EDQW	Environmental Data Quality Workgroup
		EISB	enhanced *in situ* bioremediation
		ELISA	enzyme-linked immunosorbent assay
AIChE	American Institute of Chemical Engineers	EOFA	edible oil fatty acids
		EOS®	Edible Oil Substrate
amsl	above mean sea level	ER	extraction and reinjection
AN	ammonium nitrate	ESI	electrospray ionization
ANFO	ammonium nitrate-fuel oil	ESTCP	Environmental Security Technology Certification Program
APA	American Pyrotechnics Association		
bgs	below ground surface	EVO	emulsified vegetable oil
BOD	biochemical oxygen demand	EWG	Environmental Working Group
CA	chloroethane	FBR	fluidized bed reactor
CCL	Contaminant Candidate List	ft	feet
CD	chlorite dismutase	g	gram
CDC	Center for Disease Control	g/L	grams per liter
CDHS	California Department of Health Services	GAC	granular activated carbon
		gal	gallon
CERCLA	Comprehensive Environmental Response, Compensation, and Liability Act	GCIRMS	gas chromatography-isotope ratio mass spectrometry
		GEDIT	Gaseous Electron Donor Injection Technology
cis-1,2-DCE	*cis*-1,2-dichloroethene	gpm	gallons per minute
cm	centimeter	HFTW	horizontal flow treatment well
cm/sec	centimeters per second		
COC	Contaminant of Concern	HPLC	high performance liquid chromatography
COD	chemical oxygen demand		
CSM	Conceptual Site Model		
DAP	di-ammonium phosphate	hr	hour

HRC®	Hydrogen Release Compound	mM	millimolar
HRT	hydraulic retention time	mmol	millimole
IC	ion chromatography	MNA	monitored natural attenuation
IHDIV	Indian Head Division	MPN	most probable number count
IME	Institute of Makers of Explosives	MRL	method reporting limit
in	inch	MS	mass spectrometry
IRMS	isotope ratio mass spectrometry	mS/cm	millisiemens per centimeter
		MSDS	Material Safety Data Sheet
ISE	ion specific electrode	MTBE	methyl tertiary butyl ether
ISEE	International Society of Explosives Engineers	mV	millivolt
		NAVAIR	Naval Air Systems Command
ITRC	Interstate Technology & Regulatory Council	NFESC	Naval Facilities Engineering Service Center
JPL	Jet Propulsion Laboratory	NPV	Net Present Value
K	conductivity	NRC	National Research Council
K_d	partition coefficient	NWIRP	Naval Weapons Industrial Reserve Plant
kg	kilogram		
km	kilometer	O&M	operation and maintenance
kPa	kilopascal	OB/OD	open burn, open detonation
L	liter	OM&M	operation, maintenance and monitoring
L/min	liters per minute		
lb	pound	OMB	Office of Management and Budget
LC	liquid chromatography		
LCFA	long-chain fatty acid	OMRI	Organic Materials Review Institute
LHAAP	Longhorn Army Ammunition Plant		
LLC	Limited Liability Company	ORP	oxidation reduction potential
LLNL	Lawrence Livermore National Laboratory	OSW	Office of Solid Waste
		PAH	polycyclic aromatic hydrocarbon
LPG	liquefied petroleum gas		
LUFT	Leaking Underground Fuel Tank	p-CBS	p-chlorobenzenesulfonate
		PCE	perchloroethene (also termed tetrachloroethene)
LVW	Las Vegas Wash		
m	meter	PCL	protective concentration limit
MADEP	Massachusetts Department of Environmental Protection		
		PCR	polymerase chain reaction
MCL	maximum contaminant level	ppb	part per billion
		PPE	personal protective equipment
MCT	matrix conductivity threshold		
		ppm	part per million
MDL	method detection limit	PQL	practical quantitation limit
mg	milligram	PRB	permeable reactive barrier
mg/kg	milligrams per kilogram	PRG	preliminary remediation goal
mg/L	milligrams per liter		
mi	mile	PTA	pilot test area
min	minute	PV	pore volume
mL	milliliter		

Acronyms and Abbreviations

QA/QC	quality assurance/quality control	**THPS**	tetrakis(hydroxymethyl)phosphonium sulfate or Tolcide
RAO	Remedial Action Objective	**TIC**	total inorganic carbon
RCRA	Resource Conservation and Recovery Act	**TNT**	2,4,5-trinitrotoluene
rDNA	Recombinant Deoxyribonucleic Acid	**TOC**	total organic carbon
RDX	cyclotrimethylenetrinitramine or **R**oyal **D**emolition e**X**plosive	**UCMR**	Unregulated Contaminant Monitoring Rule
RfD	reference dose	**UMD**	University of Massachusetts at Dartmouth
RFI	RCRA Facility Investigation	**USACE**	U.S. Army Corps of Engineers
RO	reverse osmosis	**USDOC**	U.S. Department of Commerce
ROD	Record of Decision	**USEPA**	U.S. Environmental Protection Agency
ROI	radius of influence	**USFDA**	U.S. Food and Drug Administration
SAMNAS	Surface Application and Mobilization of Nutrient Amendments	**USGS**	U.S. Geological Survey
SERDP	Strategic Environmental Research and Development Program	**UTC**	United Technologies Corporation
SMOC	standard mean ocean chloride	**UV**	ultraviolet
SU	standard unit	**VC**	vinyl chloride
TCE	trichloroethene	**VFA**	volatile fatty acid
TCEQ	Texas Commission on Environmental Quality	**VOC**	volatile organic compounds
TDS	total dissolved solids	**vol/vol**	volume per volume
TEAP	terminal electron accepting process	**WMP**	Waste Management Plan
		wt/vol	weight per volume
		ZVI	zero valent iron

UNIT CONVERSION TABLE

MULTIPLY	BY	TO OBTAIN
acres	0.405	hectares
acres	1.56 E-3	square miles (statute)
centimeters	0.394	inches
cubic feet	0.028	cubic meters
cubic feet	7.48	gallons (U.S. liquid)
cubic feet	28.3	liters
cubic meters	35.3	cubic feet
cubic yards	0.76	cubic meters
feet	0.305	meters
feet per year	9.66 E-7	centimeters per second
gallons (U.S. liquid)	3.79	liters
hectares	2.47	acres
inches	2.54	centimeters
kilograms	2.20	pounds (avoir)
kilograms	35.3	ounces (avoir)
kilometers	0.62	miles (statute)
liters	0.035	cubic feet
liters	0.26	gallons (U.S. liquid)
meters	3.28	feet
miles (statute)	1.61	kilometers
ounces (avoir)	0.028	kilograms
ounces (fluid)	29.6	mililiters
pounds (avoir)	0.45	kilograms
square feet	0.093	square meters
square miles	640	acres

GLOSSARY[1]

Abiotic—Occurring without the direct involvement of organisms.

Active treatment—*In situ* bioremediation approach in which water-soluble amendments are added to the subsurface intermittently, frequently, or even continuously, by pumping liquid solutions into injection wells. Extraction may also be used to recover water prior to amendment addition and/or to recirculate amendments through the target treatment zone.

Advection—Transport of molecules dissolved in water along the groundwater flow path at an average expected velocity.

Aerobic—Environmental conditions where oxygen is present.

Aerobic respiration—Process whereby microorganisms use oxygen as an electron acceptor to generate energy.

Analytical model—A mathematical model that has a closed form solution (i.e., the solution to the equations used to describe changes in a system can be expressed as a mathematical analytic function). Analytical solutions can be more exact and aesthetically pleasing than numerical models, but analytical solutions to equations describing complex systems can often become very difficult.

Anaerobic—Means "without air". It generally refers to occurring or living without oxygen present. Thus, in an anaerobic groundwater system, the chemistry is characterized by reductive conditions. Sometimes (e.g., in wastewater treatment) anaerobic is used to indicate a lack of any electron acceptors (i.e., including nitrate and sulfate). In groundwater, a dissolved oxygen concentration below 1.0 mg/L is generally considered anaerobic.

Anaerobic bioventing—Delivery of gases such as hydrogen to the subsurface to stimulate the activity of anaerobic microorganisms.

Anaerobic respiration—Process whereby microorganisms use an electron donor such as hydrogen and a chemical other than oxygen as an electron acceptor. Common "substitutes" for oxygen are nitrate, sulfate, iron, carbon dioxide, and other organic compounds (fermentation).

Anoxic—Literally means "without oxygen." For example, anoxic groundwater is groundwater that contains no dissolved oxygen.

Aquifer—An underground geological formation that stores groundwater. A confined aquifer lies beneath a confining unit of lower hydraulic conductivity. An unconfined aquifer does not have a confining unit and is defined by the water table.

Aquitard—An underground geological formation of low permeability that does not readily transmit groundwater.

---

[1] This glossary is a compilation of definitions of terms synthesized by the volume editors and chapter authors. Select definitions are reprinted from *In Situ Bioremediation: When does it work?* (National Research Council, 1993) with permission from the National Academies Press, Copyright 1993, National Academy of Sciences.

Attenuation—Reduction of contaminant concentrations over space or time. It includes both destructive (e.g., biodegradation, hydrolysis) and non-destructive (e.g., volatilization, sorption) removal processes.

Attenuation Rate—The rate of contaminant concentration reduction over time. Typical units are milligrams per liter per year (mg/L/yr).

Bacterium—A single-celled organism of microscopic size (generally 0.3 to 2.0 micrometers in diameter). As opposed to fungi and higher plants and animals ("eukaryotes"), bacteria are "prokaryotes" (i.e., they are characterized by the absence of a distinct, membrane-bound nucleus or membrane-bound organelles, and by DNA that is not organized into chromosomes).

Bedrock—The solid or fractured rock underlying surface solids and other unconsolidated material or overburden.

Bioaugmentation—The addition of microbes to the subsurface to improve the biodegradation of target contaminants. Microbes may be "seeded" from populations already present at a site, or from specially-cultivated strains of bacteria.

Biobarrier—The concept of intercepting and treating a contaminant plume as it passes through a permeable subsurface barrier. Biobarriers are created by installing wells or trenches across the width of a plume to deliver substrate to the microorganisms in the groundwater as it flows through the barrier.

Biochemical—Produced by, or involving, chemical reactions of living organisms.

Biodegradation—Biologically mediated conversion of one compound to another.

Biofouling—Impairment of the functioning of wells or other equipment as a result of the growth or activity of microorganisms.

Biomass—Total mass of microorganisms present in a given amount of water or soil.

Bioremediation—Use of microorganisms to control and destroy contaminants.

Biotransformation—Biologically catalyzed transformation of a chemical to some other product.

Biowall—A form of passive *in situ* bioremediation, in which the contaminant plume is intercepted and treated as it passes through an emplaced porous barrier (e.g., trenches filled with sand-mulch mixtures). Microorganisms growing on the wall materials remove contaminants through biodegradation processes as groundwater passes through the barrier.

Catalyst—A substance which promotes a chemical reaction, but does not itself enter into the reaction.

Chlorinated solvent—A hydrocarbon in which chlorine atoms substitute for one or more hydrogen atoms in the compound's structure. Chlorinated solvents commonly are used for grease removal in manufacturing, dry cleaning, and other operations. Examples include trichloroethene, perchloroethene and trichloroethane.

Glossary

Chlorite dismutase—An enzyme that catalyzes the disproportionation (i.e., a chemical reaction in which a single reactant breaks up to produce two different products) of chlorite to chloride and oxygen. Chlorite dismutase is present in bacteria capable of cell respiration using perchlorate or chlorate.

Conceptual site model—A hypothesis about how contaminant releases occurred at a site, the current state of the contaminant source, an idealized geochemical site type, and the current plume characteristics (plume stability).

Dechlorination—A type of dehalogenation reaction involving replacement of one or more chlorine atoms with hydrogen.

Degradation—The transformation of a compound through biological or abiotic reactions.

Dehalogenation—Replacement of one or more halogens (e.g., chlorine, fluorine, or bromine) with hydrogen atoms.

Dense nonaqueous phase liquid (DNAPL)—A liquid that is denser than water and does not dissolve or mix easily in water (it is immiscible). In the presence of water it forms a separate phase from the water. Many chlorinated solvents, such as trichloroethene, are DNAPLs.

Desorption—Opposite of sorption; the release of chemicals from solid surfaces.

Diffusion—Dispersive process resulting from the movement of molecules along a concentration gradient. Molecules move from areas of high concentration to areas of low concentration.

Dilution—The combined processes of advection and dispersion resulting in a net dilution of the molecules in the groundwater.

Dispersion—The spreading of molecules along and away from the expected groundwater flow path during advection as a result of mixing of groundwater in individual pores and channels.

Dissimilatory—A biochemical process in which an inorganic compound is used for an energy source but is not assimilated into the organism (i.e., as occurs in perchlorate reduction to chloride, the metabolites are all inorganic compounds).

Electron—A negatively charged subatomic particle that may be transferred between chemical species in chemical reactions. Every chemical molecule contains electrons and protons (positively charged particles).

Electron acceptor—Compound that receives electrons (and therefore is reduced) in the oxidation-reduction reactions that are essential for the growth of microorganisms and for bioremediation. Common electron acceptors in the subsurface are oxygen, nitrate, sulfate, iron and carbon dioxide. Chlorinated solvents (e.g., trichloroethene) can serve as electron acceptors under anaerobic conditions.

Electron donor—Compound that donates electrons (and therefore is oxidized) in the oxidation-reduction reactions that are essential for the growth of microorganisms and bioremediation. Organic compounds (e.g., lactate) generally serve as an electron donor during anaerobic bioremediation. Less chlorinated solvents (e.g., VC) can also serve as electron donors. Hydrogen generated in fermentation reactions also can serve as an electron donor.

Emulsified edible oil—A formulation in which an edible oil (such as soybean oil) is dispersed into water (e.g., through stirring or use of homogenizers) to form a mixture of oil droplets in water. Emulsifying the oil greatly improves the distribution of the oil in the subsurface.

Enzyme—A protein created by living organisms to use in transforming a specific compound. The protein serves as a catalyst in the compound's biochemical transformation.

Ex situ—Latin term referring to the removal of a substance from its natural or original position, e.g., treatment of contaminated groundwater aboveground.

Fermentation—Process whereby microorganisms use an organic compound as both electron donor and electron acceptor, converting the compound to fermentation products such as organic acids, alcohols, hydrogen and carbon dioxide.

Geochemical—Produced by, or involving, non-biochemical reactions of the subsurface.

Growth substrate—An organic compound upon which a bacteria can grow, usually as a sole carbon and energy source.

Hydraulic conductivity—A measure of the rate at which water moves through a unit area of the subsurface under a unit hydraulic gradient.

Hydraulic gradient—Change in head (i.e., water pressure) per unit distance in a given direction, typically in the principal flow direction.

Hydrophobic compound—A "water-fearing" compound, such as oil, that has low solubility in water and tends to form a separate phase.

In situ—Latin term meaning "in place"—in the natural or original position, e.g., treatment of groundwater in the subsurface.

Inorganic compound—A chemical that is not based on covalent carbon bonds. Perchlorate is an inorganic compound, as are metals, nutrients such as nitrogen and phosphorus, minerals, and carbon dioxide.

Intrinsic bioremediation—A type of *in situ* bioremediation that uses the innate capabilities of naturally occurring microbes to degrade contaminants without requiring engineering steps to enhance the process.

Intrinsic remediation—*In situ* remediation that uses naturally occurring processes to degrade or remove contaminants without using engineering steps to enhance the process.

Isotope—Any of two or more species of an element in the periodic table with the same number of protons. Isotopes have nearly identical chemical properties but different atomic masses and physical properties. For example, the isotopes chlorine 37 (^{37}Cl) and chlorine 35 (^{35}Cl) both have 17 protons, but ^{37}C has two extra neutrons and thus a greater mass.

Isotope fractionation—Selective degradation of one isotopic form of a compound over another isotopic form. For example, microorganisms degrade the ^{35}Cl isotopes of perchlorate more rapidly than the ^{37}Cl isotopes.

Glossary

Kinetics—Refers to the rate at which a reaction occurs.

Life cycle cost—The overall estimated cost for a particular remedial alternative over the time period corresponding to the life of the program including direct and indirect initial costs plus any periodic or continuing costs of operation and maintenance.

Mass balance—An accounting of the total inputs and outputs to a system. For dissolved plumes, it refers to a quantitative estimation of the mass loading to a dissolved plume and the mass attenuation capacity within the affected subsurface environment.

Mass flux—The rate of mass flow across a unit area (typically measured in grams per square meter per day [$g/m^2/day$]). Typically calculated by integrating measured groundwater contaminant concentrations across a transect. Often used interchangeably with mass discharge or mass loading (expressed in grams per day [g/day] to describe the mass emanating from a source zone or the mass passing a given transect across the plume.

Mass spectrometer—Instrument used to identify the chemical structure of a compound. Usually, the chemicals in the compound are separated beforehand by chromatography.

Mass transfer—The general term for the physical processes involving molecular and convective transport of atoms and molecules within physical systems. In this context, the term refers to the transport of solute mass from the nonaqueous phase (e.g., NAPL) into the aqueous phase. The rate of mass transfer is controlled by the differences in concentrations between the phases, as well as the interfacial tension at the NAPL:water interface.

Metabolic intermediate—A chemical produced by one step in a multistep biotransformation (e.g., chlorite produced during stepwise reduction of perchlorate to chloride).

Metabolism—The chemical reactions in living cells that convert food sources to energy and new cell mass.

Methanogen—A microorganism that exists in anaerobic environments and produces methane as the end product of its metabolism. Methanogens use carbon dioxide, or simple carbon compounds such as methanol, as an electron acceptor.

Methanogenesis—Process of producing methane gas during biological metabolism.

Microcosm—A laboratory vessel set up to resemble as closely as possible the conditions of a natural environment.

Microorganism—An organism of microscopic or submicroscopic size. Bacteria are microorganisms.

Mineralization—The complete degradation of an organic chemical to carbon dioxide, water, and possibly other inorganic compounds.

Monitored natural attenuation (MNA)—Refers to the reliance on natural attenuation processes (within the context of a carefully controlled and monitored site cleanup approach) to achieve site-specific remediation objectives within a time frame that is reasonable compared to that offered by other more active methods.

Natural attenuation—Reduction in the mass, toxicity, mobility, volume or concentration of contaminants in soil and/or groundwater caused by natural processes that act without human intervention. These *in-situ* processes include biodegradation, dispersion, dilution, sorption, volatilization, radioactive decay, and chemical or biological stabilization, transformation, or destruction of contaminants.

Nonaqueous phase liquid (NAPL)—An organic liquid that is maintained as a separate phase from water.

Numerical model—A mathematical model that uses a numerical time-stepping procedure to estimate behavior of a system over time (as opposed to an analytical model). The mathematical solution is represented by a generated table and/or graph. Numerical models require greater computing power, but they can allow more realistic simulations of complex systems.

Oxidation—Transfer (loss) of electrons from a compound, such as an organic contaminant. The oxidation can supply energy that microorganisms use for growth and reproduction. Often (but not always), oxidation results in the addition of an oxygen atom and/or the loss of a hydrogen atom.

Oxygenase—An enzyme that introduces oxygen into an organic molecule.

Passive treatment—*In situ* bioremediation approach in which amendments are added to the subsurface on a one-time, or infrequent basis. Passive treatment relies on the use of slow-release electron donors, which can be injected into the subsurface or placed in trenches or wells.

Perchlorate—An anion consisting of one chlorine atom and four oxygen atoms, with the chlorine atom present at an oxidation state of +7. Perchlorate occurs naturally, and because it is a potent oxidizer, it has also been manufactured and used for solid rocket propellants and explosives.

Permeable reactive barrier—A permeable zone containing or creating a reactive treatment area oriented to intercept and remediate a contaminant plume.

Phytoaccumulation—Plant uptake and retention of a compound within plant tissues.

Phytodegradation—Degradation of a contaminant within a plant.

Phytoremediation—The use of plants and in some cases the associated rhizosphere (root zone) microorganisms for *in situ* remediation of contaminants.

Plume—A zone of dissolved contaminants. A plume usually originates from a contaminant source zone, and extends for some distance in the direction of groundwater flow.

Primary substrates—The electron donor and electron acceptor that are essential to ensure the growth of microorganisms. These compounds can be viewed as analogous to the food and oxygen that are required for human growth and reproduction.

Radius of influence—The radial distance from the center of an injection point or well to the point where there is no significant impact from the injected material.

Reduction—Transfer of electrons to a compound such as oxygen. It occurs when another compound is oxidized.

Glossary

Reductive dechlorination—The removal of chlorine atoms from an organic compound and their replacement with hydrogen atoms (subset of reductive dehalogenation).

Reductive dehalogenation—The process by which a halogen atom (e.g., chlorine or bromine) is replaced on an organic compound with a hydrogen atom. The reactions result in the net addition of two electrons to the organic compound.

Rhizodegradation—Degradation of compounds by organisms living on or near plant roots.

Saturated zone—Part of the subsurface that is beneath the water table and in which the pores are filled with water.

Semi-passive treatment—*In situ* bioremediation approach in which amendments are added to the subsurface intermittently (i.e., at intervals of a few weeks to a few months). Generally, water-soluble compounds serve as the electron donor. The accumulation of biomass can also serve as a longer-term source of electron donors.

Sorption—Collection of a substance on the surface of a solid by physical or chemical attraction. Can refer to either absorption (in which one substance permeates another) or adsorption (surface retention of solid, liquid, or gas molecules, atoms, or ions).

Source zone—A subsurface zone that serves as a reservoir of contaminants that sustains a dissolved plume. The source includes the material that is or has been in contact with the separate phase (DNAPLs for chlorinated solvents), and the source zone mass includes the sorbed and aqueous phase contaminants as well as any residual NAPL.

Stakeholder—A person other than regulators, owners or technical personnel, who has a legitimate interest in a contaminated site.

Substrate—A compound that microorganisms can use in the chemical reactions catalyzed by their enzymes.

Sulfate reducer—A bacterium that converts sulfate to hydrogen sulfide. Because they can act without oxygen, sulfate-reducing bacteria can be important players in the oxygen limited subsurface.

Transcription—Transfer of information in DNA sequences to produce complementary messenger RNA sequences, which are then translated into functional polypeptides and proteins.

Translation—The decoding of messenger RNA to produce specific polypeptides. It occurs after transcription.

Vadose zone—Subsurface solids above the water table, where pores are partially or largely filled with air. Also called the unsaturated zone.

Volatilization—Transfer of a chemical from the liquid to the gas phase (as in evaporation).

Water table—The top of an unconfined aquifer. Indicates the level below which subsurface solids and rock are saturated with water.

TABLE OF CONTENTS

Preface vii

About the Editors xi

About the Authors xiii

External Reviewers xix

Acronyms and Abbreviations xxi

Unit Conversion Table xxv

Glossary xxvii

CHAPTER 1   *IN SITU* BIOREMEDIATION OF PERCHLORATE IN GROUNDWATER: AN OVERVIEW 1
    1.1   Introduction 1
    1.2   How Did Perchlorate Become Such a Problem? 2
            1.2.1   Perchlorate Properties and Behavior in the Subsurface 2
            1.2.2   Production and Disposal 3
            1.2.3   Regulatory History 4
            1.2.4   Evolution of Analytical Capabilities 5
            1.2.5   Evolution of Toxicological Understanding 6
    1.3   How Can *In Situ* Bioremediation Help Solve the Perchlorate Problem? 7
            1.3.1   Treatment Technology Overview 7
            1.3.2   Why Use *In Situ* Bioremediation? 9

CHAPTER 2   DEVELOPMENT OF *IN SITU* BIOREMEDIATION TECHNOLOGIES FOR PERCHLORATE 15
    2.1   Introduction 15
    2.2   Early Discoveries 15
    2.3   Analytical Methods and Pilot Programs 16
    2.4   Ubiquitous Occurrence of Perchlorate Degraders 17
    2.5   Field Demonstrations 18
    2.6   Bioremediation Strategies 19
    2.7   Remediation of Perchlorate in Soil—The New Challenge 21
    2.8   The Challenges Ahead 24

CHAPTER 3   PRINCIPLES OF PERCHLORATE TREATMENT 29
    3.1   Introduction 29
    3.2   Abiotic Remediation Processes 29
            3.2.1   Ion Exchange 29
            3.2.2   Abiotic Reduction Technologies 32
            3.2.3   Overview of Abiotic Processes 34

3.3 Biological Remediation Processes 34
    3.3.1 General Characteristics of DPRB 35
    3.3.2 Diversity of DPRB 36
    3.3.3 Environmental Factors Controlling DPRB Activity 36
    3.3.4 Summary 38
3.4 Challenges Associated with Microbial Perchlorate Reduction 39
    3.4.1 Biofouling and Electron Donor Selection 39
3.5 The Tools Available for Predicting and Monitoring Microbial Perchlorate Reduction 42
    3.5.1 Most Probable Number Counts 43
    3.5.2 Probes to Specific Groups of Perchlorate-Reducing Organisms 44
    3.5.3 Biomarkers for All DPRB 44
    3.5.4 Immunoprobes Specific for DPRB 45
    3.5.5 Use of Stable Isotopes to Identify Perchlorate Source and Monitor Degradation 45
3.6 Enrichment, Isolation, and Maintenance of DPRB 46
    3.6.1 Direct Isolation 46
    3.6.2 Culture Maintenance 47
3.7 Conclusions 47
Appendix 3.1 Medium for Freshwater Perchlorate-Reducing Microorganisms 53

CHAPTER 4 PERCHLORATE SOURCES, SOURCE IDENTIFICATION AND ANALYTICAL METHODS 55

4.1 Introduction 55
4.2 Sources of Perchlorate 55
    4.2.1 Anthropogenic Sources 56
    4.2.2 Natural Sources of Perchlorate 60
4.3 Distinguishing Synthetic from Natural Perchlorate Using Stable Isotope Analysis 62
    4.3.1 Stable Isotope Analysis 62
    4.3.2 Stable Isotope Methods for Perchlorate 63
4.4 Analytical Methods for Perchlorate Analysis 68
    4.4.1 DoD-Approved Analytical Methods 68
    4.4.2 Other Analytical Methods for Perchlorate 70
4.5 Site Characterization for Perchlorate Treatment 73
4.6 Summary 74

CHAPTER 5 ALTERNATIVES FOR *IN SITU* BIOREMEDIATION OF PERCHLORATE 79

5.1 Introduction 79
5.2 Technology Selection Process 81
    5.2.1 *In Situ* Bioremediation 81
    5.2.2 Active Treatment 82
    5.2.3 Semi-Passive Treatment 84
    5.2.4 Passive Treatment 85

Contents xxxvii

5.3	Decision Guidelines	85
	5.3.1 Ability to Meet Management Objectives	86
	5.3.2 Problematic Site Conditions	88
5.4	Summary	88

CHAPTER 6 ACTIVE BIOREMEDIATION 91

6.1	Background and General Approach	91
6.2	When to Consider an Active Treatment System	92
6.3	Treatment System Configurations	93
	6.3.1 Groundwater Extraction and Reinjection (ER)	93
	6.3.2 Horizontal Flow Treatment Wells (HFTWs)	94
6.4	System Applications	95
	6.4.1 Biobarriers	95
	6.4.2 Source Area Treatment	98
6.5	System Design, Operation and Monitoring	100
	6.5.1 Site Assessment Needs	100
	6.5.2 Modeling	101
	6.5.3 Electron Donor	105
	6.5.4 Performance Monitoring	108
	6.5.5 Operational Issues	111
6.6	Case Study: Aerojet Area 20 Groundwater Extraction – Reinjection System	113
	6.6.1 Site Description	113
	6.6.2 Site Geology and Hydrogeology	115
	6.6.3 Pilot Test Design	117
	6.6.4 PTA Installation, Instrumentation and Operation	118
	6.6.5 Baseline Geochemical Characterization	119
	6.6.6 Hydraulic Characterization (Tracer Testing)	121
	6.6.7 System Operation	123
	6.6.8 Demonstration Results	124
	6.6.9 Pilot Test Conclusions	130
6.7	Summary	131

CHAPTER 7 SEMI-PASSIVE *IN SITU* BIOREMEDIATION 135

7.1	Background	135
	7.1.1 What is a Semi-Passive Approach?	135
	7.1.2 When to Consider a Semi-Passive Approach	137
	7.1.3 Advantages and Limitations Relative to Other Approaches	137
	7.1.4 Technology Maturity	138
7.2	System Design, Operation, and Monitoring	139
	7.2.1 Typical System Design	139
	7.2.2 Site Assessment Needs	141
	7.2.3 Groundwater Modeling	142
	7.2.4 Tracer Testing	142
	7.2.5 Operation and Maintenance	143

| | | 7.2.6 | Monitoring .... 143 |
| | | 7.2.7 | Health and Safety .... 144 |

7.3 Case Study: Semi-Passive Bioremediation of Perchlorate at the Longhorn Army Ammunitions Plant... 144
    7.3.1 Demonstration Test Procedures 144
    7.3.2 Demonstration Test Results 147
    7.3.3 Conclusions of Case Study 152

7.4 Summary 153

CHAPTER 8   PASSIVE BIOREMEDIATION OF PERCHLORATE USING EMULSIFIED EDIBLE OILS 155

8.1 Introduction 155
8.2 Design of Passive Bioremediation Systems 156
    8.2.1 Treatment System Configurations 156
    8.2.2 Planning and Design of Passive Bioremediation Systems 158
    8.2.3 Site Characterization Requirements 163
    8.2.4 Monitoring 165
8.3 Case Study 166
    8.3.1 Demonstration Design 167
    8.3.2 Monitoring 169
    8.3.3 Results 169
8.4 Tools and Resources 171
8.5 Factors Controlling Cost and Performance 171
8.6 Summary 172

CHAPTER 9   PERMEABLE ORGANIC BIOWALLS FOR REMEDIATION OF PERCHLORATE IN GROUNDWATER 177

9.1 Introduction 177
    9.1.1 Applications to Date 177
    9.1.2 Technology Description 178
9.2 Site Suitability 179
    9.2.1 Land Use and Infrastructure 179
    9.2.2 Contaminant Concentration and Distribution 180
    9.2.3 Hydrogeology 181
    9.2.4 Geochemistry 181
    9.2.5 Co-Contaminants 181
9.3 Design of Permeable Biowalls 181
    9.3.1 Site-Specific Hydrogeology and Contaminant Distribution 182
    9.3.2 Dimensions, Configuration and Residence Time 182
    9.3.3 Biowall Materials 183
    9.3.4 Recharge Options and Alternative Configurations 183
    9.3.5 Regulatory Compliance 184
9.4 Biowall Installation and Construction 185
    9.4.1 Construction Methods 185
    9.4.2 Quality Assurance/Quality Control 186
    9.4.3 Waste Management Plan 186

9.5	Performance Monitoring	187
	9.5.1 Biogeochemistry	187
	9.5.2 Perchlorate Degradation	188
	9.5.3 Sustaining the Reaction Zone	188
9.6	Biowall System Costs	189
	9.6.1 Installation and Trenching Costs	189
	9.6.2 Operations and Monitoring Costs	189
	9.6.3 Summary of Life Cycle Costs	190
9.7	Case Study: Former NWIRP McGregor, McGregor, Texas	191
	9.7.1 Fast Track Cleanup and Innovative Technology Implementation	191
	9.7.2 *Ex Situ* Groundwater Treatment	192
	9.7.3 *In Situ* Groundwater Treatment	192
	9.7.4 Natural Attenuation in Groundwater	195
	9.7.5 *Ex Situ* Soil Treatment	195
	9.7.6 Operations and Maintenance	196
9.8	Summary	197

CHAPTER 10   COST ANALYSIS OF *IN SITU* PERCHLORATE BIOREMEDIATION TECHNOLOGIES 199

10.1	Background	199
10.2	Costing Methodology	200
10.3	Template Site Characteristics and Variations Considered	202
10.4	Cost Estimates for Base Case Site Characteristics	206
10.5	Impacts of Changes In Site Characteristics on Costs	211
	10.5.1 Case 2: Accelerated Clean Up	211
	10.5.2 Cases 3 and 4: Reduced and Elevated Concentrations of Perchlorate	211
	10.5.3 Cases 5 and 6: Lower and Higher Electron Acceptor Concentrations	213
	10.5.4 Cases 7 and 8: Low and High Groundwater Seepage Velocities	214
	10.5.5 Case 9: Deep Groundwater	214
	10.5.6 Cases 10 and 11: Thin and Thick Saturated Vertical Intervals	215
	10.5.7 Cases 12 and 13: Narrow and Wide Plumes	215
10.6	Summary	215

CHAPTER 11   EMERGING TECHNOLOGIES FOR PERCHLORATE BIOREMEDIATION 219

11.1	Introduction	219
11.2	Monitored Natural Attenuation	219
	11.2.1 Basis	219
	11.2.2 Advantages and Limitations	222
	11.2.3 Case Studies	222
11.3	Phytoremediation	224
	11.3.1 Basis	224
	11.3.2 Status	227
	11.3.3 Advantages and Limitations	228
	11.3.4 Case Studies	228

11.4 Vadose Zone Bioremediation 229
    11.4.1 Basis 231
    11.4.2 Status 233
    11.4.3 Advantages and Limitations 234
    11.4.4 Case Studies 234

Index **245**

LIST OF FIGURES

Figure 2.1	Perchlorate biodegradation pathway	16
Figure 2.2	(a) Layout and (b) photograph of pilot test infrastructure at Area 20, Aerojet Superfund site, Sacramento, California	19
Figure 2.3	(a) Electron donor addition regime and (b) perchlorate biodegradation results in groundwater at Area 20, Aerojet Superfund site, Sacramento, California	20
Figure 2.4	(a) Photograph and (b) results of anaerobic composting of perchlorate-impacted soil at the Aerojet Superfund site, Sacramento, California	22
Figure 2.5	(a) Photograph and (b) results of full-scale anaerobic composting of perchlorate-impacted soil in Santa Clara, California	23
Figure 4.1	Comparison of $\delta^{37}Cl$ and $\delta^{18}O$ for various sources of synthetic perchlorate and for natural perchlorate derived from the Atacama Desert of Chile	65
Figure 4.2	Comparison of $\delta^{37}Cl$ and $\delta^{18}O$ for three sources of synthetic perchlorate	66
Figure 4.3	Comparison of $\Delta^{17}O$ and $\delta^{18}O$ for synthetic perchlorate and natural perchlorate derived from the Atacama Desert of Chile	67
Figure 5.1	Alternatives for *in situ* bioremediation of perchlorate	80
Figure 6.1	Schematic of an active ER system consisting of two extraction wells and a single injection well	94
Figure 6.2	Schematic of an active HFTW system consisting of two treatment wells	95
Figure 6.3	Schematic showing different configurations of active biobarrier systems: (a) upgradient extraction and downgradient injection wells; (b) upgradient injection and downgradient extraction wells; (c) alternating injection and extraction wells in a line perpendicular to groundwater flow	96
Figure 6.4	Photograph of the HFTW system layout at Aerojet during construction	97
Figure 6.5	Photograph of the field pilot system installed to treat perchlorate in a source area at the Indian Head Division, Naval Surface Warfare Center (IHDIV), Indian Head, MD	99
Figure 6.6	Perchlorate levels in groundwater during the IHDIV pilot test: (a) test plot monitoring wells receiving electron donor and buffer; (b) control plot monitoring wells receiving no amendments	100
Figure 6.7	(a) Schematic showing the conceptual geologic layering pattern used in the fate and transport model developed for the Aerojet HFTW system. (b) Comparison of model-simulated and measured water elevations in Well 3633 during pump testing at Aerojet	104
Figure 6.8	Plan view of streamlines showing the simulated capture of groundwater in layer 3 of the conceptual site model at a pumping rate of 26.5 L/min and 10 m HFTW spacing	105

Figure 6.9	Perchlorate biodegradation in laboratory microcosms amended with different electron donors	107
Figure 6.10	Map of the Pilot Test Area (PTA)	114
Figure 6.11(a)	Geologic cross sections of the PTA oriented perpendicular to groundwater flow	115
Figure 6.11(b)	Geologic cross sections of the PTA oriented parallel to groundwater flow	116
Figure 6.12	Groundwater flow simulation at PTA scale	118
Figure 6.13	Plan view of the groundwater extraction, injection and monitoring wells in the PTA	119
Figure 6.14	Schematic of the active treatment system	120
Figure 6.15	Bromide concentrations in downgradient monitoring wells 3600, 3617 and 100 (11, 15, and 20 m from the injection well, respectively) with time	122
Figure 6.16	(a) Oxidation-reduction potential (ORP) and (b) dissolved oxygen (DO) concentrations in the influent water and in monitoring wells 3600, 3617, and 100 over the duration of the pilot test	125
Figure 6.17	Perchlorate concentrations in the influent water and in monitoring wells 3600, 3617, and 100 during the pilot test	126
Figure 6.18	Concentrations of trichloroethene and daughter products *cis*-1,2-dichloroethene, vinyl chloride and ethene as a function of time in the influent water and in monitoring wells 3601, 3600, 100, 3618, and transgradient well 3617	127
Figure 6.19	Acetate concentrations in the influent water and in monitoring wells 3600, 3617, and 100 during the pilot test	129
Figure 6.20	Dissolved iron (Fe) concentrations in the influent water and in monitoring wells 3600, 3617, and 100 during the pilot test	129
Figure 6.21	Dissolved manganese (Mn) concentrations in the influent water and in monitoring wells 3600, 3617, and 100 during the pilot test	130
Figure 7.1	Plan view of groundwater flow for semi-passive biobarrier system	136
Figure 7.2	Layout of demonstration test area at LHAAP	145
Figure 7.3	Perchlorate concentrations over time in monitoring transect one	149
Figure 7.4	Perchlorate concentrations over time in monitoring transect four	150
Figure 7.5	ORP over time in monitoring transect one	151
Figure 7.6	ORP over time in monitoring transect four	152
Figure 8.1	Use of emulsified oils to treat contaminated groundwater in (a) source areas using grid injection and (b) plumes using permeable reactive barriers	157
Figure 8.2	Cross-section showing zones of higher permeability	157
Figure 8.3	Example cost analysis for a PRB with various injection well spacings (1 m = 3.28 ft)	162

List of Figures xliii

Figure 8.4	Comparison of perchlorate concentrations in microcosms after 2 and 14 days illustrating the effect of different treatments on perchlorate biodegradation	168
Figure 8.5	Reductive dechlorination of 1,1,1-TCA to 1,1-Dichloroethane (1,1-DCA) and Chloroethane (CA) in microcosms constructed with aquifer material, groundwater and emulsified soybean oil	168
Figure 8.6	Change in perchlorate concentration versus time in monitoring wells upgradient and downgradient of the emulsified oil barrier	169
Figure 8.7	Change in 1,1,1-TCA and degradation products versus time in Monitoring Well SMW-6 showing effect of emulsified oil addition in stimulating reductive dechlorination	170
Figure 9.1	Biowall conceptual design	179
Figure 9.2	Continuous one-pass trencher in operation at Ellsworth Air Force Base, South Dakota	185
Figure 9.3	*Ex situ* bioreactor at NWIRP McGregor, McGregor, Texas	192
Figure 9.4	Construction of biowalls at NWIRP McGregor	193
Figure 9.5	Typical biowall construction specifications at NWIRP McGregor, McGregor, Texas	194
Figure 10.1	Base case plume and biobarrier or biowall configuration	204
Figure 11.1	Perchlorate concentrations in microcosms containing groundwater collected from Naval Surface Warfare Center, Indian Head, Maryland	223
Figure 11.2	Processes that may occur during phytoremediation of perchlorate	225
Figure 11.3	Concentrations of perchlorate, chlorate and chloride in growth solutions from hydroponics bioreactors	226
Figure 11.4	Effectiveness of stimulation and enhancement of rhizodegradation of perchlorate during phytoremediation	227
Figure 11.5	Full-scale *ex situ* soil bioremediation in (a) Ag-Bag® and (b) concrete containment cells	230
Figure 11.6	Breakthrough curve of ethanol transport through soil column	232
Figure 11.7	Schematics comparing gaseous electron donor injection and aerobic bioventing	232
Figure 11.8	Preliminary results for *in situ* reduction of nitrate concentration in vadose zone soil	233

LIST OF TABLES

Table 2.1	Perchlorate Biodegradation in Microcosms for SERDP Test Sites	18
Table 4.1	Blasting Agents and Explosives Containing Perchlorate	58
Table 4.2	DoD Recommended Methods for Perchlorate Analysis	69
Table 4.3	Other Available Methods for Perchlorate Analysis	70
Table 4.4	Additional Chemical and Geochemical Parameters to Measure during Site Characterization	73
Table 5.1	Characteristics of Alternative Approaches for Implementing *In Situ* Bioremediation for Perchlorate in Groundwater	80
Table 5.2	Advantages and Limitations of Perchlorate *In Situ* Bioremediation Approaches	83
Table 5.3	Comparisons of Ability of Perchlorate *In Situ* Bioremediation Approaches to Meet Typical Management Objectives	86
Table 6.1	Cost Comparison of Three Common Soluble Electron Donors Used for Perchlorate Treatment	108
Table 6.2	Typical Performance Monitoring Activities for an Active Treatment System	109
Table 6.3	Typical Monitoring Parameters for an Active Perchlorate Treatment System Utilizing Ethanol as an Electron Donor and with One or More VOCs as Co-Contaminants	110
Table 6.4	Baseline Geochemical Parameters and Contaminant Concentrations at the Aerojet Pilot Test Area	121
Table 7.1	Summary of Groundwater Monitoring Results at Site 16 Landfill, LHAAP, Karnack, Texas	148
Table 8.1	Observed Oil Droplet Retention by Aquifer Solids	159
Table 8.2	Estimated Volumes of Oil and Water Required for Treatment of 30-m x 30-m (100-ft x 100-ft) Area	161
Table 9.1	Suitability of Site Characteristics for Biowalls	180
Table 9.2	Characteristic of Materials Used for Permeable Biowalls	184
Table 9.3	Biowall Technology Costs, Site BG05, Ellsworth AFB, South Dakota	190
Table 10.1	Summary of Site Characteristics and Design Parameters for Biological Treatment of Perchlorate-Impacted Groundwater	203
Table 10.2	Cost Components for Active Biobarrier Treatment of Perchlorate-Impacted Groundwater	207
Table 10.3	Cost Components for Semi-Passive Biobarrier Treatment of Perchlorate-Impacted Groundwater	207

Table 10.4	Cost Components for Passive Injection Biobarrier Treatment of Perchlorate-Impacted Groundwater	208
Table 10.5	Cost Components for Passive Trench Biowall Treatment of Perchlorate-Impacted Groundwater	208
Table 10.6	Cost Components for Extraction and Treatment of Perchlorate-Impacted Groundwater	208
Table 10.7	Summary of Capital Costs and NPV of Costs for Operation, Maintenance and Monitoring for Biological Treatment of Perchlorate-Impacted Groundwater	209
Table 10.8	Summary of Impact of Site Characteristics and Design Parameters on Costs for Biological Treatment of Perchlorate-Impacted Groundwater	212
Table 11.1	Summary of Vadose Zone Bioremediation Case Studies by Location	234

CHAPTER 1

IN SITU BIOREMEDIATION OF PERCHLORATE IN GROUNDWATER: AN OVERVIEW

Hans F. Stroo,[1] Raymond C. Loehr[2] and C. Herb Ward[3]

[1]HydroGeoLogic, Inc., Ashland, OR 97520; [2]University of Texas, Austin, TX 78705; [3]Rice University, Houston, TX 77005

1.1 INTRODUCTION

Perchlorate contamination is an explosive issue in many ways. It burst into prominence as a pollutant very rapidly in the late 1990s, and it has remained a hot topic for the past decade. It has also ignited an enormous amount of controversy and public interest (Cheremisinoff, 2001; Cunniff et al., 2006). And of course, its primary use has been as a component of solid rocket propellants (ITRC, 2005), and the public understandably gets concerned about "rocket fuel in drinking water" (EWG, 2007).

The reasons for the explosion of interest in perchlorate include recent advances in both analytical chemistry and in our understanding of perchlorate's health impacts. The advances in chemistry have allowed detection at low part-per-billion (microgram per liter [µg/L]) concentrations (Urbansky, 2000), and the toxicological research has suggested that such concentrations may be of concern, particularly to developing fetuses and infants (USEPA, 2002). However, there has been a great deal of uncertainty, and controversy, regarding perchlorate's impacts on human health, and on other organisms (Kendall and Smith, 2006; NRC, 2005).

In addition, there have been several high-profile cases of perchlorate contamination of surface waters and drinking water supplies in major metropolitan areas (Gullick et al., 2001). The responsible parties, notably the U.S. Department of Defense (DoD), have had to respond quickly to regulatory and public pressure to prevent further exposures and clean up contaminated sites.

The chemical properties of perchlorate make it an extremely difficult contaminant to contain and treat. In particular, it is highly mobile in groundwater, and it is resistant to biodegradation under aerobic conditions. As a result, many contaminant plumes have spread over large areas, potentially impacting large numbers of people. Perchlorate in groundwater is also very expensive to treat, particularly with the technologies that were initially available in the late 1990s (USEPA, 2005; Roote, 2001).

Public concerns over groundwater contamination cases, and the enormous potential financial liability for the DoD, has led to a rapid and focused research effort on perchlorate by the DoD's Strategic Environmental Research and Development Program (SERDP), and the related Environmental Security Technology Certification Program (ESTCP). This research has been focused on three areas: (1) the development and testing of cost-effective remedial technologies for drinking water, soils, and groundwater; (2) improved methods to characterize perchlorate-contaminated sites; and (3) an improved understanding of the human and ecological health effects due to perchlorate exposure.

Research on treatment technologies has emphasized more cost-effective *ex situ* treatment, as well as *in situ* treatment approaches that can eliminate the need for extraction and aboveground treatment (Logan, 2001). The potential value of *in situ* bioremediation for treating perchlorate in groundwater was recognized early (Urbansky, 1998), and research has proven that it can help reduce further exposures to perchlorate, and also save DoD significant resources in dealing with this issue. Several approaches for implementing *in situ* bioremediation have been successfully demonstrated and accepted for full-scale cleanups. Other approaches are still in development, but seem very promising.

This chapter provides information on how perchlorate contamination of groundwater has become a significant problem, and why *in situ* bioremediation can be a promising technology for solving this problem. It is intended to introduce readers to the fundamental issues involved in perchlorate contamination and remediation, as well as to provide a historical perspective on the evolution of our present knowledge and capabilities.

1.2 HOW DID PERCHLORATE BECOME SUCH A PROBLEM?

1.2.1 Perchlorate Properties and Behavior in the Subsurface

Perchlorate (ClO_4^-) is the oxidation product of chlorate (ClO_3^-). It is a negatively charged ion that consists of one chlorine atom and four oxygen atoms that form a weak association with a positively charged ion such as ammonium. As a result, perchlorate salts are extremely soluble in water and in polar organic solvents. The order of solubility of the common perchlorate salts is sodium>lithium>ammonium>potassium. Because perchlorate salts are so soluble, the health risks associated with them are considered equivalent to those associated with perchlorate itself (NRC, 2005).

The high solubility of perchlorate salts, and limited sorption to solids, results in very low retardation factors in the subsurface. As a result, when perchlorate is released to groundwater, it can spread over large distances and impact large volumes of groundwater (Xu et al., 2003).

Perchlorate has excellent oxidizing ability under some conditions. However, the activation energy required to initiate the chemical reaction is very high. The high required activation energy for oxidation and the high solubility of perchlorate salts results in both the stability and mobility of perchlorate in the environment. The high activation energy also leads to the nonreactivity of perchlorate in the human body, from which it is excreted virtually unchanged as indicated by absorption, distribution, metabolism and elimination studies (NRC, 2005).

The primary exposure pathway of concern for perchlorate is ingestion because of its rapid uptake from the gastrointestinal tract. Dermal uptake is minimal. In addition, perchlorate sorbs only weakly to mineral and organic subsurface materials. Perchlorate salts are considered non-volatile and their low vapor pressure results in negligible inhalation of perchlorate vapors (NRC, 2005).

1.2.2 Production and Disposal

1.2.2.1 History of Use

The earliest uses of perchlorate were for the production of fireworks. The first fireworks used black powder, which contained saltpeter (potassium nitrate). Shortly after 1800, potassium chlorate was substituted for some or all of the potassium nitrate in the explosive mixture for economic reasons. Although it is not clear when perchlorate salts were first deliberately used, perchlorate was not manufactured industrially until the 1890s, and commercial production in the United States started in the early 1900s (ITRC, 2005). Chlorate and perchlorate salts are still a key part of many fireworks mixtures (Aziz et al., 2006).

Perchlorate was also used unknowingly for several decades, because it is a constituent of natural caliche deposits that are high in nitrate content, particularly nitrate ores found in the deserts of Chile. We now know that perchlorate is produced naturally, and it can persist for thousands of years under arid conditions (Dasgupta et al., 2005). The Chilean nitrate ores were first imported into the United States in 1857, and were used for black powder production (ITRC, 2005). They also were widely used as fertilizers, starting in the early 1900s (Aziz et al., 2006).

The high oxidizing ability of perchlorate eventually led to its use as a component of propellants and explosives, with a rapid increase in its use during the 1940s. Before the 1940s, annual global production of perchlorate was estimated to be 1,800 tons. In the mid 1940s, annual perchlorate production increased to 18,000 tons because of demand by the military and aerospace industry (NRC, 2005). Similar amounts were produced into the 1990s, based on the total estimated production capacity (Mendiratta et al., 1996). It is still produced in the United States, but current production amounts are difficult to estimate because ammonium perchlorate is classified as a strategic compound.

Perchlorates have been used for propellants and explosives because they are more stable and result in increased performance compared to nitroglycerin and nitrocellulose propellants (Cunniff et al., 2006). They also increase the range of tactical and strategic missiles. Approximately 90% of all perchlorate salts are manufactured as ammonium perchlorate for use in rocket and missile propellants (Xu et al., 2003). The fuel for rockets can contain as much as 70% ammonium perchlorate, as finely ground crystalline particles distributed in a polymer matrix. Ammonium perchlorate is the oxidizer of choice for solid propellants because of its performance, safety and ease of manufacturing and handling, although the increased environmental concern regarding perchlorate in water supplies has led to efforts to find alternatives (Dewey, 2007).

Though the dominant use of the perchlorate produced in the United States has been for rocket fuel production, perchlorate also is used in the production of explosives, pyrotechnics, and blasting formulations (ITRC, 2005). It also has been used medically in the past to treat patients with hyperthyroidism (ITRC, 2005). Chapter 4 of this volume provides further information on perchlorate uses and source identification techniques.

1.2.2.2 Disposal Practices

The majority of the anthropogenic perchlorate found in ground and surface waters appears to result from historical disposal practices used by the aerospace and ordnance industries, the military and chemical manufacturers. During the period of the 1950s through the mid 1970s, solid perchlorate-containing fuels requiring disposal often were burned in open pits and open detonation areas. Also during this period, wastewaters containing perchlorate were released to surface soils or discharged into earthen lagoons or evaporation ponds (Cheremisinoff, 2001; Hatzinger, 2005).

Such wastes and wastewaters result from the preparation and processing of the solid rocket fuel and the cleanup of wastes from processing equipment and rocket or ordnance firings. The periodic replacement and use of solid propellant has resulted in the discharge of more than 15.9 million kilograms (kg) (35 million pounds [lb]) of perchlorate salts into the environment since the 1950s (Xu et al., 2003).

1.2.3 Regulatory History

The regulatory history related to control and treatment of perchlorate in groundwater (up to 2004) has been summarized in the National Research Council (NRC) report that dealt with the health implications of perchlorate (NRC, 2005). The following brief chronology of major events and regulatory history is excerpted from that NRC report.

In 1985, the U.S. Environmental Protection Agency (USEPA) Regional Office in California (Region 9) raised concern about potential perchlorate contamination at Superfund sites in the San Gabriel Valley (Takata, 1985). At the time, no validated analytical method was available to measure low perchlorate concentrations and little information on the possible health effects of perchlorate was available. As a result, attention was focused on other chemicals at these Superfund sites.

In the early 1990s, detection of part-per-million levels of perchlorate in a drinking water supply aquifer near Sacramento, California, led the USEPA's Superfund team to request the country's first evaluation of perchlorate's toxicity (USEPA, 2002). In 1992, a provisional perchlorate reference dose (RfD) was issued by the USEPA Superfund Technical Support Center, and in 1995, a revised provisional RfD was released by the USEPA (USEPA, 2002). These RfDs were considered provisional because they had not had rigorous internal or external peer review. However, these provisional RfDs were used to derive guidance levels for groundwater remediation goals.

In March 1997, the USEPA convened an independent peer review of the provisional RfD. That peer review concluded that the then available scientific data base was insufficient to conduct a credible quantitative risk analysis. In May 1997, another independent peer review panel met and developed a testing strategy to address data gaps and reduce uncertainties regarding possible health effects of low concentration perchlorate ingestion (USEPA, 2002).

In December 1998, the USEPA released its first formal draft risk assessment of perchlorate (USEPA, 1998b). A USEPA-sponsored peer review of that draft risk assessment occurred in February 1999 (USEPA, 2002). The peer review panel made a number of suggestions that included completion of studies recommended earlier, additional studies to evaluate the effects of perchlorate on fetal development and a review of existing thyroid histopathology data.

Also in 1998, perchlorate was placed on the USEPA final version of the Contaminant Candidate List (CCL) (USEPA, 1998a). The CCL identifies unregulated contaminants that may pose a public health concern in drinking water. Chemicals on the CCL are then considered for regulation. To determine the extent of perchlorate contamination of the national drinking water supply, in 2001, monitoring of perchlorate in all large public water systems and in a representative sample of small public water systems became mandatory (USEPA, 2004).

In 2002, the USEPA issued a revised draft risk assessment that incorporated revisions suggested by the 1999 peer review panel as well as new data that had been generated as of fall 2001 (USEPA, 2002). The USEPA also convened a new risk assessment panel in March 2002 to review the revised draft risk assessment (USEPA, 2002). However, as of 2008, no national drinking water standard for perchlorate exists and the concentration at which a standard should be set continues to be hotly debated (Sass, 2004). In view of the controversy surrounding the concentration at which perchlorate should be regulated, the NRC was asked by the USEPA, DoD, the U.S. Department of Energy, and the National Aeronautics and Space Administration to independently assess the adverse health effects of perchlorate ingestion from clinical, toxicological and public health perspectives.

In completing its study, the NRC Committee did comment on the previous perchlorate risk assessments conducted. The Panel also suggested specific scientific research that could reduce the uncertainty in the understanding of human health effects associated with ingestion of low concentrations of perchlorate (NRC, 2005).

After reviewing the available scientific information, the NRC Committee concluded that a perchlorate RfD of 0.0007 milligrams per kilogram (mg/kg) body weight (0.7 micrograms per kilogram [µg/kg]) per day should be protective of the health of even the most sensitive populations. The Committee also acknowledged that the RfD may need to be adjusted upward or downward on the basis of future research (NRC, 2005).

The RfD of 0.7 µg/kg/day corresponds to a "Drinking Water Equivalent Level" of 24.5 µg/L, after accounting for average body weight and water consumption rate, and incorporating a safety factor. Although to date (2008) no federal drinking water standard for perchlorate has been established, several states have set their own advisory levels. These advisory levels have ranged from 1 to18 µg/L (Hatzinger, 2005). Massachusetts promulgated the first state drinking water standard in 2006, at 2 µg/L (MADEP, 2006), and California recently established a drinking water standard of 6 µg/L (CDHS, 2007).

1.2.4 Evolution of Analytical Capabilities

The extent of the perchlorate problem was not recognized when the initial regulatory criteria were being developed, because of the limitations of the available analytical methods. Perchlorate can be detected by many methods, such as ion-selective electrodes, ion chromatography, capillary electrophoresis, high performance liquid chromatography and spectrophotometry (Xu et al., 2003). But until the mid 1990s, the analytical methods available were limited to detection limits of 400 µg/L, even though the proposed allowable concentrations were 100 times lower (Urbansky, 2000). To solve this problem, the California Department of Health Services developed an ion chromatography (IC) method for drinking water in 1997, and showed that it was capable of detecting perchlorate concentrations as low as 6 µg/L (CDHS, 1997).

Ion chromatography has become the standard method for analysis of perchlorate in water. The original IC method was revised by the USEPA (USEPA, 1999) and published as EPA Method 314.0. Method 314.0 has a widely achievable detection limit of 4 µg/L. The use of this method, and a concerted effort to identify potential exposures, has led to a rapid increase in the number of known perchlorate sites.

However, there have long been concerns regarding this method, and several other methods have been developed to improve perchlorate analysis (ITRC, 2005). Notably, there have been reports of interferences leading to both false positives and false negatives. As a result of the recognized deficiencies in the standard method, research is underway to develop reliable field tests and analyses that could provide greater specificity and lower detection limits for the presence of perchlorate (Mosier-Boss, 2006).

1.2.5 Evolution of Toxicological Understanding

Most of the medical knowledge available on perchlorate toxicity has resulted from its use as an antithyroid agent in the treatment of hyperthyroidism. Perchlorate had been used for this purpose because it can act to reduce thyroid iodide uptake, thus decreasing the production of thyroid hormones. However, due to its potential toxicity and side effects, perchlorate use for this purpose has been replaced by other drug treatments (Herman and Frankenberger, 1998; NRC, 2005).

Although perchlorate is rapidly eliminated from the body, environmental concerns exist because its ionic radius and charge are similar to that of iodide. As a result, perchlorate can competitively block thyroid iodide uptake (Clewell et al., 2004). Iodine is necessary for the production of thyroid hormones, which regulate metabolism in all cells. Hence perchlorate can interfere with the normal functions of the thyroid, and alter the concentrations of key hormones. Exposure to perchlorate can result in varying degrees of cognitive impairment, though some have proposed other health effects, including carcinogenicity (NRC, 2005; Mattie et al., 2006).

Thyroid hormones are critical determinants of growth and development in fetuses, infants and young children. These groups are considered sensitive populations for thyroid deficiency problems. Adolescents and adults who have compromised thyroid function and people who are iodine-deficient are also potentially sensitive populations (NRC, 2005). In addition, perchlorate impacts the normal development of other animal species, so there has been considerable research on its ecotoxicity (Kendall and Smith, 2006). Many animals appear more susceptible to perchlorate than humans, partly because humans have relatively high stores of hormones within the thyroid, and humans may also bind the hormones in the blood more tightly, so that they are lost less rapidly (Smith, 2006).

It has been difficult to develop protective criteria for perchlorate for several reasons. Perchlorate's effects, particularly at relatively low doses, are subtle, complex and difficult to detect (NRC, 2005). Perchlorate is flushed from the system relatively quickly, and its effects are not only dependent upon the dose and duration of exposure, but on the organism's iodide status (Merrill et al., 2005). Databases are not available to allow robust epidemiological studies of humans, and there is little information available on perchlorate effects on potential ecological receptors.

Also, there has been considerable uncertainty and controversy regarding the available underlying toxicological studies. The quality and validity of the available data have been repeatedly questioned, as well as the interpretation of published results. It has proven difficult to even define what constitutes an adverse effect. Finally, the uncertainty factors that should be applied to the data when developing risk-based criteria have been hotly debated. Although the NRC report resolved many of the issues, the controversy is not completely over, and surprising results are still being observed (Blount et al., 2006).

1.2.5.1 Magnitude of the Problem

Although perchlorate contamination of groundwater was discovered in wells at California Superfund sites in 1985, nationwide perchlorate contamination of water sources was not recognized until 1997 (NRC, 2005). Based on sampling conducted by the USEPA, as of 2004, over 11 million people in the United States had perchlorate in their public drinking water supplies at concentrations of 4 µg/L or higher. In addition, as of September 2004, environmental releases of perchlorate have been confirmed in 35 states (NRC, 2005).

More specifically, as of late 2004, 361 out of approximately 6,800 public drinking water sources in California have tested positive for perchlorate. In addition, the Lower Colorado River, which supplies drinking water to about 15 million people, contains measurable concentrations of perchlorate at certain times of the year (Hatzinger, 2005).

Elevated perchlorate concentrations have been found in monitoring wells associated with Superfund sites and groundwater and surface water not directly associated with drinking water supplies. Once considered to be mainly a water contaminant, perchlorate also has been detected in soils, plants, foods and human breast milk (NRC, 2005; Gu and Coates, 2006). Recent evidence suggests that the primary exposure to perchlorate in the United States is through the consumption of food (USFDA, 2007). Blount et al. (2007) have conducted the most extensive characterization of background exposure to perchlorate to date. Using a subset of the National Health and Nutrition Examination Survey, Blount et al. (2007) found that perchlorate was present in all of the urine samples tested (a total of 2,820 United States residents), and they estimated median and 95th percentile exposures of 0.064 and 0.234 µg/kg body weight/day.

Natural sources of perchlorate have been recently recognized, and researchers have developed the ability to distinguish among natural and anthropogenic sources associated with propellant manufacturing and use (Böhlke et al., 2005). In particular, the long-term use of Chilean nitrate ores (discussed earlier) led the USEPA to assess the potential contribution of fertilizers to perchlorate contamination. That study concluded that fertilizer use is not a major source of perchlorate contamination of water (NRC, 2005).

1.3 HOW CAN *IN SITU* BIOREMEDIATION HELP SOLVE THE PERCHLORATE PROBLEM?

The reasons for the rapid development and application of *in situ* bioremediation for perchlorate include the advantages of the technology itself, as well as the deficiencies of other available treatment technologies. Briefly, available technologies have proven to be relatively expensive, and generally require extraction of the groundwater or treatment at a point of exposure (e.g., at a wellhead). However, *in situ* bioremediation can provide a cost-effective method to treat contamination before it spreads to drinking water wells or other receptors, and it can greatly decrease the life-cycle costs for managing a perchlorate contaminated site. The following sections briefly describe the available technologies, and then summarize the rationale for using *in situ* bioremediation.

1.3.1 Treatment Technology Overview

Several technologies have been used to treat perchlorate contamination of soils and waters (USEPA, 2005; ITRC, 2008; Logan, 2001). Initially, *ex situ* technology at the wellhead was used to treat perchlorate contamination of drinking water wells. Perchlorate was removed from potable water by pumping through granular activated carbon (GAC) or ion-exchange resin developed for removal of anions.

There are two types of conventional ion exchange systems: (1) single-pass systems, in which the resin is loaded with perchlorate (and other anions such as nitrate and sulfate) and then properly disposed; and (2) regenerable systems, in which the perchlorate is stripped from the resin and the resin reused. Regeneration yields a brine solution that must be disposed, or treated by physical, chemical or biological technologies. The most common regeneration technology is thermal treatment, although anaerobic bioremediation also has been tested and is potentially more cost-effective (Gingras and Batista, 2002).

All conventional ion exchange technologies have proven to be extraordinarily costly, partly because other anions are also removed, often at much higher concentrations than the perchlorate. As the extent of perchlorate contamination was realized, efforts were initiated to develop less expensive approaches. One notable improvement in ion exchange technology has been the development of highly perchlorate-specific resins. These selective resins have included so-called strong base and weak base resins (ITRC, 2008). The former have been more widely used to date, although the latter may prove to be more cost-competitive.

GAC has proven to be expensive as well, because perchlorate is highly water-soluble and adsorbs poorly to GAC, so that it is not efficiently removed. Hence, GAC columns have a short useful lifetime. However, GAC can be "tailored" to increase the positive charge and greatly increase the efficiency of perchlorate removal (Parette and Cannon, 2006). As with ion exchange resins, disposal or regeneration is required after the material is exhausted.

Membrane treatment technologies can also remove perchlorate from water. Reverse osmosis (RO) has been used in some cases. RO is a physical separation technology that utilizes pressure gradients to drive water through semi-permeable membranes. RO also is not selective for perchlorate. Nanofiltration and ultrafiltration have been tested, and can work for some waters, but have proven ineffective in the presence of significant amounts of other ions. Electrodialysis is a membrane technology that has been pilot-tested, and can remove perchlorate at low concentrations. However, electrodialysis has not been used at full-scale to date (Zhou et al., 2006).

The high cost for physical-chemical treatment technologies has prompted research and development of alternative technologies, notably biological treatment. Anaerobic bioreactors, including both fixed-bed and fluidized-bed reactors, have been tested and employed commercially for *ex situ* treatment of groundwaters. However, biological treatment of perchlorate for drinking water use has been slow to gain regulatory acceptance, largely due to concerns about the robustness and reliability of the technology (ITRC, 2005).

Ex situ treatment is generally far more costly than *in situ* treatment, for several reasons. Perchlorate contaminated groundwater must be extracted via wells, requiring considerable capital and operational costs. Water extraction may also have to be continued for very long periods of time to slowly flush perchlorate from the subsurface. Residual sources in the vadose zone or areas of high concentration within the saturated zone can continue to feed contaminants to the plume for long periods, and even highly soluble contaminants such as perchlorate can be retarded during transport. Finally, *ex situ* treatment often requires expensive disposal of water and treatment residuals.

The high costs for the available *ex situ* technologies, and the large volumes of water that may require treatment, have led to the interest in developing *in situ* treatment technologies for perchlorate in soil and groundwater. As described below, bioremediation has been by far the most commonly used *in situ* remedial approach, because it has proven to be effective and economical, and also because it is a flexible technology that can be adapted for use under a wide variety of site conditions.

1.3.2 Why Use *In Situ* Bioremediation?

Bioremediation is the use of organisms to destroy or transform contaminants (Norris et al., 1994). The ability of organisms to degrade environmental contaminants has been recognized for over a century, and research has continued to demonstrate the remarkable capabilities of microorganisms to degrade a wide variety of pollutants to innocuous end products.

The concept of biodegrading groundwater contaminants in place became accepted in the late 1970s to mid 1980s (Raymond et al., 1976; Thomas and Ward, 1989). *In situ* bioremediation was first used to treat aerobically biodegradable contaminants such as petroleum-based fuels, and it proved to be effective, reliable and economically attractive compared to other alternatives (NRC, 1993). Enhanced aerobic biodegradation generally relies on the injection of oxygen as an electron acceptor to promote biodegradation of reduced organic compounds that act as electron donors.

More recently, practitioners have exploited the ability of anaerobic microorganisms to degrade contaminants that were previously considered recalcitrant in the environment. For example, the most prevalent groundwater contaminants, chlorinated solvents such as perchloroethene (also termed tetrachloroethene) and trichloroethene (TCE), are relatively oxidized compounds and are resistant to conventional aerobic biodegradation. However, during the 1990s, enhanced anaerobic bioremediation techniques were developed to completely degrade these solvents, as well as other oxidized compounds. In this case, it is the electron donor that is added, to promote the growth and activities of organisms that can use the target contaminant as the electron acceptor.

Several electron donors and application methods have been developed and tested for enhancing anaerobic biodegradation, and protocols for *in situ* anaerobic bioremediation have been developed (Parsons, 2004). Regulatory acceptance of the approach has increased as carefully-controlled demonstrations showed its potential efficacy and safety (ITRC, 1998). Anaerobic biological treatment also has been used for both *ex situ* and *in situ* treatment of nitrate, a compound that is very similar to perchlorate, and in fact both can be degraded by the same enzyme (Herman and Frankenberger, 1998). It was therefore reasonable to investigate whether *in situ* bioremediation could be used for perchlorate as well (Logan, 2001).

The ability of microorganisms to degrade perchlorate has been known since the 1960s (Hackenthal et al., 1964; Romanenko et al., 1976), and biological treatment of perchlorate in wastewater was demonstrated in the early 1990s (Attaway and Smith, 1993; Attaway, 1994). By the late 1990s, it was clear that numerous organisms were capable of reducing perchlorate to chloride, that they were near-ubiquitous in the subsurface and that the basic mechanism for biological perchlorate reduction was well understood (Coates et al., 1999; Rikken et al., 1996; Wallace et al., 1996).

Hence, the basic knowledge and related practical experience was largely in place when the magnitude of the perchlorate problem became evident. Laboratory and field tests were quickly initiated, and the results confirmed that perchlorate, and its intermediates, could be biologically removed to below detectable levels. Also, bioremediation technology could be implemented with relatively low capital and operating costs (Cox et al., 2000; Xu et al., 2003). Further, biological processes could effectively treat common co-contaminants found at perchlorate sites, including TCE, nitrate and the explosive compound cyclotrimethylenetrinitramine (RDX or Royal Demolition eXplosive).

In situ bioremediation of perchlorate has been rapidly adopted and used at field sites, and the flexibility of the technology has led to many variations that can be used to adapt to site-specific conditions. *In situ* bioremediation has already significantly reduced the costs of managing perchlorate contamination, and it will continue to be an important remedial technology for perchlorate-contaminated sites, alone or in combination with other technologies.

REFERENCES

Attaway H. 1994. Propellant wastewater treatment process. U.S. Patent No. 5,302,285.

Attaway H, Smith M. 1993. Reduction of perchlorate by an anaerobic enrichment culture. J Ind Microbiol 12:408–412.

Aziz C, Borch R, Nicholson P, Cox E. 2006. Alternative causes of wide-spread, low concentration perchlorate impacts to groundwater. In Gu B, Coates JD, eds, Perchlorate: Environmental Occurrence, Interactions and Treatment. Springer Science+Business Media, Inc., New York, NY, USA, pp 71–91.

Blount BC, Pirkle JL, Osterloh JD, Vaentin-Blasini L, Caldwell KL. 2006. Urinary perchlorate and thyroid hormone levels in adolescent and adult men and women living in the United States. Environ Health Perspectives 114:1865–1871.

Blount BC, Valentin-Blasini L, Osterloh JD, Mauldin JP, Pirkle JL. 2007. Perchlorate exposure of the U.S. population, 2001–2002. J Expo Sci and Environ Epidemiology 17:400–407.

Böhlke JK, Sturchio NC, Gu B, Horita J, Brown GM, Jackson WA, Batista J, Hatzinger PB. 2005. Perchlorate isotope forensics. Anal Chem 77:7838–7842.

CDHS (California Department of Health Services). 1997. Determination of Perchlorate by Ion Chromatography, Rev. 0. CDHS, Sacramento, CA, USA. June 3.

CDHS. 2007. Maximum Contaminant Levels – Inorganic Chemicals. 22 California Code of Regulations §64431.

Cheremisinoff NP. 2001. National defense programs lead to groundwater contamination: The perchlorate story. Pollut Eng (August):38–43.

Clewell RA, Merrill EA, Narayanan L, Gearhart JM, Robinson PJ. 2004. Evidence for competitive inhibition of iodide uptake by perchlorate and translocation of perchlorate into the thyroid. Int J Toxicol 23:17–23.

Coates JD, Michaelidou U, Bruce R, O'Connor S, Crespi J, Achenbach L. 1999. Ubiquity and diversity of dissimilatory (per)chlorate—reducing bacteria. Appl Environ Microbiol 65:5234–5241.

Cox EE, Edwards E, Neville S. 2000. In situ bioremediation of perchlorate in groundwater. In Urbansky ET, ed, Perchlorate in the Environment. Kluwer/Academic/Plenum Publishers, New York, NY, USA, pp 231–240.

Cunniff SE, Cramer RJ, Maupin HE. 2006. Perchlorate: Challenges and lessons. In Gu B, Coates JD, eds, Perchlorate: Environmental Occurrence, Interactions and Treatment. Springer Science+Business Media, Inc., New York, NY, USA, pp 1–15.

Dasgupta PK, Martinelango PK, Jackson WA, Anderson TA, Tian K, Tock RW, Rajagopalan S. 2005. The origin of naturally occurring perchlorate. Environ Sci Technol 39:1569–1575.

Dewey MA. 2007. Synthesis, evaluation, and formulation studies on new oxidizers as alternatives to ammonium perchlorate in DoD missile propulsion applications. Project WP1403 Final Report. Department of Defense (DoD) Strategic Environmental Research and

Development Program (SERDP), Arlington, VA, USA. http://www.serdp.org/Research/upload/WP-1403-FR.pdf. Accessed February 1, 2008.

EWG (Environmental Working Group). 2007. Rocket fuel in drinking water: Perchlorate pollution spreading nationwide. http://www.ewg.org/node/8203. Accessed February 1, 2008.

Gingras TM, Batista JR. 2002. Biological reduction of perchlorate in ion exchange regenerant solutions containing high salinity and ammonium levels. J Environ Monit 4:96–101.

Gu B, Coates JD. 2006. Preface. In Gu B, Coates JD, eds, Perchlorate: Environmental Occurrence, Interactions and Treatment. Springer Science+Business Media, Inc., New York, NY, USA, pp xi–xii.

Gullick RW, LeChevallier M, Barhorst T. 2001. Occurrence of perchlorate in drinking water sources. J Am Water Works Assoc 93:66–77.

Hackenthal E, Mannheim W, Hackenthal R, Becher R. 1964. The reduction of perchlorate by bacteria - I. Studies with whole cells. Biochem Pharm 13:195–206.

Hatzinger PB. 2005. Perchlorate biodegradation for water treatment. Environ Sci Technol 39:239A–247A.

Herman DC, Frankenberger WT. 1998. Microbial-mediated reduction of perchlorate in groundwater. J Environ Qual 27:750–754.

ITRC (Interstate Technology & Regulatory Council). 1998. Technical and regulatory requirements for enhanced in situ bioremediation of chlorinated solvents in groundwater. ITRC, Washington, DC, USA. http://www.itrcweb.org/Documents/ISB-6.pdf. Accessed April 22, 2008.

ITRC. 2005. Perchlorate: Overview of issues, status, and remedial options. ITRC, Washington, DC, USA. http://www.itrcweb.org/Documents/PERC-1.pdf. Accessed February 1, 2008.

ITRC. 2008. Remediation technologies for perchlorate contamination in water and soil. PERC-2. ITRC Perchlorate Team, Washington DC, USA. http://www.itrcweb.org/Documents/PERC-2.pdf. Accessed April 20, 2008.

Kendall RJ, Smith PN, eds. 2006. Perchlorate Ecotoxicology. Society of Environmental Toxicology and Chemistry Press, Boca Raton, FL, USA. 288p.

Logan BE. 2001. Assessing the outlook for perchlorate remediation. Environ Sci Technol 35:482A–487A.

MADEP (Massachusetts Department of Environmental Protection). 2006. Inorganic Chemical Maximum Contaminant Levels, Monitoring Requirements and Analytical Methods. 310 Code Massachusetts Regulations §22.06.

Mattie DR, Strawson J, Zhao J. 2006. Perchlorate toxicity and risk assessment. In Gu B, Coates JD, eds, Perchlorate: Environmental Occurrence, Interactions and Treatment. Springer Science+Business Media, Inc., New York, NY, USA, pp 169–196.

Mendiratta SK, Dotson RL, Brooker RT. 1996. Perchloric acid and perchlorates. In Kroschwitz JI, Howe-Grant M, eds, Kirk-Othmer Encyclopedia of Chemical Technology, Vol 18, 4th ed. John Wiley & Sons, Inc, New York, NY, USA, pp 157–170.

Merrill EA, Clewell RA, Robinson JP, Jarabek AM, Gearhart JM, Sterner TR, Yu KO, Fisher JW. 2005. PBPK model for radioactive iodide and perchlorate kinetics and the perchlorate induced inhibition of iodide uptake in humans. Toxicol Sci 83:25–43.

Mosier-Boss PA. 2006. Recent developments in perchlorate detection. In Gu B, Coates JD, eds, Perchlorate: Environmental Occurrence, Interactions and Treatment. Springer Science+Business Media, Inc., New York, NY, USA, pp 111–152.

NRC (National Research Council). 1993: In Situ Bioremediation: When Does It Work? National Academies Press, Washington, DC, USA, 184 p.

NRC. 2005. Health Implications of Perchlorate Ingestion. National Academies Press, Washington, DC, USA, 276 p.

Norris RD, Hinchee RE, Brown R, McCarty PL, Semprini L, Wilson JT, Kampbell DH, Reinhard M, Bouwer EJ, Borden RC, Vogel TM, Thomas JM, Ward CH. 1994. Handbook of Bioremediation. Lewis Publishers, CRC Press, Boca Raton, FL, USA, 257 p.

Parette R, Cannon FS. 2006. Perchlorate removal by modified activated carbon. In Gu B, Coates JD, eds, Perchlorate: Environmental Occurrence, Interactions and Treatment. Springer Science+Business Media, Inc., New York, NY, USA, pp 343–372.

Parsons. 2004. Principles and Practices of Enhanced Anaerobic Bioremediation of Chlorinated Solvents. Prepared for Air Force Center for Environmental Excellence, Brooks City-Base, TX, USA; Naval Facilities Engineering Service Center, Port Hueneme, CA, USA; DoD Environmental Security Technology Certification Program, Arlington, VA, USA. http://www.afcee.af.mil/shared/media/document/AFD-071130-020.pdf. Accessed September 21, 2008.

Raymond RL, Jamison VW, Hudson JO. 1976. Beneficial stimulation of bacterial activity in groundwater containing petroleum products. American Institute of Chemical Engineers (AIChE) Symposium Series 73:390–404.

Rikken GB, Kroon AGM, van Ginkel CG. 1996. Transformation of perchlorate into chloride by a newly isolated bacterium: Reduction and dismutation. Appl Microbiol Biotechnol 45:420–426.

Romanenko VI, Korenkov VN, Kuznetsov SI. 1976. Bacterial decomposition of ammonium perchlorate. Mikrobiologiya 61:347–356.

Roote DS. 2001. Technology status report: Perchlorate treatment technologies. Ground-Water Remediation Technology Analysis Center, Pittsburgh, PA, USA. http://cluin.org/download/toolkit/thirdednew/gwrtacperchlorate.pdf. Accessed February 1, 2008.

Sass J. 2004. U.S. Department of Defense and White House working together to avoid cleanup and liability for perchlorate pollution. Int J Occup and Environ Health 10:330–334.

Smith PN. 2006. The ecotoxicology of perchlorate in the environment. In Gu B, Coates JD, eds, Perchlorate: Environmental Occurrence, Interactions and Treatment. Springer Science+Business Media, Inc., New York, NY, USA, pp 153–168.

Takata K. 1985. Letter from Chief, Superfund Programs Branch, to Don Hawkins, Center for Disease Control (CDC) Regional Representative. Request for CDC assistance regarding potential health effects of perchlorate contamination at the San Gabriel Valley Superfund sites. Doc. AR0128. December 23.

Thomas JM, Ward CH. 1989. *In situ* biorestoration of organic contaminants in the subsurface. Environ Sci Technol 23:760–766.

Urbansky ET. 1998. Perchlorate chemistry: Implications for analysis and remediation. Bioremediation J 2:81–95.

Urbansky ET. 2000. Quantitation of perchlorate ion: Practices and advances applied to the analysis of common matrices. Crit Rev Anal Chem 30:311–343.

USEPA (U.S. Environmental Protection Agency). 1998a. Announcement of the Drinking Water Contaminant Candidate List: Notice. Federal Register 63, No. 40 (March 2): 10273–10287.

USEPA. 1998b. Perchlorate Environmental Contamination: Toxicological Review and Risk Characterization Based on Emerging Information. NCEA-1-0503, Review Draft. Office of Research and Development, USEPA, Washington, DC, USA.

USEPA. 1999. Method 314.0 Determination of Perchlorate in Drinking Water Using Ion Chromatography, Rev.1. EPA 815/B/99/003. USEPA, Washington, DC, USA.

USEPA. 2002. Perchlorate Environmental Contamination: Toxicological Review and Risk Characterization. NCEA-1-0503, External Review Draft. National Center for Environmental Assessment, Office of Research and Development, USEPA, Washington, DC, USA. http://cfpub.epa.gov/ncea/cfm/recordisplay.cfm?deid=24002. Accessed February 1, 2008.

USEPA. 2004. Revisions to the Unregulated Contaminant Monitoring Rule Fact Sheet. 815-F-01-008. Office of Water, USEPA, Washington, DC, USA. http://www.epa.gov/safewater/standard/ucmr/ucmrfact.html. Accessed February 1, 2008.

USEPA. 2005. Perchlorate Treatment Technology Update: Federal Facilities Forum Issue Paper. 542-R-05-015. USEPA, Solid Waste and Emergency Response, Washington, DC, USA. http://www.clu-in.org/download/remed/542-r-05-015.pdf. Accessed February 1, 2008.

USFDA (U.S. Food and Drug Administration). 2007. 2004–2005 Exploratory survey data on perchlorate in food. Posted May 2007. http://www.cfsan.fda.gov/~dms/clo4data.html. Accessed February 1, 2008.

Wallace W, Ward T, Breen A, Attaway H. 1996. Identification of an anaerobic bacterium which reduces perchlorate and chlorate as *Wolinella succinogenes*. J Ind Microbiol 16:68–72.

Xu J, Song Y, Min B, Steinberg L, Logan BE. 2003. Microbial degradation of perchlorate: Principles and applications. Environ Eng Sci 20:405–422.

Zhou P, Brown GM, Gu B. 2006. Membrane and other treatment technologies – pros and cons. In Gu B, Coates JD, eds, Perchlorate: Environmental Occurrence, Interactions and Treatment. Springer Science+Business Media, Inc., New York, NY, USA, pp 389–404.

CHAPTER 2

DEVELOPMENT OF *IN SITU* BIOREMEDIATION TECHNOLOGIES FOR PERCHLORATE

Evan E. Cox[1]

[1]Geosyntec Consultants, Inc., Guelph, ON, Canada

2.1 INTRODUCTION

In the early 1960s, research at the University of Heidelberg, Germany, by Eberhard Hackenthal and others revealed that various heterotrophic bacteria containing nitrate reductase enzymes were capable of reducing perchlorate to chloride (Hackenthal et al., 1964). The importance of this finding to the U.S. Department of Defense (DoD) and the environmental remediation community in North America would not be recognized for more than two decades, when research by Attaway et al. (1989) and later Attaway and Smith (1993) reported the reduction of perchlorate by an anaerobic enrichment culture. This research, prompted by the desire to develop a treatment process for solid rocket propellant waste streams, also spurred several lines of further research, initially including the development of *ex situ* biotreatment processes targeting industrial waste streams at solid rocket manufacturing facilities, and later the development of *ex situ* and *in situ* biotreatment techniques for perchlorate in groundwater and drinking water supplies.

Today, a wide variety of *ex situ* and *in situ* biological treatment approaches are available to remediate perchlorate in groundwater and soil, and remediation tools and techniques are available from a collection of technology vendors and environmental consultants. While the path to commercialization of these technologies and tools has been relatively short (less than 10 years), many lessons have been learned along the way, and challenges still lie ahead before the technology can be used to its full potential. The path traveled and the challenges ahead are chronicled herein.

2.2 EARLY DISCOVERIES

The early perchlorate biodegradation literature consists of a handful of papers identifying the potential for perchlorate reduction linked to the nitrate reductase enzyme. As previously indicated, the first of these papers by Hackenthal et al. (1964) and Hackenthal (1965) revealed the ability of various heterotrophic bacteria and later *Bacillus cereus* containing nitrate reductase enzymes to reduce perchlorate to chloride. A decade later in the former Soviet Union, Romanenko et al. (1976) reported on the ability of *Vibrio dechloraticans Cuznesove* B-1168 to degrade perchlorate at a concentration of ~300 milligrams per liter (mg/L) to chloride in industrial wastewater when grown anaerobically on acetate or ethanol. In these studies, perchlorate reduction was linked to nitrate reductase activity, and nitrate inhibited perchlorate reduction. Based on the initial work of Romanenko et al. (1976), United States Patent 3,943,055 was issued to Korenkov et al. in March 1976 for a process for purification of industrial waste waters containing perchlorates (including ammonium perchlorate) and chlorates using the aforementioned organism, although no reason for the presence of ammonium perchlorate in industrial waste waters was cited in these documents.

In the late 1980s, an Air Force Office of Scientific Research funded initiative demonstrated the reduction of perchlorate by an anaerobic enrichment culture from municipal digester sludge (Attaway et al., 1989; Attaway and Smith, 1993). This research established that perchlorate reduction is not mediated by nitrate reductase, but rather by its own enzymes. The enrichment culture was capable of degrading high concentrations (in excess of 6,000 mg/L) of perchlorate in the presence of protein-based carbon substrates. Interestingly, the enrichment culture could not degrade perchlorate in the presence of many of the simple carbon substrates (sugars, organic acids and alcohols) that are used today for perchlorate bioremediation. Attempts to isolate the organism(s) responsible for perchlorate reduction in this enrichment culture were unsuccessful during the initial study, although in later studies, Wallace et al. (1996) isolated an organism, *Wolinella succinogenes*, which they identified as responsible for the observed perchlorate reduction. Despite being a rather rare and finicky organism for use in engineered perchlorate reduction, the discovery of this organism led to significant research and development of a bioreactor treatment process for the treatment of perchlorate at elevated concentrations in waste streams at several rocket manufacturing sites.

In 1996, research conducted by Rikken et al. at the Akzo Nobel Central Research Center in the Netherlands resulted in the isolation of a bacterial strain, GR-1, capable of perchlorate reduction via chlorate to chlorite, which disproportionates into molecular oxygen and chloride, using acetate as electron donor. Through this work, Rikken et al. (1996) proposed the perchlorate reduction pathway that is now commonly understood (Figure 2.1).

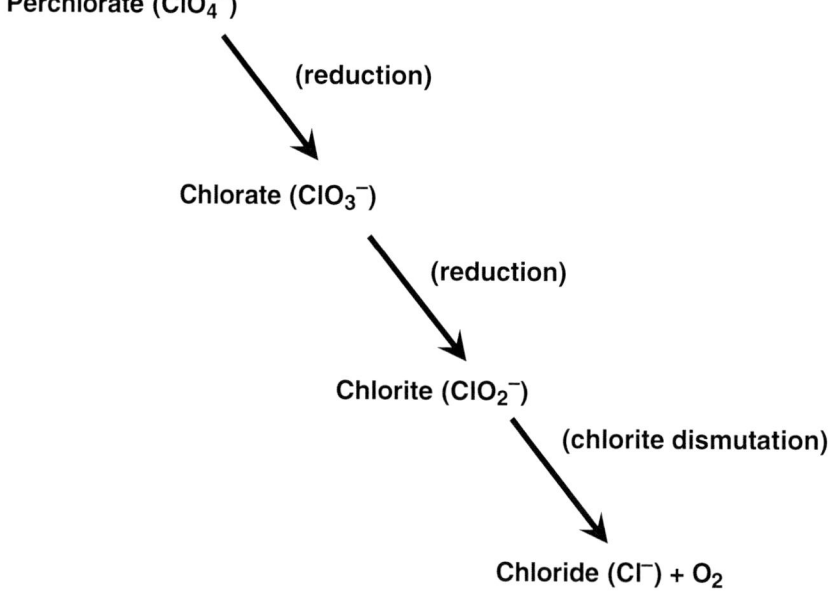

Figure 2.1. Perchlorate biodegradation pathway (after Rikken et al., 1996)

2.3 ANALYTICAL METHODS AND PILOT PROGRAMS

The first tests conducted to demonstrate *in situ* bioremediation of perchlorate were performed at a site in Rancho Cordova (Sacramento), California owned by Aerojet-General Corporation (Aerojet). Significant quantities of perchlorate were handled at the facility over nearly six decades, and perchlorate was released to the environment in various areas of the

3,440-hectare (8,500-acre) site (USEPA, 2000). While environmental investigations at the facility contemplated or targeted perchlorate impacts since at least the 1980s, the magnitude of the perchlorate impacts and the distribution of perchlorate in groundwater at the site were not fully known until the advent of analytical techniques that could detect perchlorate at low part per billion concentrations (ppb, equivalent to micrograms per liter, or µg/L). In the early 1990s, ion specific electrode (ISE) methods were commonly used to detect perchlorate in groundwater samples. The detection limits of the ISE methods were generally in the part per million (ppm, or mg/L) range, and as such, only the higher concentration perchlorate source areas could be delineated. While refinement of the ISE techniques in the mid 1990s reduced the detection limit to about 400 to 700 µg/L, it wasn't until the State of California and Dionex Corporation developed ion chromatographic techniques capable of reliable perchlorate detection to a level of 4 µg/L that the magnitude of perchlorate impacts at the facility, including off-site areas, were truly known.

To address their cleanup needs, Aerojet joined the DoD in the role of technology development pioneer, funding much of the initial research related to both *ex situ* biological reduction and enhanced *in situ* bioremediation. In the late 1980s, Aerojet initiated research to develop an *ex situ* biological process to treat perchlorate, including design and operation of an interim pilot treatment plant consisting of a packed bed, recycle reactor (Andrews, 1989). Approximately ten years later, Aerojet designed and built the world's first full-scale aboveground treatment facility to remove perchlorate from groundwater using a fluidized bed reactor (FBR). Microbes isolated from strawberry jam manufacturing waste were used to seed the FBRs, and ethanol was used as the electron donor. The initial FBRs are still in use at the facility today, and *ex situ* biological reduction remains the primary technology for perchlorate treatment at the site, cleaning nearly 38 million liters (L) (10 million gallons [gal]) of groundwater daily.

Aerojet also funded research on the potential of microbes from the soil and groundwater of the Sacramento site to degrade perchlorate. In microcosms containing materials from one area of the site, degradation of more than 100 mg/L of perchlorate was observed after an acclimation period of 30 to 40 days, using either ethanol or manure as electron donor. In microcosms from a second area, degradation of more than 80 mg/L perchlorate was observed within several weeks using ethanol or molasses as electron donors (Cox et al., 1999; Cox et al., 2000). These studies showed that the microbial communities naturally present in at least some aquifer materials might be capable of addressing what was generally thought to be a recalcitrant groundwater contaminant. However, it took some time to identify the key organisms involved and to realize that perchlorate degraders were virtually ubiquitous at contaminated sites.

2.4 UBIQUITOUS OCCURRENCE OF PERCHLORATE DEGRADERS

As information on the widespread occurrence of perchlorate in groundwater related to military activities grew in the late 1990s (e.g., Renner, 1998), the DoD responded with a request for proposal under the Strategic Environmental Research and Development Program (SERDP) to conduct fundamental research on the use of *in situ* bioremediation as a potential technology for remediation of perchlorate-impacted groundwater. The three projects funded through SERDP were designed to: (1) assess the ubiquity of perchlorate reducing bacteria at DoD sites; (2) determine the geochemical tolerances and electron donor preferences of perchlorate reducers from diverse environments; and (3) elucidate the organisms and enzymes involved in perchlorate reduction and develop tools/assays for identification of these organisms/enzymes.

The SERDP research projects produced a wealth of information regarding the ubiquity of perchlorate-reducers in subsurface environments, the wide range of electron donors that could be used to stimulate perchlorate reduction, and the relative ease with which perchlorate could be degraded. The final reports for each of these projects are available at www.serdp.org/research/er-perchlorate.cfm (Projects ER-1162, ER-1163 and ER-1164). Table 2.1 provides a summary of the DoD installations from which subsurface materials were tested by the three research groups.

In all cases, perchlorate reduction could be stimulated through the addition of acetate, and in the few cases where perchlorate reduction could not be stimulated, the inhibition was related to geochemical factors such as low pH. Perhaps most surprising was the discovery that dozens of different organisms could degrade perchlorate (Coates et al., 1999; Waller et al., 2004), dispelling notions from the early (pre-1998) research that perchlorate biodegradation was a unique microbial capability.

Table 2.1. Perchlorate Biodegradation in Microcosms for SERDP Test Sites

Electron Donor	Aerojet (CA)	Edwards AFB (CA)	Navy (CA)	Navy (WV)	Navy (MD)	Industrial (NV)	Boeing (CA)	JPL (CA)	UTC (CA)
Acetate	✓	✓	✓	✓	✓	✓	✓	✓	✓
Lactate	✓	—	—	—	—	—	—	✓	—
Oleate	—	✓	—	—	—	✓	—	—	—
Molasses	✓	✓	✓	✓	✓	✓	✓	✓	✓
Canola Oil	—	—	—	—	—	—	✓	—	—
Methanol	—	—	—	—	—	—	—	✓	—
Ethanol	✓	—	—	—	—	—	—	✓	—
Hydrogen	—	—	—	—	✓	—	—	✓	—
Propane	—	—	—	—	—	—	—	✓	—

✓ = biodegradation tested and observed
— = biodegradation not tested
AFB = Air Force Base
JPL = Jet Propulsion Laboratory
UTC = United Technologies Corporation

2.5 FIELD DEMONSTRATIONS

The first rigorous demonstration of *in situ* perchlorate biodegradation was conducted at Aerojet's Area 20. The surface infrastructure for this pilot test was minimal. It consisted of a control shed, a single extraction-injection well pair separated by 20 meters (m) (65 feet [ft]), and several monitoring wells (Figure 2.2). Oddly enough, this is a key advantage of *in situ* bioremediation: little to no aboveground infrastructure is required to achieve the desired end result of perchlorate treatment. The pilot test employed an active recirculation approach, whereby groundwater containing approximately 15,000 µg/L of perchlorate was extracted, amended with acetate as electron donor, and reinjected via an upgradient injection well (McMaster et al., 2001; Cox et al., 2001).

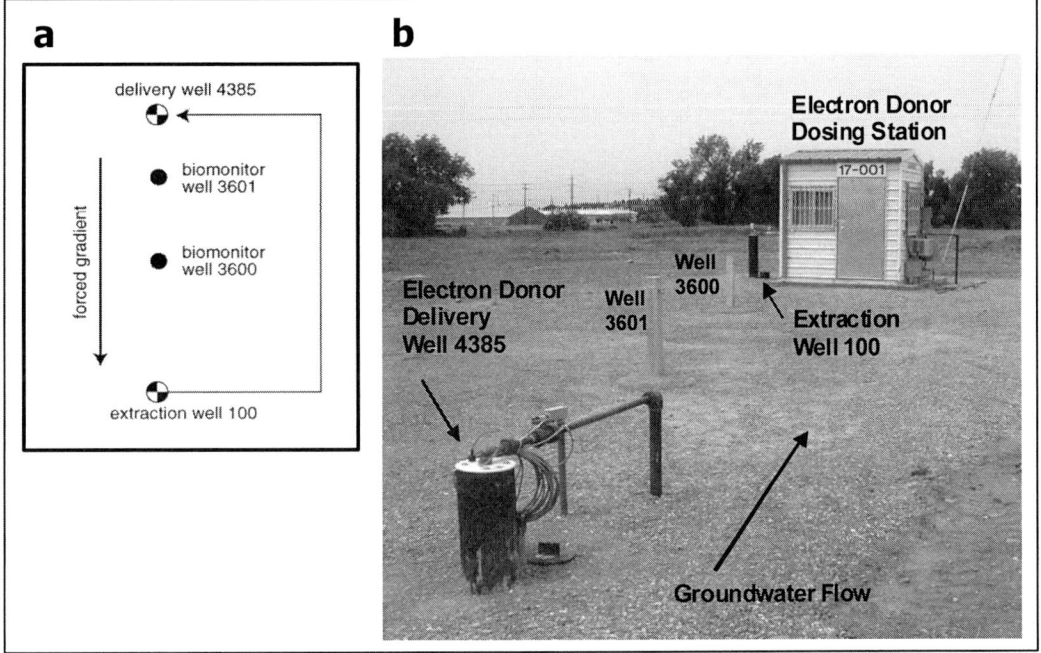

Figure 2.2. (a) Layout and (b) photograph of pilot test infrastructure at Area 20, Aerojet Superfund site, Sacramento, California (Geosyntec Consultants, 2002)

Within weeks of startup in July 2000, perchlorate concentrations in the groundwater at the monitoring wells located 4.6 and 10.7 m (15 and 35 ft) downgradient from the electron donor delivery well began to decline, eventually dropping to less than the method detection limit (MDL) of 4 µg/L (Figure 2.3). After years of theorizing that *in situ* bioremediation of perchlorate might be possible, the proof that this technology could completely eliminate perchlorate from groundwater was now available, sparking implementation of *in situ* bioremediation projects at many other sites, in a wide variety of configurations.

2.6 BIOREMEDIATION STRATEGIES

The SERDP-funded research projects clearly showed that perchlorate reduction can be readily stimulated in subsurface environments through the addition of a wide range of carbon-based electron donors. The key to successful *in situ* bioremediation is effective delivery of appropriate quantities of the electron donor, to promote reduction of perchlorate and competing electron acceptors (primarily oxygen and nitrate, and at some sites chlorate). Electron donor delivery can be accomplished in many ways, from continuous or periodic delivery and recirculation of soluble electron donors—such as alcohols, organic acids (acetate, lactate, citrate) or sugars—to direct injection of soluble or insoluble (or at least less soluble) substrates such as vegetable oils or chitin.

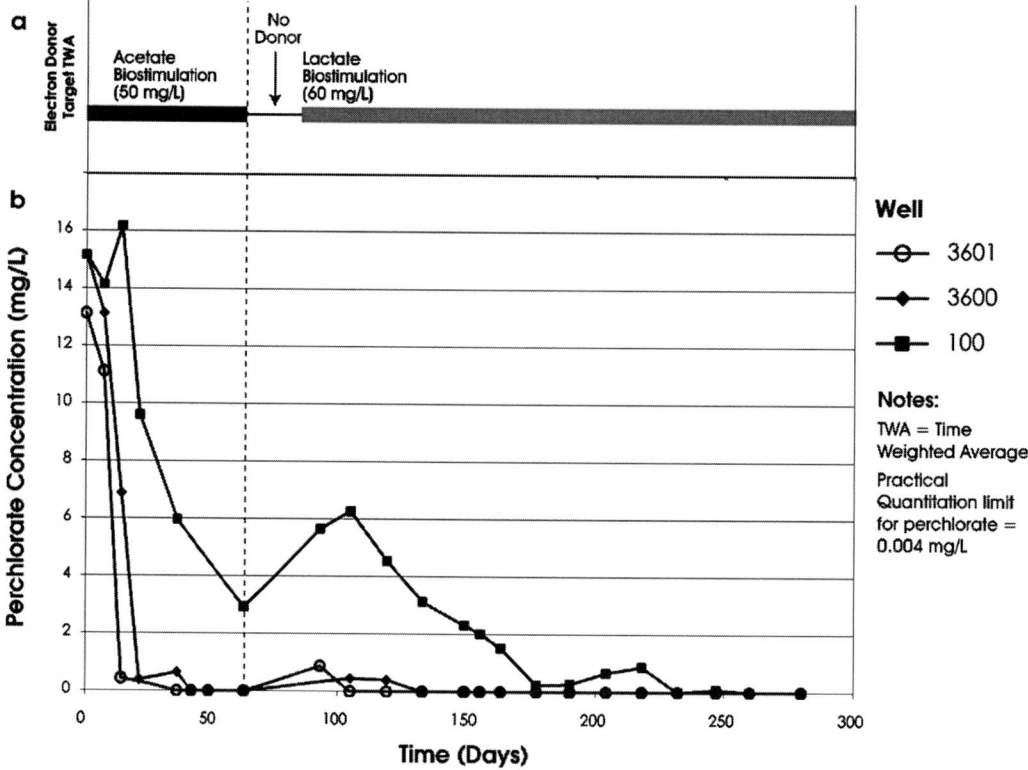

Figure 2.3. (a) Electron donor addition regime and (b) perchlorate biodegradation results in groundwater at Area 20, Aerojet Superfund site, Sacramento, California (Geosyntec Consultants, 2002)

Since the initial demonstration of perchlorate reduction *in situ*, many *in situ* bioremediation applications have been attempted and/or completed at perchlorate sites using varying approaches. These approaches generally group into several application categories, depending on the frequency of electron donor addition and the delivery infrastructure. The common bioremediation approaches can be grouped as follows:

- **Active Bioremediation**: Active bioremediation systems typically employ conventional extraction wells and infrastructure to continuously mix and distribute soluble electron donors into the perchlorate-impacted groundwater.

- **Passive Bioremediation**: Passive bioremediation systems typically inject slow-release electron donors to create biologically active zones in the subsurface to either treat source areas or to create biological treatment zones (biobarriers) through which perchlorate-impacted groundwater must flow.

- **Semi-Passive Bioremediation**: This is a hybrid bioremediation approach that attempts to balance the benefits of the active and passive systems, namely more effective electron donor mixing and distribution in the subsurface through a periodic active phase, followed by low operations and maintenance (O&M) requirements during the passive phase.

Each of these electron donor delivery approaches has specific benefits and limitations that are discussed in greater detail in Chapter 5, with the experiences largely gained through the design and execution of three large field demonstration/validation projects funded (starting in 2002) by the DoD Environmental Security Technology Certification Program (ESTCP). These projects included: 1) a passive biobarrier project at an active rocket manufacturing facility in Maryland (IES Solutions); 2) a semi-passive biobarrier project at Longhorn Army Ammunition Plant (LHAAP) in Texas (Geosyntec); and 3) an active biobarrier project at Aerojet in California (Shaw). Information for each of these projects is available at http://www.estcp.org/technology/er-perchlorate.cfm (Projects ER-0221, ER-0219 and ER-0224, respectively).

The results of these and other demonstrations, presented in detail within this book (Chapters 6, 7, 8 and 9), provide compelling evidence that *in situ* bioremediation is an effective technique to remediate perchlorate in groundwater. In the passive biobarrier demonstration (ER-0221), perchlorate concentrations were reduced from more than 10,000 µg/L to less than 4 µg/L within days to weeks after injection of Edible Oil Substrate (EOS®), a patented slow-release electron donor. During this project, reduction of 1,1,1-trichloroethane (1,1,1-TCA), a common chlorinated solvent present in groundwater at rocket manufacturing facilities, was also observed. While largely successful, the injection of large quantities of slow-release electron donor promoted significant mobilization of iron and manganese in the groundwater, a consequence that would need to be considered during design of bioremediation applications for other sites. In the semi-passive (ER-0219) and active (ER-0224) demonstrations, perchlorate concentrations were also reduced to less than 4 µg/L in groundwater. Implementation and operation of these projects required a higher degree of effort as compared to the passive approach, but the benefit of the higher activity (and lower electron donor dosage) was seen in lower impacts to secondary water quality. The choice of bioremediation configuration for a given site thus depends on finding the acceptable balance between infrastructure and operational demands, and tolerance to water quality impacts that result from adding large quantities of electron donor at any one event.

While final data are not yet available, ESTCP has also funded two other perchlorate–related demonstrations. The first of these projects involves a demonstration of passive bioremediation using a mulch biowall at Redstone Arsenal in Alabama (Parsons) (ER-0427); see http://www.estcp.org/technology/er-perchlorate.cfm. The approach is simple in design, and the cost of the mulch and installation are expected to be modest compared to other bioremediation configurations. However, uncertainties exist as to the longevity of the mulch as an effective electron donor, and the results of this and similar demonstrations will need to be evaluated before conclusions can be drawn regarding the long-term effectiveness of the approach. The second project involves an evaluation of monitored natural attenuation of perchlorate at Indian Head Naval Weapons Reserve in Maryland (ER-0428); see http://www.estcp.org/technology/er-perchlorate.cfm. Given the propensity for perchlorate to degrade whenever/wherever carbon-based electron donors and anaerobic/reducing redox conditions are present, one would anticipate that natural degradation and attenuation of perchlorate occurs far more commonly in the environment than we currently realize. The results of this demonstration, which will attempt to quantify natural attenuation of perchlorate in groundwater at multiple DoD sites, should be enlightening.

2.7 REMEDIATION OF PERCHLORATE IN SOIL—THE NEW CHALLENGE

For the past 10 years, the focus of technology development for perchlorate remediation has understandably been on groundwater, due to the need to preserve and/or clean our nation's drinking water supplies. In the past few years, however, greater emphasis and effort have been placed on the remediation of perchlorate-impacted soils, due to the realization that soils at many sites represent persistent sources of long-term impacts to groundwater. Interestingly, while perchlorate is extremely soluble and does not sorb to soil particles, it is still present in soil at concentrations of concern (10s to 100s of mg/kg) at a significant number of sites where perchlorate handling ceased decades ago. The continued presence of perchlorate at these sites can largely be correlated to soil lithology, with the highest concentrations present in lower permeability materials.

Low permeability materials may beneficially delay the transfer of perchlorate to groundwater. However, the degree of control afforded is often insufficient to protect groundwater from continuing impacts in excess of regulatory action or notification levels, which are typically in the low µg/L range. Furthermore, while the U.S. Environmental Protection Agency's (USEPA's) residential and commercial/industrial preliminary remediation goals (PRGs) for soil are relatively high (7,800 and 100,000 µg/kg, respectively), soil remediation at many sites is being driven by the potential for contaminated soils to impact groundwater (i.e., the soil to groundwater pathway). Calculations of site-specific soil cleanup levels following USEPA guidance have resulted in soil cleanup levels for many sites that are orders of magnitude lower than the PRGs, and in some cases as low as 20 to 50 µg/kg (Cox et al., 2006), necessitating remediation of large volumes of soil.

Several biotreatment techniques have emerged over the past decade to address perchlorate in soils. The first of these was anaerobic composting, which was first demonstrated at the Aerojet site in California by Cox et al. (1999). Shallow soils containing elevated concentrations of perchlorate (up to 4,200 mg/kg) were excavated, mixed with manure, alfalfa and water, and piled to promote biological reduction. The results were encouraging, demonstrating a decline in perchlorate concentrations from 25 mg/kg to less than 0.1 mg/kg (the detection limit at the time) within several weeks (Figure 2.4).

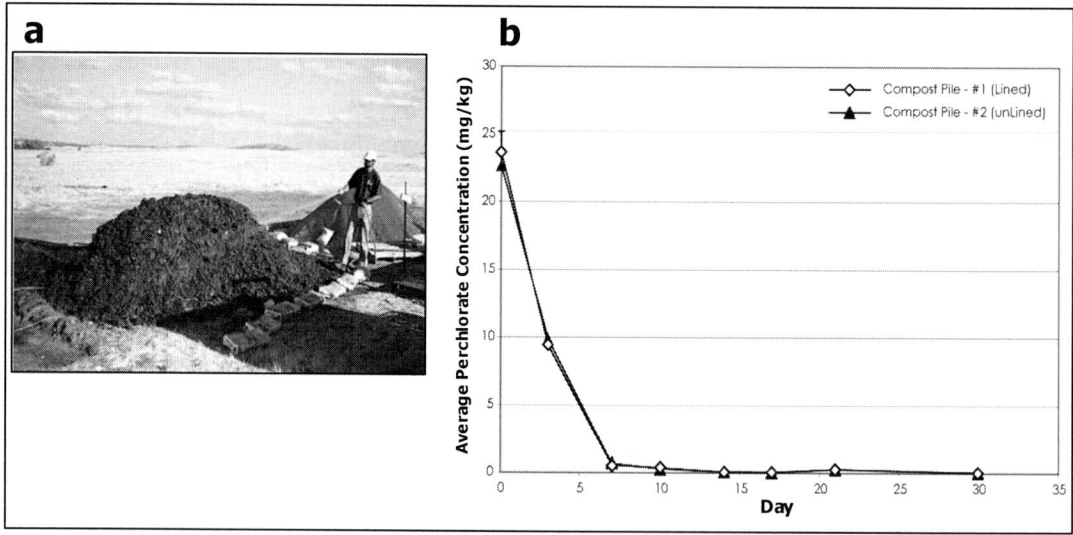

Figure 2.4. (a) Photograph and (b) results of anaerobic composting of perchlorate-impacted soil at the Aerojet Superfund site, Sacramento, California

This technology has now been used at many sites nationwide to treat perchlorate impacted soils. While simple to implement, it is limited by the depth which can be feasibly excavated, and for large sites, by materials handling constraints. Recent improvements to the process include the use of electron donors, such as acetate and citrate instead of manure, to reduce materials handling and to reduce analytical interferences so that low detection limits can be achieved. For example, Griffin et al. (2007) reported on the treatment of approximately 765 cubic meters (m^3) (1,000 cubic yards [yd^3]) of soil containing an average perchlorate concentration of 7,000 µg/kg. Using calcium-magnesium acetate and citric acid as electron donors, perchlorate concentrations were successfully reduced to an average of 12 µg/kg within eight months (Figure 2.5).

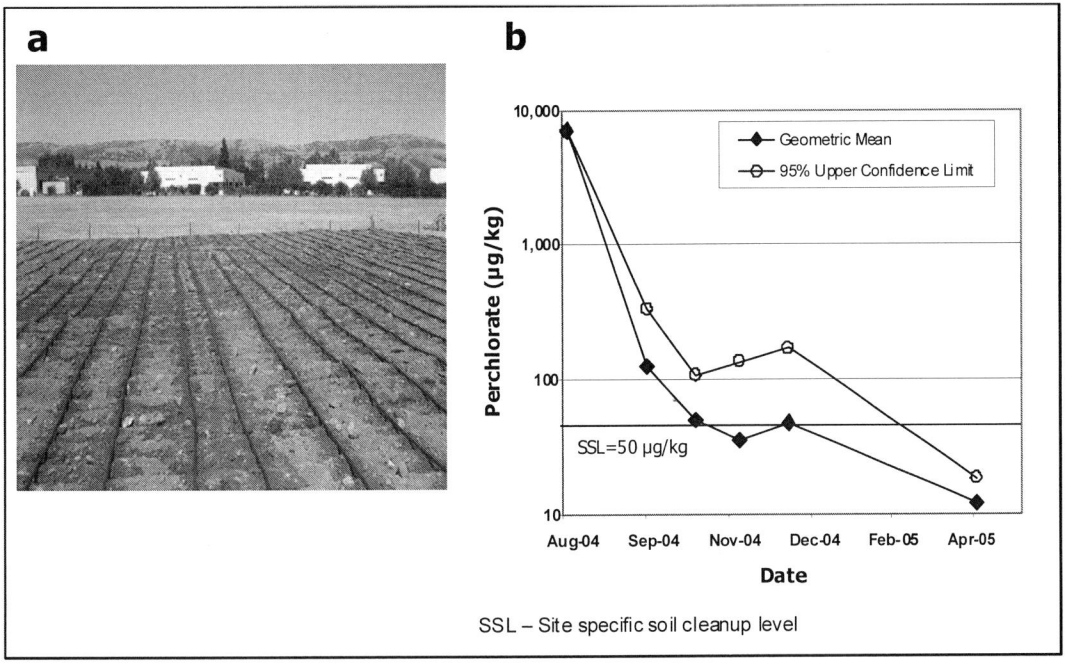

Figure 2.5. (a) Photograph and (b) results of full-scale anaerobic composting of perchlorate-impacted soil in Santa Clara, California

Several alternative approaches to excavation and *ex situ* biotreatment of perchlorate-impacted soils have also emerged in recent years, including surface infiltration of water containing electron donor (Borch, 2001; Kastner et al., 2001; Cox et al., 2006), direct subsurface injection of electron donor (Wuerl et al., 2004), and gas-phase electron donor addition (Evans, 2004). The first attempt at surface infiltration involved the application of manure and water to surface soils containing perchlorate at high concentrations at the Aerojet site in California (Borch, 2001). The approach successfully reduced perchlorate concentrations in the site soils by an average of 96% across the 65 high concentration hot spots where this technique was applied. Following this work, Kastner et al. (2001) demonstrated that perchlorate in the shallow vadose zone (generally less than 3 m or 10 ft deep) could be readily treated through surface infiltration of water containing a variety of carbon substrates such as ethanol and acetate.

More recently, the surface infiltration approach was used to successfully treat more than 30,580 m^3 (40,000 yd^3) of perchlorate-impacted soil at a former road-flare manufacturing facility in California. In this full-scale application, approximately 150 L/min (40 gpm) of treated groundwater (from an ion exchange treatment system) was amended with citric acid and applied to the ground surface via agricultural irrigation methods (drip tape) to treat perchlorate in the unsaturated zone to a depth of 5 m (16 ft) below groundwater surface. Using this method, perchlorate concentrations were effectively reduced from an average of

215 μg/kg to an average of 15 μg/kg within eight months of operation, achieving regulatory approval for closure (Griffin et al., 2007).

While surface infiltration has been shown to be an effective soil remediation technique at shallow sites, there is significant debate as to the feasibility of this approach for deeper sites. Many practitioners believe that it will be impractical to deliver electron donor from ground surface to the depths that require remediation (in some cases in excess of approximately 60 m, or 225 ft), and the experiences of surface infiltration projects at shallower sites confirm that electron donor is rapidly consumed during infiltration. While it may be possible to flush the perchlorate from the vadose zone to groundwater, where it can be captured and treated *ex situ* by a variety of technologies (e.g., ion exchange), as was recently demonstrated by Battey et al. (2006) at Edwards Air Force Base in California, many regulatory jurisdictions are not supportive of the flushing approach.

The use of gas-phase electron donors to promote perchlorate reduction in deeper vadose zone environments may provide an effective alternative to electron donor delivery via surface infiltration or direct injection (see Chapter 11). The approach, as yet tested in the field, would involve injection or circulation of gas-phase electron donors, such as hydrogen, propane or methane, to promote perchlorate reduction. As an advantage of this approach, the addition process would not displace the perchlorate, facilitating treatment. ESTCP is currently funding a field demonstration of vadose zone treatment by gas-phase electron donor addition (ER-0511); see http://www.estcp.org/technology/er-perchlorate.cfm.

2.8 THE CHALLENGES AHEAD

After nearly 10 years of research, technology development and application, it would seem that there should be few hurdles left to overcome for widespread use of *in situ* bioremediation at perchlorate sites. Unfortunately, this is not the case. Selection of *in situ* bioremediation as the full-scale remedy for groundwater sites is still somewhat rare, although its selection and use is increasing. Pump and treat by ion exchange remains the preferred remedy at most sites, due largely to the wider range of disposal/use options available for the treated water, and also due to greater understanding of the process by site owners and stakeholders as compared to bioremediation. The high value placed on drinking water supplies largely drives the decision to respond to perchlorate contamination at this time. Hence, treatment systems designed for direct public water supply application may be more highly represented in the present universe of treatment sites. Presently, ion exchange systems are generally preferred for public water supply treatment.

More effective technology transfer is needed to educate technology vendors and users on the appropriate application of bioremediation technology for perchlorate remediation. Unfortunately, some technology vendors and environmental consultants who adopt a cookie-cutter, "one solution fits all sites" approach, have applied *in situ* bioremediation at sites where use of the technology has been inappropriate. Such application failures have, in some cases, eroded stakeholder confidence in the technology. Other key challenges that lie ahead include:

- Greater understanding of the secondary impacts to water quality that are caused by the varying *in situ* bioremediation application approaches. As previously indicated, the addition of excess electron donor to groundwater will result in the mobilization of dissolved metals, primarily iron and manganese, and in extreme cases, the production of sulfide and methane. The acceptability of these impacts for a given site will depend on the site setting, water quality and use in the area, and the regulatory setting. At some sites, the mobilization of metals or production of sulfide or methane may be an acceptable tradeoff for perchlorate treatment; however, even at sites where water is classified as non-potable, most regulatory jurisdictions are unsupportive of such tradeoffs. In such cases, the amount of electron donor added will need to be tightly balanced with the amount required to degrade perchlorate and competing electron acceptors, namely oxygen and nitrate (and in some cases chlorate). Improved understanding of secondary water quality impacts is an issue that has been ignored by many bioremediation vendors and consultants, but is an area that warrants greater attention and understanding.

- Development of cost-effective electron donor delivery to large plumes. Despite decades of use of *in situ* bioremediation projects for varying contaminants, there remains a limited number of ways that electron donor can be effectively introduced to the subsurface. Case studies of the most common approaches are contained within this book. Better electron donor delivery techniques will always be welcome, as will be the creation of slow-release electron donors that can persist for longer timeframes but minimize the impacts to other water quality parameters.

- Remediation of perchlorate in the deep vadose zone. The lifespan of remediation at many sites, particularly those in the arid southwest, will depend on our ability to develop and commercialize effective techniques to treat perchlorate sources in vadose zone systems. While some advances have been made in this area in recent years, this remains a high priority for additional research.

- Site-specific cleanup levels – how low can we go? There is significant debate as to what levels are achievable for soil and vadose zone cleanup, although several recent large full-scale applications have successfully reduced perchlorate levels to below 20 µg/kg (Griffin et al., 2007). The results of these cleanups will, over time, help us to understand whether cleanup to such low levels was necessary in the first place, and will hopefully provide guidance on the perchlorate levels that can be safely left in the subsurface while still being protective of our drinking water supplies.

Much progress has been made in developing *in situ* and *ex situ* remediation techniques for perchlorate. When the timeframe is examined, there are few other contaminants that have received so much attention over such a short period. With continued application and improved understanding, *in situ* bioremediation will undoubtedly see increased use for groundwater, soil, and vadose zone cleanups, at both military and non-military sites. From simple discoveries about perchlorate reduction four decades ago to the large-scale commercial treatment of perchlorate-impacted water today, it has certainly been an interesting and encouraging path forward in environmental remediation technology development.

ACKNOWLEDGEMENTS

The author wishes to thank the following individuals and organizations for their contribution to the research cited herein: Scott Neville of Aerojet; Robert Borch, Jim Deitsch, Leslie Griffin, Thomas Krug and Michaye McMaster of Geosyntec Consultants; Paul Hatzinger of Shaw Environmental, Inc.; John Coates of the University of California at Berkeley; Robert Borden of the North Carolina State University; Richard McClure of Olin Corporation; and SERDP and ESTCP.

REFERENCES

Andrews S. 1989. Final Report on Anaerobic Treatment of Ammonium Perchlorate Interim Pilot Plant Operation. Aerojet Contract Reference Number 061787. Manville RemedTech, Denver, CO, USA. May 1989.

Attaway H, Smith M. 1993. Reduction of perchlorate by an anaerobic enrichment culture. J Ind Microbiol 12:408–412.

Attaway HS, Baca S, Williams B. 1989. Reduction of perchlorate to chloride by an anaerobic microbial consortium. Proceedings, 89th Annual Meeting of the American Society of Microbiology, New Orleans, LA, USA, May 14-18, 1989, Abstract Q-191.

Battey T, Shepard A, Curtis K. 2006. Soil flushing to remove perchlorate from a thick vadose zone in the arid southwest. Proceedings, Groundwater Resources Association of California Perchlorate 2006: Progress Toward Understanding and Cleanup. Glendale, CA, USA. January 26, 2006.

Borch R. 2001. Case study: Bioremediation of perchlorate in soils by the surface application of wet cow manure. Abstracts, Association for Environmental Health and Sciences, 11th Annual West Coast Conference on Contaminated Soils, Sediments and Water. San Diego, CA, USA. March 21, 2001.

Coates JD, Michaelidou U, Bruce R, O'Connor S, Crespi J, Achenbach L. 1999. Ubiquity and diversity of dissimilatory (per)chlorate-reducing bacteria. Appl Environ Microbiol 65:5234–5241.

Cox EE, Allan JP, Neville S. 1999. Rapid bioremediation of perchlorate in soil and groundwater. Proceedings, American Chemical Society, New Orleans, LA, USA, August 1999.

Cox EE, Edwards E, Neville S. 2000. In situ bioremediation of perchlorate in groundwater. In Urbansky ET, ed, Perchlorate in the Environment, Kluwer Academic/Plenum Publishers, New York, NY, USA, pp 231-240.

Cox EE, McMaster M, Neville S. 2001. Perchlorate in groundwater: Scope of the problem and emerging remedial solutions. Proceedings, Symposium on Engineering Geology and Geotechnical Engineering. Las Vegas, NV, USA, March 2001.

Cox EE, Borch R, Deitsch J, Griffin L, McClure R. 2006. Full-scale remediation of perchlorate in vadose zone soils using innovative in situ and ex situ bioremediation techniques. Proceedings, Groundwater Resources Association of California Perchlorate 2006: Progress Toward Understanding and Cleanup. Glendale, CA, USA, January 26, 2006.

Evans P. 2004. In situ bioremediation of perchlorate in vadose zone soil. Proceedings, Groundwater Resources Association of California, Perchlorate in California's Groundwater. Glendale, CA, USA, August 4, 2004.

GeoSyntec Consultants. 2002. In Situ Bioremediation of Perchlorate Impacted Groundwater. Final Technical Report for Project CU-1164. Submitted to the DoD Strategic Environmental Research and Development Program (SERDP), Arlington, VA, USA. http://www.serdp.org/Research/upload/CU-1164-FR-01.pdf. Accessed May 5, 2008.

Griffin LM, Deitsch J, Borch R, Cox E, McClure R. 2007. Successful full-scale remediation of perchlorate in vadose zone soils using innovative in situ & ex situ bioremediation techniques. Proceedings, In Situ and On-Site Bioremediation Symposium, Baltimore, MD, USA. May 7–10, 2007.

Hackenthal E. 1965. The reduction of perchlorate by bacteria – II. The identity of the nitrate reductase and the perchlorate-reducing enzyme of *B. cereus*. Biochem Pharm 14:1313–1324.

Hackenthal E, Mannheim W, Hackenthal R, Becher R. 1964. The reduction of perchlorate by bacteria - I. Studies with whole cells. Biochem Pharm 13:195–206.

Kastner JR, Das KC, Nzengung VA, Dowd J, Fields J. 2001. In-situ bioremediation of perchlorate-contaminated soils. In Leeson A, Peyton BM, Means JL, Magar VS, eds, Bioremediation of Inorganic Compounds. Battelle Press, Columbus, OH, USA, pp 289-295.

Korenkov VN, Romanenko VI, Kuznetsov SI, Voronov JV. 1976. Process for purification of industrial waste waters from perchlorates and chlorates. United States Patent 3,943,055.

McMaster M, Neville S, Bonsack L, Cox EE. 2001. Successful demonstration of in situ bioremediation of perchlorate in groundwater. In Leeson A, Peyton BM, Means JL, Magar VS, eds, Bioremediation of Inorganic Compounds, Battelle Press, Columbus, OH, USA, pp 297-302.

Renner R. 1998. Perchlorate tainted wells spur government action. Environ Sci Technol News 33:210A.

Rikken GB, Kroon AGM, van Ginkel CG. 1996. Transformation of (per)chlorate into chloride by a newly isolated bacterium: reduction and dismutation. Appl Microbiol Biotechnol 45:420–426.

Romanenko VI, Korenkov VN, Kuznetsov SI. 1976. Bacterial decomposition of ammonium perchlorate. Mikrobiologiya 61:347–356.

USEPA (U.S. Environmental Protection Agency). 2000. Aerojet Superfund Site Proposed Plan, USEPA Region IX. November 2000.

Wallace W, Ward T, Breen A, Attaway H. 1996. Identification of an anaerobic bacterium which reduces perchlorate and chlorate as Wolinella succinogenes. J Ind Microbiol 16:68–72.

Waller AD, Cox EE, Edwards E. 2004. Perchlorate-reducing organisms isolated from contaminated sites. Environ Microbiol 6:517–527.

Wuerl BJ, Owsianiak LM, Frankel AJ, Molnaa B. 2004. In-situ anaerobic bioremediation of perchlorate-impacted vadose zone soil. Proceedings, Groundwater Resources Association of California, Perchlorate in California's Groundwater, Glendale, CA, USA, August 4, 2004.

CHAPTER 3

PRINCIPLES OF PERCHLORATE TREATMENT

John D. Coates[1] and W. Andrew Jackson[2]

[1]University of California, Berkeley, Berkeley, CA 94720; [2]Texas Tech University, Lubbock, TX 79409

Dedication:
The authors dedicate this chapter to the memory of David C. White M.D. Ph.D. for his many paradigm shifting contributions to the fields of microbiology and microbial ecology.

3.1 INTRODUCTION

Perchlorate remediation has progressed significantly in an extremely short period. Prior to 1997 relatively few technologies were available, and there was not a substantial body of research to support the development of new technologies. With the recognition of perchlorate as a groundwater contaminant at a number of high profile sites, an enormous effort and concomitant success has been made in developing a variety of technologies capable of perchlorate remediation. These technologies include both biological- and chemical-based systems. In general, most *ex situ* remediation systems in use at drinking water treatment facilities have utilized abiotic technologies while there are numerous examples of both biotic and abiotic processes for *ex situ* remediation facilities. In contrast to the *ex situ* treatment facilities, *in situ* remediation efforts are solely based on biotic treatment processes. This chapter presents the technical foundation for perchlorate remediation technologies with a specific emphasis on those capable of *in situ* applications.

3.2 ABIOTIC REMEDIATION PROCESSES

Abiotic perchlorate remediation can generally be separated into sequestration and transformation reactions. Sequestration reactions include sorption, precipitation, ion exchange reactions and various membrane-based separations. Abiotic transformation reactions include chemical reduction and electrochemical reduction. Examples of all the above processes exist and indeed some (e.g., ion exchange) have been commonly implemented in *ex situ* clean-up efforts or drinking water treatment. However, many of the processes above have very limited or no application even in *ex situ* strategies and none, as far as the authors could discern, have been used in an *in situ* process. Regardless, a brief summary of the relevant abiotic reactions is given below including a discussion of their future potential for effective application.

3.2.1 Ion Exchange

Ion exchange is one of the most successful technologies for *ex situ* treatment of perchlorate-contaminated groundwater, especially when the water is used as a drinking water source. Ion exchange reactions can be conducted using traditional resins or activated carbon either modified or non-modified. There are numerous examples of both large- and small-scale installations and accompanying performance data. The effectiveness of ion exchange for perchlorate remediation is primarily a function of the ion exchange matrix used and, to some extent, of the concentration of other ionic constituents in water. The overall efficiency of ion exchange as a remediation option is a function of both the exchanger and the regeneration method selected, as will be discussed later. While at this point there are no known applications of *in situ* ion exchange remediation, the technology is covered here both due to its importance in *ex situ* remediation and because there is the *potential* for *in situ* application in select situations. A full discussion of the science and engineering underlying various ion exchange technologies for perchlorate remediation is beyond the scope of this chapter but can be found in Gu and Brown (2006) and accompanying references.

3.2.1.1 Non-Selective Resins

Ion exchange resins can be classified as selective or non-selective. Non-selective resins sorb a variety of ions and have higher affinities for ions with multiple charges and ions at higher concentrations. These resins are of little use for perchlorate treatment (because of low charge density) except in cases where perchlorate is at very high concentrations and the exchange process is used for product recovery, or the water has an extremely low concentration of other ions. While these resins can be regenerated with simple brine solutions, the low sorption potential necessitates frequent regeneration, producing large quantities of contaminated brine solution.

3.2.1.2 Selective Resins

Selective resins are highly specific for perchlorate, even in the presence of other ions that may be multiple orders of magnitude higher in concentration. A number of selective resins are available with varying characteristics optimized for specific remediation applications. The specificity of these resins is related to the use of quaternary ammonium groups attached to a polystyrene divinylbenzene support matrix. This matrix is inherently hydrophobic and naturally selects for poorly hydrated anions. Further selectivity is achieved by increasing the trialkyl chain length, which also decreases selectivity for hydrated multivalent anions (e.g., SO_4^{-2}). This selectivity allows the removal of perchlorate, even at low concentrations, from water with extremely high concentrations of other anions.

However, this increased selectivity is accompanied by a decrease in the rate of sorption. Formulations are based on the potentially antagonistic requirements of selective sorption versus rate of sorption. In either case, the end result of decreasing either the sorption capacity or sorption rate is an increase in the volume of resin required. One solution has been the use of bi-functional resins, that combine two types of exchange groups, to provide a fast acting exchanger with a highly selective large capacity exchanger to meet both requirements.

3.2.1.3 Advanced Regeneration Technologies

Selective resins cannot be regenerated to any significant extent using sodium chloride (NaCl) brine solutions but can be regenerated using specialized regenerating solutions. The use of specialized regeneration solutions can, in some cases, allow for the nearly complete regeneration of the resin in a very small volume of solution. In other cases, the lifetime sorption capacity of the resin is sufficiently large that the selective resin is not regenerated but simply sent for final disposal. In general, all selective resins rely on the use of pH dependent ionic compounds. Over a specific pH range, these compounds have a higher affinity than perchlorate for the resin exchange sites and a very low affinity outside this pH range. The addition of highly concentrated solutions allows for the rapid and nearly complete desorption of perchlorate in a small volume of regenerate solution (less than 1 to 3 bed volumes). Once the perchlorate has been desorbed, the column is flushed with a solution outside the optimum pH range, which causes the resin compound to reverse charge, release perchlorate and complete the regeneration cycle.

Examples of this type of process include the use of $FeCl_4^-$ and salicylic acid (Gu and Brown, 2006). $FeCl_4^-$ is applied to the spent column in a concentrated solution of hydrochloric acid in which it is stable and it preferentially replaces perchlorate from exchange sites on the resin. When the bound $FeCl_4^-$ is exposed to a neutral pH solution it spontaneously dechlorinates, producing a net positive or neutral charge, and is quickly desorbed by charge repulsion. As only the acid stabilized eluent contains perchlorate, the neutral or slightly acidic pH rinse water can be easily disposed. Use of salicylic acid is similar except in this case its ionic form occurs at neutral or alkaline pH in which it displaces perchlorate. The salicylic anion is then protonated by application of an acidic solution (e.g., HCl) and perchlorate is again repelled from the resin. Both of these examples also have very clever methods for regenerating or processing the eluent solution containing high concentrations of perchlorate. These methods include recovery of perchlorate by precipitation and destruction of perchlorate by chemical reduction or biological reactions.

The key points to the success of selective ion exchange resins are their ability to rapidly remove perchlorate, even at low concentrations from water, almost regardless of the presence of other ionic constituents and their ability to desorb the accumulated perchlorate in a very small volume (3 to 10 bed volumes) compared to the total water treated (greater than 100,000 bed volumes) prior to resin exhaustion.

3.2.1.4 Activated Carbon

Another related technology is the use of activated carbon for perchlorate removal, summarized here and extensively reviewed by Parette and Cannon (2006). Unlike the traditional use of activated carbon, which focuses on the removal of organics by adsorption, the use of activated carbon for perchlorate removal is really another form of anion exchange utilizing either natural exchange sites on the activated carbon or those added to the carbon by modification. Regardless, the removal generally conforms to the basic theory of ion exchange. Virgin activated carbon, while effective at removing low-level perchlorate, has a substantially lower capacity than the selective resins discussed above or even that of non-selective resins.

Numerous modifications to activated carbon have been investigated for their ability to increase the bed lifetime or overall sorption capacity. These include preloading with iron and oxalic acid, cationic polymer and cationic surfactants. Of these, the most effective appears to be preloading with cationic surfactants, one of which was able to increase bed life 30 times that of virgin granular activated carbon. Interestingly, if spent (in relation to perchlorate) unmodified activated carbon is loaded with cationic surfactants, 60 percent of the capacity of pre-loaded activated carbon capacity can be regained. All forms (virgin and pre-loaded) of the activated carbon appear to be highly impacted by the presence of other anions in solution, which could severely impact their performance in groundwater with moderate to high total dissolved solids (TDS).

While even modified activated carbon does not appear to be nearly as efficient as selective resins, it does have a few potential advantages. Activated carbon can be regenerated thermally with the complete decomposition of sorbed perchlorate, thus eliminating brine or regenerate stream disposal and treatment. This advantage may be less compelling if specialized regenerating solutions are used as previously discussed. Another potential advantage to the use of activated carbon is its capacity to sorb multiple contaminants simultaneously. In cases where other organic co-contaminants (e.g., chlorinated solvents) exist, the ability to remove both organic compounds and perchlorate simultaneously could be advantageous.

3.2.1.5 Potential *In Situ* Applications

Both the use of ion exchange and exchange resins have been completely confined to *ex situ* applications. However, a case could be made in specialized circumstances for the *in situ* application of the highly selective resins or even the modified activated carbon. Given the extraordinary capacity of the exchange resin, it could be argued that in situations involving very low groundwater velocities, low-level perchlorate, and a relatively narrow shallow plume that an interception permeable wall composed of resin and a support material could be economical. In cases of low permeability, where pump and treat technologies can take decades and produce extraordinarily high operating and maintenance costs, even passive biological technologies such as slow release biobarriers or substrate injections could be costly if numerous re-applications of substrate were required.

For instance, given a plume 100 meters (m) (328 feet [ft]) wide and 10 m (33 ft) deep, a reactive permeable wall (50% resin) of similar dimensions and 0.1 m (0.3 ft) thick could theoretically completely remediate a plume 5,000 m (16,404 ft) long, even assuming a low range for the partition coefficient (K_d). Significantly smaller (cross sectional area) walls could be achieved by the use of hydraulic wall and gate scenarios to funnel the plume through a small cross sectional area. While no example of this application is known and numerous issues would impact its overall cost effectiveness (long-term permeability, loss of capacity due to biofouling or scaling, etc.), application is perhaps worth considering in very specialized circumstances due to the potential for almost zero operation and maintenance costs, excluding resin replacement that may or may not be required.

3.2.2 Abiotic Reduction Technologies

3.2.2.1 Chemical Reduction

The chemical reduction of perchlorate has been extensively studied and a number of metals are known to reduce perchlorate to chloride. None of the typical abiotic remediation reductants (e.g., zero valent iron) are capable of rapid perchlorate transformation in the typical range of ambient *in situ* conditions. With one exception (Fe^{2+}), even those metals or metal complexes (e.g., Ti (III), Ru (II)) that are known to reduce perchlorate are exceedingly slow when compared to typical reduction rates for other environmental contaminants. More success has been reported for the *ex situ* application of abiotic perchlorate reduction. In these cases, the pH is typically significantly below natural pH values and/or the temperature is normally elevated. This section summarizes the most salient features of abiotic chemical reduction with an emphasis on its potential application. A full review of the mechanistic aspects and kinetics is included in Brown and Gu (2006).

Iron is one of the more commonly studied metal reductants. Rapid perchlorate reduction has been reported using Fe^{2+} at elevated temperatures, pressures and low pH (Gu et al., 2003). Perchlorate also has been shown to be slowly reduced by iron surfaces under ambient conditions (Cao et al., 2005). The use of zero valent iron, elevated temperatures (75 degrees Celsius [°C]), or ultraviolet (UV) irradiation greatly increases the reaction rates (Gurol and Kim, 2000). While *ex situ* applications of this technology may exist, such as in the regeneration of spent ion exchange regenerant solutions, it is difficult to see how they could be effectively applied *in situ*. The presence of oxygen and other constituents capable of competing with perchlorate on the iron surface would greatly decrease the already slow reaction rates.

Titanium is another commonly studied metal for remediation. The use of Ti (III) is similar to that reported for Fe. Numerous studies have shown its potential for perchlorate reduction at elevated temperatures or in solutions of low pH (Gu and Brown, 2006). Other modifications (use of ethanol or addition of ligands) have been shown to further increase the reaction rate (Earley et al., 2000). These technologies also have been proposed for regeneration of spent brine solutions where the active metal complex can be regenerated and the reaction environment controlled, but no application is envisioned for *in situ* treatment.

Other transition metals, V (III or II), Mo (III), Re (V) and Ru (II), are similarly capable of reducing perchlorate (Gu and Brown, 2006). Again the reactions at ambient conditions for the aqua metal ions are quite slow (half-life greater than 87 hours [hr]) even for the most rapid metal ions (Ru II). Much faster rates have been reported for organometallics, such as methylrhenium, and other oxorhenium (V) complexes (Abu-Omar et al., 1996; Espenson, 1999; Abu-Omar et al., 2000). The organic oxorhenium complexes are currently the most promising as they have sufficiently fast reaction rates even at typical environmental conditions, are stable to air and moisture, can be regenerated using organic thioesters and potentially could be tethered to supports. The impact of other reducible species or even other typical aqueous species has not been investigated, nor has any pilot study been reported. Improvements in chemical reduction technologies may greatly increase their application for *ex situ* perchlorate remediation in the near future. However, certainly in the near term, it does not appear that the *in situ* application of chemical reductants is likely to be practical.

3.2.2.2 Electrochemical Reduction

The difficulty of using electrochemical processes for the reduction of perchlorate is highlighted by its historic use as an inert electrolyte in corrosion and electrochemical studies. Perchlorate can be reduced on a number of electrode materials including noble (Pt, Ir, Rh, Ru), non-noble (WC, Re, TC) and in conjunction with metal corrosion (passive or induced electrode destruction) (Co, Fe, Ti) (Brown, 1986). Again, at ambient conditions, these reactions are quite slow, would require very large surface areas, and would be impacted by the presence of other species (reactive or non-reactive) in solution. No *in situ* application is envisioned but use of electrochemical processes for concentrated waste streams or for regeneration of spent brines may be feasible.

3.2.2.3 Other Abiotic Technologies

Other processes capable of perchlorate remediation include various membrane separation systems (nanofiltration, reverse osmosis, electrodialysis and precipitation). None of these technologies have any *in situ* application, but at least in the case of the membrane separation systems, they are useful for some applications (e.g., household use). Perchlorate, like any other dissolved constituent, will be rejected by membranes based on steric hindrance (pore size) and charge exclusion. While membrane systems have shown various removal levels, all will produce concentrated waste streams with respect to perchlorate and other dissolved constituents that will still require some type of treatment. Precipitation is another possible technology for some highly concentrated waste streams. However, perchlorate salts have very high solubility constants and therefore precipitation is unlikely to be useful except in cases of product recovery.

3.2.3 Overview of Abiotic Processes

Currently the outlook for the *in situ* use of abiotic processes is poor at best. While numerous abiotic processes can and are being successfully used for the remediation of perchlorate, they are all confined to *ex situ* applications. Fortuitously, the biological reduction of perchlorate is very applicable to perchlorate treatment processes and is currently being used extensively in both *in situ* and *ex situ* applications. As is often the case, reactions which are seemingly difficult to promote abiotically can readily be accomplished by the use of microorganisms.

3.3 BIOLOGICAL REMEDIATION PROCESSES

Biological remediation of perchlorate relies upon stimulating the activity of dissimilatory perchlorate-reducing bacteria (DPRB), which in the absence of oxygen, utilize the anion as a respiratory terminal electron acceptor, completely reducing it to innocuous chloride (Coates and Achenbach, 2004). These bacteria sequentially reduce perchlorate to chlorite with small amounts of chlorate sometimes being produced as a transient intermediate (Dudley et al. 2008). Chlorite is disproportionated by chlorite dismutase in a non-energy yielding reaction to produce O_2 and chloride. The oxygen is further reduced by DPRB to water. Generally, the initial reductive steps of perchlorate to chlorite are rate limiting for this metabolism and, as such, neither chlorite nor molecular oxygen are ever detectable as metabolic intermediates. The large reduction potential of perchlorate (ClO_4^-/Cl^- E^o = 1.287 V) makes it an ideal electron acceptor for microbial metabolism (Coates et al., 2000). DPRB are readily isolated from most pristine environmental sources as well as those contaminated with perchlorate, indicating that bioaugmentation is not normally a prerequisite for most sites (Coates et al., 1999b; Coates and Achenbach, 2004; Waller et al., 2004). However, little is known of this metabolism under the more extreme conditions of aridity, pH, temperature or salinity that may exist at some sites where perchlorate is known to persist.

Initial studies published on the reduction of chlorine oxyanions indicated that microorganisms rapidly reduced chlorate that was used for thistle control (Aslander, 1928). These early studies suggested that this reductive process was mediated by nitrate-respiring organisms that were using chlorate as a competitive substrate for their nitrate reductase enzymes (Hackenthal et al., 1964; Hackenthal, 1965; de Groot and Stouthamer, 1969). In support of this, many known nitrate respiring organisms, including *Escherichia coli, Proteus mirabilis, Rhodobacter capsulatus* and *Rhodobacter sphaeroides*, were shown to be capable of this metabolism (de Groot and Stouthamer, 1969; Roldan et al., 1994). However, in all cases chlorite (ClO_2^-) was produced as the final metabolic product that resulted in rapid death of the respective organisms.

Over the last decade a unique group of organisms has been identified that evolved to grow by the anaerobic reductive dissimilation of perchlorate and produce chloride as the final end product (Coates and Achenbach, 2004). Many DPRB are now available as pure cultures (Romanenko et al., 1976; Stepanyuk, 1992; Malmqvist et al., 1994; Rikken et al., 1996; Wallace et al., 1996; Bruce et al., 1999; Coates et al., 1999b; Herman and Frankenberger, 1999; Michaelidou et al., 2000; Coates et al., 2001; Okeke et al., 2002; Zhang et al., 2002) and much has been revealed of the nature of this unique metabolism and the organisms involved (Coates and Achenbach, 2004). These organisms appear to be ubiquitous in nature and have been isolated from numerous environments including both pristine and contaminated soils, waters and sediments (Romanenko et al., 1976; Stepanyuk, 1992; Malmqvist et al., 1994; Rikken et al., 1996; Wallace et al., 1996; Bruce et al., 1999; Coates et al., 1999b; Michaelidou et al., 2000; Waller et al., 2004). The environmental prevalence of these organisms was unexpected given the accepted paradigm that perchlorate in the environment is the sole result of anthropogenic activities of the armed forces as well as the manufacturing, munitions, and agricultural industries. In addition to their use as oxidizing agents in rocket propellants and other munitions, perchlorate salts have many industrial applications ranging from pyrotechnics to lubricating oils (Motzer, 2001). Perchlorate contaminantion also has been associated with the use of Chilean nitrate-based fertilizers that have been known to naturally contain perchlorate for over a century. However, the ubiquity of DPRB may be explained by recent studies that demonstrated atmospheric production of perchlorate (Dasgupta et al., 2005) and the finding of the existence of large natural perchlorate reservoirs in arid and semi-arid areas (Rajagopalan et al., 2006; Rao et al., 2007).

3.3.1 General Characteristics of DPRB

The known DPRB exhibit a broad range of metabolic capabilities including the utilization of hydrogen, simple organic acids (acetate, propionate, butyrate), alcohols (ethanol, propanol), aromatic hydrocarbons (benzene, toluene, ethylbenzene), reduced quinones (2,6-anthrahydroquinone disulfonate), both soluble and insoluble ferrous iron and hydrogen sulfide (Coates and Achenbach, 2004; Coates and Achenbach, 2006, and references therein). No DPRB are known to utilize complex substrates such as methyl soyate, molasses, or various edible oils, compounds that are often utilized as electron donors for *in situ* bioremediation of chlorinated solvents.

All known dissimilatory perchlorate-reducing bacteria are facultatively anaerobic or microaerophilic, which is understandable given that molecular oxygen is produced as a transient intermediate during the microbial reduction of perchlorate (Rikken et al., 1996; Wallace et al., 1996; Bruce et al., 1999; Coates et al., 1999b; Michaelidou et al., 2000). Most, but not all, DPRB can also respire nitrate, usually in favor of perchlorate (Chaudhuri et al., 2002; Coates et al., 1999b; Coates and Achenbach, 2004). Generally, these organisms use either chlorate or perchlorate as terminal electron acceptors, although this has only been demonstrated in a few cases. Interestingly, several pure culture chlorate-reducing bacteria, including the well-characterized *Ideonella dechloratans* and *Pseudomonas chloritidismutans* strain AW-1, are incapable of utilizing perchlorate, indicating that the dual metabolic capability is not implicit (Malmqvist et al., 1994; Wolterink et al., 2002;

Danielsson-Thorell et al., 2003; Wolterink, 2003; Coates and Achenbach, 2004; Bender et al., 2005).

3.3.2 Diversity of DPRB

The known DPRB are dominated by two novel genera, the *Dechloromonas* species and the *Azospira* (formerly *Dechlorosoma*) species (Coates and Achenbach, 2004; Coates and Achenbach, 2006). Using both culture-based and culture-independent methods, these genera have been identified and isolated from almost all environments examined, including both pristine and contaminated field sites and, as such, are considered to represent the most environmentally relevant perchlorate-reducing bacteria (Coates et al., 1999b; Achenbach et al., 2001; Coates, 2004; Coates and Achenbach, 2006). The *Dechlorospirillum* species (Coates et al., 2000; Michaelidou, 2005; Thrash et al., 2007) represent a third important group of DPRB that are underrepresented in pure culture and are closely related to the magnetotatic *Magnetospirillum* species. This group can be found in sediments but is more often identified in bioreactors treating groundwater contaminated with perchlorate (Thrash et al., 2007; Coates and Achenbach, 2006). The selective pressure for this group in bioreactors remains to be determined. The type strain and best described of this group is *Dechlorospirillum anomalous* strain WD, an organism isolated from a swine waste lagoon (Michaelidou et al., 2000). *D. anomalous* shows almost 97% 16S rDNA sequence identity to *Magnetospirillum gryphiswaldense* (Michaelidou et al., 2000) and, similarly to *Magnetospirillum* species, is also a microaerophile (Michaelidou et al., 2000). The *Magnetospirillum* genus is characterized by its ability to form magnetosomes when growing micro-aerophilically on iron-based media, which confers a unique magnetotactic characteristic on these microorganisms. In contrast to this unique characteristic of *Magnetospirillum* species, none of the *Dechlorspirillus* species tested are capable of magnetosome production.

3.3.3 Environmental Factors Controlling DPRB Activity

The nutritional requirements of all of the phylogenetically diverse perchlorate reducing bacteria remain to be determined. However, the few studies that have been performed (Bruce et al., 1999; Chaudhuri et al., 2001) indicate that the environmentally dominant *Dechloromonas* and *Azospira* DPRB have simple nutritional requirements readily found in most environments. Similarly to all organisms, they require some form of available carbon (either as organic or inorganic depending on the specific species), nitrogen, phosphorous and iron for growth (Coates and Achenbach, 2004). Molybdenum is also a required trace element for perchlorate reduction due to its functional role in the biochemistry of the perchlorate reductase enzyme (Chaudhuri et al., 2002). In acidic soil environments, bioremediation efforts may be hindered by the decreased bioavailability of molybdenum due to its enhanced adsorption to soil particulates. While pure-culture studies have normally been performed in media with a defined or undefined vitamin source, these organisms have been shown to grow and metabolize robustly in media devoid of any vitamin supplementation, suggesting that they can synthesize their own vitamin requirements (Bruce et al., 1999).

In general, DPRB prefer neutral or near neutral pH environments (Bruce et al., 1999; Coates et al., 1999b; Michaelidou et al., 2000). However, more recent field studies suggest that some related deep-branching members of the *Dechloromonas* and *Azospira* genera are common at sites with unfavorable pH or salinity, including certain species capable of growth and perchlorate reduction at pH values as low as 5. These results suggest that pH buffering may not be required at most sites (Pollock, 2005), although it may impact rates of degradation.

To date, no bacterial isolate has been demonstrated to reduce perchlorate in salinities greater than 2%. One putative perchlorate reducer, *Citrobacter* sp. strain IsoCock1 (Okeke et al., 2002), was reported to partially reduce perchlorate in salt concentrations as high as 7.5%; however, neither growth coupled to perchlorate reduction nor complete reduction of perchlorate to chloride was demonstrated for this microorganism. This presents a problem for the biological treatment of the waste brine concentrated with perchlorate collected by ion-exchange processes. Enrichment cultures, however, have been shown to reduce perchlorate at higher salinities, although nothing is known of the organisms involved or the metabolisms being utilized (Cang et al., 2004; Logan et al., 2001). Enrichment cultures were obtained from the Great Salt Lake, seawater, biofilm sludge, and marine sediments. These cultures were shown to reduce perchlorate in solutions containing up to 11% salinity. In most cases these enrichment cultures have not been rigorously studied but in general growth rates were significantly reduced at higher salinities.

Oxygen is an inhibitor of microbial perchlorate reduction at even modest concentrations (Chaudhuri et al., 2002; O'Connor and Coates, 2002; Coates and Achenbach, 2004). Perchlorate reduction by *Azospira suillum* occurred only under anaerobic conditions, and required enzymes that were only induced in the presence of perchlorate or chlorate (Chaudhuri et al., 2002; O'Connor and Coates, 2002). The absence of oxygen alone is not enough to induce enzymes required for perchlorate metabolism by DPRB, suggesting a more complex genetic regulation than simple anoxia (Chaudhuri et al., 2002). Undefined mixed perchlorate reducing consortia exposed to oxygen for limited periods of time can quickly re-establish perchlorate reduction after the added oxygen is consumed; however, longer exposures can significantly increase recovery times (Song and Logan, 2004; Shrout and Parkin, 2006). Remediation processes can operate at elevated oxygen concentrations as long as the system, whether natural (e.g., groundwater) or a biological reactor, has sufficient retention time and available substrate to biologically deplete the oxygen in addition to the perchlorate. However, these systems are heterogenous and likely contain biological flocs or biofilms in which bulk liquid oxygen or elevated redox may not be indicative of actual conditions at the point of perchlorate reduction. These systems all exhibit rapid reduction of free oxygen in solution. Regardless, it is clear from both published research and active remediation processes that systems which have steady state free oxygen will not reduce perchlorate similar to systems designed to reduce nitrate.

Nitrate also can negatively impact the production of the active enzymes involved in perchlorate reduction (Chaudhuri et al., 2002). This effect, where nitrate is preferentially used even if the cultures had previously been grown on perchlorate, has been observed with both pure cultures and some environmental samples (Chaudhuri et al., 2001; Thrash et al., 2007). However, such preferential use is not universal as several notable exceptions are known to exist (Herman and Frankenberger, 1999; Chaudhuri et al., 2002). Nitrate had no significant effect on perchlorate reduction by *Dechloromonas agitata* strain CKB, the only perchlorate-reducer described that is incapable of growth by dissimilatory nitrate reduction (Bruce et al., 1999; Coates, 2004). Interestingly, during perchlorate reduction by *D. agitata* the nitrate in the culture medium was concomitantly reduced to nitrite which accumulated in solution, suggesting that the nitrate is co-reduced by the perchlorate reductase enzyme in the organism (Chaudhuri et al., 2002). This was also shown for the DPRB strain perc1ace (Herman and Frankenberger, 1999). Further work has shown that perc1ace uses separate reductases and that there was no impact on nitrate or perchlorate reduction even if cells were grown in medium containing the opposite electron acceptor (Giblin et al., 2000). However, in contrast to *D. agitata*, strain perc1ace can grow by nitrate reduction with no nitrite accumulation (Herman and Frankenberger, 1999). *Citrobacter* strains can also reduce perchlorate in the presence of nitrate (Bardiya and Bae, 2004). In whole, these studies generally suggest that nitrate inhibition is species specific at least under the conditions examined in the previous cited research.

Studies that have investigated perchlorate reduction using undefined mixed cultures or environmental samples indicate that nitrate is typically preferentially reduced (Tan et al., 2003; Thrash et al., 2007), although in cases where electron donors are in excess, this impact may be minimized. In contrast to simple bottle studies, natural environments or complex bioreactor systems often exhibit simultaneous perchlorate and nitrate reduction, at least with respect to the overall spatial gradient (e.g., Tan et al., 2004; Tan et al., 2005; Thrash et al., 2007). However, this does not necessarily mean that both electron acceptors are simultaneously reduced at a specific point as biofilms can produce secondary concentration gradients. As such, the effects of the presence of nitrate on the overall removal of perchlorate at a particular site will be determined by the dominant perchlorate-reducing species, concentration ratio of the two electron acceptors and the availability of electron donor present. However, in general, it is reasonable to assume that both the oxygen and nitrate content of groundwater have to be depleted prior to the onset of robust biological removal of perchlorate.

3.3.4 Summary

Although biodegradation of perchlorate was recognized over 40 years ago, there was little known about the process until the late 1990s. However, the last decade has seen a significant increase in our understanding, and has laid the basis for implementation of full-scale *in situ* bioremediation processes for perchlorate. The key points for practitioners from the microbiological research to date are:

1. Several bacterial species, from a variety of genera, are capable of completely reducing perchlorate, through chlorate and chlorite, to innocuous chloride.

2. Perchlorate reducers are virtually ubiquitous in groundwaters and soils, and bioaugmentation should not be needed for most applications.

3. Perchlorate reducers are "generalists", with a broad range of metabolic capabilities, but they do not directly use complex organic carbon sources, including those often used for enhanced anaerobic bioremediation (e.g., edible oils or molasses).

4. Perchlorate reducers are facultative anaerobes (anaerobic bacteria that are able to use oxygen) or microaerophilic bacteria, and the presence of oxygen will strongly inhibit perchlorate reduction.

5. Most perchlorate reducers are also able to respire nitrate and, in fact, prefer nitrate over perchlorate as an electron acceptor, so the presence of nitrate will generally inhibit perchlorate reduction.

6. Perchlorate reducers grow over a broad range of conditions, though little is known about perchlorate reduction under extreme environmental conditions.

7. Perchlorate reduction is most rapid near neutral pH values, though perchlorate reducers are able to tolerate slightly acidic conditions (pH 5.0). Little is known about perchlorate under extreme pH conditions.

8. Perchlorate reduction is generally limited to low-salt conditions (i.e., <2% NaCl), though there is some evidence for reduction at higher salinities (<11% NaCl).

3.4 CHALLENGES ASSOCIATED WITH MICROBIAL PERCHLORATE REDUCTION

3.4.1 Biofouling and Electron Donor Selection

DPRB are ubiquitous and capable of growth under a wide range of environmental conditions utilizing a diverse range of substrates (Coates et al., 1999b; Coates and Achenbach, 2004). Generally, bioaugmentation is not required for *in situ* bioremediation of perchlorate. *In situ* groundwater bioremediation technologies are based on stimulating the activity of indigenous DPRB, normally by addition of an electron donor. The electron donor or substrate can either be directly used (such as acetate, ethanol, or H_2), or be a more complex organic substrate that must first be partially degraded before producing organic metabolites that can be used by DPRB.

Substrate application rates are generally in excess of perchlorate concentrations or fluxes, creating an elevated concentration of the electron donor in the reactive zone. As microbial perchlorate reduction is inhibited by the presence of O_2 and to some extent nitrate, excess substrate must be added to biologically remove these components prior to initiation of perchlorate reduction. In addition, diffusion and mixing are utilized to increase the reactive zone beyond that solely attributable to the active application technology (e.g., *in situ* injection, infiltration, *in situ* generation). However, improper application of readily degradable organic substrates into the subsurface can also stimulate other undesirable non-perchlorate-reducing microorganisms within the substrate impacted area. Over the long-term, this competition for electron donors can result in ineffective treatment of perchlorate, due to the loss of added electron donor to other microbial processes (e.g., Fe (III) or SO_4^{2-} reduction) and potential plugging of the aquifer matrix (biofouling). This also can have secondary impacts on the physical-chemical nature of the aquifer matrix such as mineral content, hydraulic conductivity and pH, and reduce overall water quality through the direct or indirect release of undesirable end-products (Fe (II), HS^-, CH_4, and mobilized heavy metals).

There are two key biological issues that impact the choice of electron donor in order to stimulate microbial perchlorate reduction as outlined below.

3.4.1.1 Stimulation of Undesirable Organisms

As mentioned, many substrates or electron donors can be used to stimulate the activity of indigenous DPRB. These substrates can either be used directly or indirectly after transformation by non-DPRB. Substrates that must first be biotransformed often cause rapid increases in fermentative microorganism populations which obtain energy during the biotransformation. These fermentative bacteria do not directly impact perchlorate. Examples of complex substrates from which this will occur include: methyl soyate, molasses and various edible oils. However, even some relatively simple organic compounds like citrate must first be fermented into simpler compounds (e.g., acetate, propionate and lactate) before microbial perchlorate reduction is stimulated. DPRB are unable to utilize complex substrates and even the ubiquitous and metabolically versatile *Dechloromonas* and *Azospira* genera are only capable of biodegrading low molecular weight organic compounds including monocarboxylic and dicarboxylic acids, simple alcohols and monoaromatic compounds (Coates and Achenbach, 2004).

3.4.1.2 Establishment of Nonproductive TEAPs

In the natural environment, the population structure of microbial communities is controlled primarily by the dominant terminal electron accepting process (TEAP). In any given environment multiple TEAPs may exist, but they will be stratified spatially with respect to the flux of electron acceptor. This stratification can occur over large distances (meters) such as in aquifers or over very small distances (micrometers) in the case of biofilms. The sequence of redox zones is based on the preferential use of more thermodynamically favorable electron acceptor-donor pairs. For instance, microbial perchlorate reduction is less energetically favorable than oxygen reduction. Thus, microbial populations will utilize oxygen prior to perchlorate as it produces a greater benefit (available energy). The energetic gain of an organism that utilizes perchlorate as an electron acceptor is similar to the use of nitrate as an electron acceptor and more favorable than the use of Fe (III), sulfate, or CO_2 as electron acceptors (Coates et al., 2000; Coates and Achenbach, 2004).

In any given location, as the most energetically favorable electron acceptor is depleted, the microbial community will evolve to take advantage of the next most thermodynamically favorable electron acceptor available (Champ et al., 1979; Lyngkilde and Christensen, 1992; Lovley and Chapelle, 1995; Anderson and Lovley, 1997; Christensen et al., 2000; Coates and Achenbach, 2001). As mentioned, if oxygen is present, an electron donor will be used by bacteria to consume oxygen. If sufficient donor is available to allow for the complete consumption of the oxygen, then NO_3^- consumption will occur next. This process will continue, as long as sufficient substrate is available, through the complete sequential process of TEAPs (ClO_4^-, Fe (III), SO_4^{-2}, CO_2). However, because CO_2 will always be present in a eutrophic environment, methanogenesis can occur as long as electron donor is continually supplied.

The particular electron acceptor being used at any specific location is a function of the rate of substrate and TEAP supply, and the rate of substrate and TEAP utilization. While the rate of TEAP and substrate utilization is largely dependent, significantly different processes can control the rate of substrate and TEAP supply. For the removal of perchlorate, both oxygen and nitrate must first be consumed at any given location and only enough electron donor must be added to reduce these plus the perchlorate. In practice, this is difficult to achieve because the total electron accepting capacity of the perchlorate is generally minor relative to the other electron acceptors available in the natural environment. If substrate is applied at a rate in excess of the consumption rate of O_2, NO_3, and ClO_4, then less thermodynamically favorable TEAPs (Fe (III), SO_4, CO_2) will be utilized.

It should be noted, as mentioned above, that spatial TEAP stratification can occur on many scales. In groundwater, TEAPs are generally organized in the direction of groundwater flow. Depending on the rate of substrate consumption, a given TEAP can be quite large. However in cases where substrate fluxes are more dominant, TEAPs can exist separated by very small spatial differences. In some cases, the differences are so small that they will appear to co-occur. In some environments it is even possible to have a bi-directional spatial gradient—one that is organized in relation to the overall flux of substrate (generally in the case of groundwater the direction of flow) and one that is controlled by diffusion of substrate into biofilms or microsites on or in the aquifer media. Again, this can cause the appearance of simultaneous TEAP consumption relative to the larger spatial domain (e.g., flow direction).

In reality, the spatial stratifications are quite complex and depend not only on the substrate flux but also on the electron acceptor flux or fluxes. Regardless, the addition of complex substrates that cannot be directly utilized by perchlorate reducing populations and/or the addition of substrates at rates in excess of perchlorate consumption rates can cause undesirable side-effects, including reduced perchlorate removal effectiveness. Typical effects could include:

1. Stimulation of substantial non-perchlorate-reducing microbial communities, resulting in a significant loss of the added electron donor to metabolisms other than perchlorate reduction.
2. Biofouling or loss of hydraulic conductivity especially near the point of electron donor addition (e.g., injection well) due to excessive microbial growth.
3. Increased activity of microbial communities, such as Fe (III) reducers, sulfate reducers, and methanogens, causing secondary water quality impacts such as increases in soluble Mn (II) and Fe (II), releases of adsorbed metals and phosphates and production of sulfides and methane gas.
4. Solubilization and mobilization of normally immobile toxic metals (e.g., copper, zinc, or chromium) through complexation with certain electron donors or their metabolites such as citrate or oxalate.

The extent of each of these effects is partially a function of the manner of injection (e.g., continuous versus pulsed) and concentration of substrate versus electron acceptors, and partially a function of the nature of the electron donor selected and its chemical and biological reactivity. While the manner of injection and concentration of substrate are dependent on site characteristics, the selection of electron donor that will minimize deleterious impacts is largely site independent. As such, the biogeochemical characteristics of an ideal electron donor can be identified:

1. Donor should be non-fermentable. The electron donor should be directly utilizable by the indigenous DPRB. This will reduce the impact of fermentative microbial populations and reduce biofouling potentials.
2. Donor should be biocidal at elevated concentrations and thus inhibit the growth of all microorganisms at elevated concentrations. This will reduce near-well biofouling and increase the zone of impact as well as reduce/remove the need for additional biocidal compounds to prevent well plugging.
3. Donor should, if possible, only be utilized by microorganisms capable of perchlorate reduction. It should be non-biodegraded by Mn(VI)-reducing, Fe(III)-reducing, sulfate-reducing or methanogenic bacteria which consume the substrate and increase the overall substrate demand and the potential for biofouling, and decrease water quality through the production of undesirable end products.
4. Donor should not readily complex and solubilize insoluble metals, thus mobilizing them in the groundwater until the donor is biodegraded, at which point the metal re-precipitates back out of solution (Ehrlich, 1990). Such mobilization and re-precipitation may result in localized mineral formation, causing irreversible reduction in porosity.
5. Donor should have a large electron donating capacity per unit cost. For example, perchlorate reduction can be stimulated by the addition of formate or acetate according to the molar ratios outlined in Reactions 3.1 and 3.2 below.

$$CH_3COO^- + ClO_4^- + H^+ \rightarrow 2CO_2 + Cl^- + 2H_2O \qquad (Rx.\ 3.1)$$

$$4HCOOH + ClO_4^- \rightarrow 4CO_2 + Cl^- + 4H_2O \qquad (Rx.\ 3.2)$$

Oxidation of acetate reduces one mole of perchlorate per mole of acetate oxidized, whereas four moles of formate are required to reduce one mole of perchlorate. This assumes that all of the available reducing equivalents are being directed into reduction of perchlorate by bacteria rather than into carbon assimilation and biomass production. In general, for heterotrophic bacteria, phenotypic studies have indicated that molar ratios of 1.2 to 1.5 times the theoretical concentrations are required to account for biomass production and effective removal of perchlorate (Bruce et al., 1999; Chaudhuri et al., 2002) Autotrophic bacteria utilizing H_2 as an energy source but growing on a separate carbon substrate may have substantially different molar ratios.

Although no one potential electron donor matches all of the requirements outlined, several requirements can be satisfied through the use of individual compounds or mixtures of non-fermentable electron donors. As an example, sodium benzoate is highly soluble, biocidal at high concentrations, non-fermentable and can be utilized directly as an electron donor by DPRB such as *Dechloromonas aromatica* (Chakraborty and Coates, 2005). Benzoate is completely oxidized to CO_2 (Rx. 3.3) and has a high electron donating capacity (30 reducing equivalents per molecule).

$$4C_6H_5COO^- + 15ClO_4^- + 4H^+ \rightarrow 28CO_2 + 15Cl^- + 12H_2O \qquad (Rx.\ 3.3)$$

Benzoate is approved as a food additive and is, at present, priced competitively with acetate, ethanol, or citrate. Complex substrates, such as molasses, or edible vegetable oils should be carefully considered. Although such compounds will initially stimulate successful microbial removal of perchlorate *in situ*, their long-term application may result in several undesirable biogeochemical and biofouling effects. However, the relative cost of these bulk compounds and their ease of use may effectively outweigh any potential disadvantages of their application, especially in short-term treatment processes.

3.5 THE TOOLS AVAILABLE FOR PREDICTING AND MONITORING MICROBIAL PERCHLORATE REDUCTION

Because current *in situ* perchlorate remediation efforts are entirely based on the activity of perchlorate reducing bacteria, it is important to not only identify the presence of these organisms prior to a process design, but also to monitor the health and activity of these organisms throughout the course of the treatment. As outlined above, in the past few years phenotypic characterization studies have demonstrated that the known perchlorate-reducing bacteria exhibit a broad range of metabolic capabilities and can thrive in adverse environments. Similarly, significant advances have been made in the biochemistry and genetic systems involved in microbial perchlorate reduction and the environmental factors that affect their activity (Bender et al., 2002; Chaudhuri et al., 2002; Okeke and Frankenberger, 2003; Bender, et al., 2004; Coates, 2004). As such, the applicability of this metabolism offers great potential for the bioremediation of perchlorate-contaminated environments.

Several tools based on unique signature molecules (biomarkers) characteristic of DPRB and novel metabolic capabilities are now available through commercial laboratories.[1] Such molecular biological tools can be used to determine the potential for *in situ* bioremediation of perchlorate as well as monitor its effectiveness in field environments. Although many field treatments have been performed without the application of these tools, in general these have been short-term projects lasting less than one year. In more long-term remediation efforts, application of these tools will provide an inexpensive preventative maintenance screen for the operator to ensure continued successful remediation and will help predict the potential for catastrophic failures before they arise. The application of these tools will also define the zone of impact of the remediation efforts to ensure design optimization and cost minimization. Because of their complexity, many of these analyses are beyond the scope of most general laboratories; however, for a specialized laboratory the techniques are relatively straightforward and a rapid sample turnaround can be achieved.

3.5.1 Most Probable Number Counts

Most probable number counts (MPNs) (Halvorson and Ziegler, 1933) for perchlorate-reducing populations (Coates et al., 1999a; 1999b) in freshwater environments can be performed with a slight modification to the medium outlined in Appendix 3.1. Sodium or ammonium perchlorate at a final concentration of 5 millimolar (mM) is optimum. Higher concentrations may result in false negatives due to toxicity, while lower concentrations may not allow for non-ambiguous results. A non-fermentable electron donor, such as H_2 (101 kilopascals [kPa]), acetate (2 mM), or ethanol (5 mM), should be used. If H_2 is being used, yeast extract (0.1 grams per liter [g/L]) should be added as a carbon source.

Media are dispensed in 9 milliliter (mL) aliquots into 30 mL glass pressure tubes and degassed individually with N_2-CO_2 (80-20, volume per volume [vol/vol]) as outlined below. The prepared media tubes are heat sterilized by autoclaving at 121°C for 15 minutes (min). Just prior to the addition of the environmental sample, 0.1 mL sodium pyrophosphate from a sterile anaerobic 10% (weight per volume [wt/vol]) aqueous stock solution is added to the initial dilution tubes. This will serve to release any cells adsorbed onto the soil/sediment particles and significantly improve the counts obtained. Tubes should be incubated at temperatures suitable to the original sample environment. MPN culture tubes should be checked after 60 days incubation. Positives can be identified unambiguously by measuring the removal of perchlorate relative to the uninoculated medium control. An initial quick screening of the tubes can be done by visual observation of development of an optically dense suspension (white cloudy suspension in colorless medium) of cells in the culture tubes. The number of DPRB in the original sample can then be calculated using a standardized formula (Halvorson and Ziegler, 1933) and tables or through application of a standardized MPN calculator such as the Most Probable Number Calculator version 4.04 ©1996 available at http://www.epa.gov/nerlcwww/other.htm.

---

[1] The only laboratory known to the authors to perform these analyses is BioInsite, LLC, Illinois, (www.bioinsite.com), but there will undoubtedly be others in the future.

3.5.2 Probes to Specific Groups of Perchlorate-Reducing Organisms

It has long been recognized that comparison of the gene sequence encoding the small subunit of the ribosomal RNA (16S rRNA) can be used to measure the relationship between any two microorganisms. From this information, certain limited conclusions can be drawn regarding the metabolic capabilities of unknown microorganisms. However, such genetic comparisons cannot be used to categorically identify the metabolic capability of a microorganism, particularly perchlorate reduction, because of the high 16S rRNA gene sequence similarity between many of the known DPRB and their closest non-perchlorate-reducing relatives. Even so, molecular probes specific to the 16S rRNA genes of the *Dechloromonas*, *Azospira*, and *Dechlorospirillum* genera have been designed and proven to be of use for the rapid prescreening of environmental samples for the presence of these bacteria or to monitor the health of a known perchlorate-reducing population in soils or bioreactors (Coates et al., 1999a).

When used in conjunction with enumeration techniques for DPRB, such as MPNs or real-time polymerase chain reaction (PCR), specific 16S rDNA molecular probes can be used to monitor population shifts in response to particular stimuli introduced as part of a bioremediation process, from which the effectiveness of the strategy can be inferred[2].

3.5.3 Biomarkers for All DPRB

The identification of several genes involved in the reduction of perchlorate and chlorate now makes it possible to use several different molecular approaches to assist bioremediation efforts. For example, because chlorite dismutase is a highly conserved enzyme unique to organisms capable of perchlorate or chlorate reduction (Coates et al., 1999b; O'Connor and Coates, 2002), the gene encoding this protein is an ideal target for detecting the presence of any perchlorate-reducing bacteria in the environment, regardless of their phylogenetic affiliation (Bender et al., 2002; Bender et al., 2004). Using this approach, detection of the chlorite dismutase gene from an environmental sample can be accomplished in a short time using specific molecular probes and can be used to determine if the indigenous bacterial population is capable of perchlorate reduction (Bender et al., 2004).

A method for quickly enumerating perchlorate-reducing bacteria using the chlorite dismutase gene is achieved by combining the traditional MPN technique with a PCR DNA amplification (Holmes et al., 2002). MPN-PCR is a technique involving extraction and dilution of DNA prior to PCR amplification for quantification of target molecules. Quantification is based on the statistical analysis of a triplicate series in which the template is diluted to extinction. The actual number of target cells is calculated using an algorithm based on sample dilution and probability. MPN-PCR can be readily and rapidly performed using standard PCR reagents and equipment. This technique is very robust and rapid, and is sensitive for DPRB populations as low as 10^2 cells per gram (or per mL) of sample. It can be readily applied to soil/sediment and groundwater samples with results being achievable within 48 hrs.

---

[2] A recent study in the author's laboratory using this approach to investigate the perchlorate-reducing population associated with an active permeable barrier treating perchlorate- and radionuclide-contaminated surface waters in Los Alamos, New Mexico, indicated that the perchlorate population was dominated by species of the *Dechloromonas* genus and that the relative size of this population responded directly to perchlorate concentrations and water volume treated within a six-month period.

3.5.4 Immunoprobes Specific for DPRB

An alternative probe for DPRB was recently developed based on the ability of antibodies to target and attach to specific antigenic structures within a compound. Because of the highly conserved nature of the chlorite dismutase (CD) enzyme at the amino acid level among all DPRB, regardless of their phylogenetic affiliation and the uniqueness of this enzyme to these microorganisms, the CD protein represents an ideal target for a DPRB-specific immunoprobe. In addition, this probe is unaffected by non-perchlorate-reducing microorganisms, such as *M. magnetotacticum*, which carry the *cld* gene but do not produce an active chlorite dismutase enzyme.

A recent study demonstrated the effectiveness of this approach by raising polyclonal antisera against the purified chlorite dismutase from *Dechloromonas agitata* strain CKB (O'Connor and Coates, 2002). Characterization studies indicated that the anitsera had a high affinity for the CD enzyme and activity was observed in dilutions as low as 1×10^{-6} of the original antisera. The antisera was active against both cell lysates and whole cells of all DPRB tested, regardless of phylogenetic affiliation but only if the cells were grown on perchlorate. Little or no cross reactivity was observed with closely related non-perchlorate-reducing relatives (O'Connor and Coates, 2002). With this immunoprobe as a basis, a rapid enzyme-linked immunosorbent assay (ELISA) was developed that is specific for DPRB actively metabolizing perchlorate. Cell populations as low as 200 DPRB cells per mL can readily be detected and enumerated in aqueous samples colorimetrically within 45 minutes. This assay allows the rapid screening of environmental samples for actively metabolizing perchlorate-reducing bacteria. However, in contrast to the molecular approach outlined above, ELISA assays are unsuitable for most solid phase samples and are better suited for rapid analysis of groundwater.

3.5.5 Use of Stable Isotopes to Identify Perchlorate Source and Monitor Degradation

Although both molecular and immunological tools based on unique signature molecules are now available to monitor the microbial populations associated with perchlorate reduction in the environment, monitoring the effectiveness of perchlorate bioremediation in field environments is often difficult owing to the complex nature of environmental samples. Results can often be tainted by many abiotic factors including adsorption, dilution or chemical reactivity of the target contaminant. One potential strategy for overcoming these shortcomings with many compounds is to follow the changes in stable isotope composition of the molecule of interest. Variations of the stable isotope ratios of many elements have been used for a long time to give valuable information about elemental sources and biogeochemical processes occurring in the environment (Nissenbaum et al., 1972; Bailey et al., 1973; Ku et al., 1999).

The stable isotopic signature of a molecule can be used as a means of fingerprinting and locating the source of a compound. Many atoms can exist in two or more forms, chemically identical but differing in mass. The relative abundances of the stable (non-radioactive) isotopes are effectively constant for each element. Chlorine has two stable isotopes, ^{35}Cl or and ^{37}Cl with a natural abundance of approximately 75% and 25%, respectively.

There are relatively few examples of major physical or chemical fractionating processes operating naturally for chlorine, although some do exist. Probably the largest fractionation effect is attributable to aqueous diffusion of dissolved chloride in marine pore-waters in low permeability rocks (originally ~ 0‰), which results in relative depletion of ^{37}Cl from the brines and an isotopic ratio of ~ –0.9‰ at the diffusion front.

By contrast, significantly larger changes in isotopic content can result from chemical manufacturing processes where, for example, chlorinated hydrocarbon solvents produced from natural sodium chloride (~0‰), can show a range of isotopic signature values from –3‰ to +4‰ depending on the manufacturing processes used (Jendrzejewski et al., 2001). As such, the stable isotopic content of anthropogenic perchlorate will be dependent on both the original source of chloride in the perchlorate and the manufacturing process used and may be distinguishable from that of naturally occurring perchlorate. As outlined by Böhlke and co-workers (Böhlke et al., 2005), this approach has been used to develop a fingerprinting technique for source identification of perchlorate contamination.

However, such fingerprinting should be viewed with caution. Microbial processes are known to make significant changes to the isotopic compositions of many elements, such as carbon or sulfur, by preferentially utilizing the lighter isotope. In general, perchlorate-reducing bacteria are able to distinguish between light and heavy isotopes in the chlorine (Coleman et al., 2003) and oxygen content (Sturchio et al., 2007) of perchlorate. Recent studies demonstrated that one of the environmentally dominant perchlorate-reducing bacteria, *Azospira suillum*, preferentially utilizes perchlorate containing the lighter isotope (^{35}Cl), resulting in a significant fractionation (-15‰) of the isotopic content of the perchlorate as the organism grows in pure culture (Coleman et al., 2003). A subsequent study demonstrated similar isotopic fractionation of the chlorine content of perchlorate when DPRB were grown in natural sediments (Sturchio et al., 2003).

The results of these studies suggest that isotope-signature tracing can be successfully applied to monitor the microbial reduction and removal of perchlorate in environments being treated for perchlorate contamination and distinguish this from abiotic reactions (dilution or absorption). However, the results also suggest that care must be taken when using isotope fingerprinting to identify perchlorate sources, because any intrinsic microbial reduction occurring in the environment may alter the isotope fingerprint obtained, resulting in false identifications. Thus, indigenous DPRB population sizes and intrinsic activity must be accounted for to ensure reliable source identification.

3.6 ENRICHMENT, ISOLATION, AND MAINTENANCE OF DPRB

Perchlorate reducing bacteria are relatively non-fastidious organisms making them easy to isolate, culture, and maintain in a general laboratory with little specialized equipment. Some proven successful approaches are outlined below.

3.6.1 Direct Isolation

Dissimilatory perchlorate-reducing bacteria can be selectively enriched from diverse habitats in an anoxic basal medium (Appendix 3.1) using various non-fermentative alternative electron donors. Samples collected from the field should completely fill any vessel in which they are collected to exclude air in the headspace. These should be sealed, transported back to the laboratory at controlled refrigerated (4°C) temperatures (on ice is appropriate as long as freezing of the sample is avoided) and used immediately. If not used immediately, samples should not be frozen but may be stored at 4°C for short periods (<48 hr). Enrichments should be initiated with inoculum sizes of 10% by weight of the culture volume. Incubations should be carried out at environmental temperatures depending on the source of the sample.

Dissimilatory perchlorate-reducing bacteria can be directly isolated from a broad diversity of environments using a modified shake tube method. The medium of choice (Appendix 3.1) uses sodium perchlorate (5 mM) as the sole electron acceptor and a non-fermentable electron donor such as H_2 (101 kPa), acetate (10 mM), or ethanol (10 mM). The medium should be prepared using standard anaerobic techniques as outlined below. Freshly collected environmental samples are serially diluted to 10^{-9} in this medium. Aliquots (7 mL) of the respective dilutions are transferred anaerobically into glass pressure tubes containing 3 mL of sterile molten noble agar (Difco) (4% wt/vol) at 55°C under a gas phase of N_2-CO_2 (80-20, vol/vol). The sample is mixed by inverting the tube several times and then solidified by plunging into an ice bath. The solidified dilutions are incubated inverted. Colonies of perchlorate-reducing bacteria should be visible in the lower dilutions (10^{-1} - 10^{-3}) after two weeks incubation. These can easily be recognized as small (0.5 to 1 millimeter diameter) pink colonies in the translucent white-colored agar. In an anaerobic glove bag, colonies can be picked as plugs, using a sterile pasteur-pipette, and transferred into fresh anaerobic medium (5 mL) with sodium perchlorate (5 mM) and a suitable electron donor. Active cultures can easily be recognized after two to four weeks incubation by the development of an optically dense suspension of cells in the perchlorate-medium. Active cultures should be transferred through a second dilution shake tube series to ensure isolation.

3.6.2 Culture Maintenance

All mesophilic perchlorate-reducing cultures can be maintained as frozen stocks at -70°C. The most reliable technique is to grow the culture in a medium amended with a soluble electron donor and acceptor respectively such as acetate (10 mM) and perchlorate or chlorate (8 mM). Once a dense culture has been obtained, aliquots (1 mL) should be anaerobically transferred into small serum vials (10 mL) that have previously been gassed out with N_2-CO_2 (80-20; vol/vol) and heat sterilized. The vials should be amended with an anaerobic sterile aqueous glycerol solution (100 microliter) (25% vol/vol), and then mixed and frozen at -70°C. Frozen stocks should be checked regularly to ensure viability.

3.7 CONCLUSIONS

Although abiotic processes for the treatment of perchlorate contamination are of limited applicability *in situ*, the application of biologically-based systems has proven to be quite robust, at least over the short-term. The field of microbial perchlorate reduction and its application to the *in situ* treatment of perchlorate contamination has clearly advanced significantly in a very short period from a poorly understood metabolism in 1997 to a burgeoning scientific field of discovery today. As outlined above, there is now a much greater understanding of the microbiology involved and the application of that understanding to the successful treatment of contaminated environments.

Overall, the future is promising, even though the application of perchlorate bioremediation in the field is still in its infancy. As more of these treatments come online, information and experience will be gathered that will allow for better predictive models of successful treatment strategies and the identification of potential pitfalls. With the development of this technology comes a better understanding of the ideal electron donors available and the individual factors which truly control the activity of these organisms, allowing for the design of more effective and robust enhanced *in situ* bioremediation technologies.

REFERENCES

Abu-Omar MM, Appleman EH, Espenson JH. 1996. Oxygen-transfer reactions of methylrhenium oxides. Inorg Chem 35:7751–7757.

Abu-Omar MM, McPherson LD, Arias J, Bereau VM. 2000. Clean and efficient catalytic reduction of perchlorate. Angew Chem Int Ed Engl 39:4310–4313.

Achenbach LA, Bruce RA, Michaelidou U, Coates JD. 2001. *Dechloromonas agitata* N.N. gen., sp. nov. and *Dechlorosoma suillum* N.N. gen., sp. nov. two novel environmentally dominant (per)chlorate-reducing bacteria and their phylogenetic position. Int J Syst Evol Microbiol 51:527–533.

Anderson RT, Lovley DR. 1997. Ecology and biogeochemistry of *in situ* groundwater bioremediation. Adv Microbial Ecol 15:289–350.

Aslander A. 1928. Experiments on the eradication of Canada Thistle, *Cirsium arvense*, with chlorates and other herbicides. J Agric Res 36:915.

Bailey NJL, Krouse HR, Evens CR, Rogers MA. 1973. Alteration of crude oil by waters and bacteria - evidence from geochemical and isotope studies. Am Assoc Petrol Geol Bull 57:1276.

Bardiya N, Bae JH. 2004. Role of *Citrobacter amalonaticus* and *Citrobacter farmeri* in dissimilatory perchlorate reduction. J Bas Microbiol 44:88–97.

Bender KS, O'Connor SM, Chakraborty R, Coates JD, Achenbach LA. 2002. The chlorite dismutase gene of *Dechloromonas agitata* strain CKB: Sequencing, transcriptional analysis and its use as a metabolic probe. Appl Environ Microbiol 68:4820–4826.

Bender KS, Rice MR, Fugate WH, Coates JD, Achenbach LA. 2004. Metabolic primers for detection of (per)chlorate-reducing bacteria in the environment and phylogenetic analysis of *cld* gene sequences. Appl Environ Microbiol 70:5651–5658.

Bender KS, Shang C, Chakraborty R, Belchik SM, Coates JD, Achenbach LA. 2005. Identification, characterization, and classification of genes encoding perchlorate reductase. J Bacteriol 187:5090–5096.

Böhlke JK, Sturchio NC, Gu B, Horita J, Brown GM, Jackson WA, Batista JR, Hatzinger PB. 2005. Perchlorate isotope forensics. Anal Chem 77:7838–7842.

Brown GM. 1986. The reduction of chlorate and perchlorate ions at an active titanium electrode. J Electroanal Chem 198:319–330.

Brown GM, Gu B. 2006. The chemistry of perchlorate in the environment. In Gu B, Coates JD, eds, Perchlorate Environmental Occurrence, Interactions and Treatment, Springer, New York, NY, USA, pp 17–47.

Bruce RA, Achenbach LA, Coates JD. 1999. Reduction of (per)chlorate by a novel organism isolated from a paper mill waste. Environ Microbiol 1:319–331.

Cang Y, Roberts DJ, Clifford DA. 2004. Development of cultures capable of reducing perchlorate and nitrate in high salt solutions. Water Res 38:3322–3330.

Cao JS, Elliott D, Zhang WX. 2005. Perchlorate reduction by nanoscale iron particles. J Nanopart Res 7:499–506.

Chakraborty R, Coates JD. 2005. Hydroxylation and carboxylation - two crucial steps of anaerobic benzene degradation by *Dechloromonas* strain RCB. Appl Environ Microbiol 71:5427–5432.

Champ DR, Gulens J, Jackson RE. 1979. Oxidation-reduction sequences in ground water flow systems. Can J Earth Sci 16:12–23.

Chaudhuri SK, Lack JG, Coates, JD. 2001. Biogenic magnetite formation through anaerobic biooxidation of Fe(II). Appl Environ Microbiol 67:2844–2848.

Chaudhuri SK, O'Connor SM, Gustavson RL, Achenbach LA, Coates JD. 2002. Environmental factors that control microbial perchlorate reduction. Appl Environ Microbiol 68:4425–4430.

Christensen TH, Bjerg PL, Banwart SA, Jakobsen R, Heron G, Albrechtsen H. 2000. Characterization of redox conditions in groundwater contaminant plumes. J Contam Hydrol 45:165–241.

Coates JD. 2004. Bacteria that respire oxyanions of chlorine. In Brenner D, Krieg N, Staley J, Garrity G, eds, Bergey's Manual of Sytematic Bacteriology, 2nd ed, Vol 2. Springer-Verlag, New York, NY, USA.

Coates JD, Achenbach LA. 2001. The Biogeochemistry of Aquifer Systems. In Hurst CJ, Knudsen GR, McInerney MJ, Stetzenbach LD, Walter MW, eds, Manual of Environmental Microbiology, 2nd ed. ASM Press, Washington, DC, USA, pp 719–727.

Coates JD, Achenbach LA. 2004. Microbial perchlorate reduction: Rocket-fuelled metabolism. Nat Rev Microbiol 2:569–580.

Coates JD, Achenbach LA. 2006. The microbiology of perchlorate reduction and its bioremediative application. In Gu B, Coates JD, eds, Perchlorate, Environmental Occurrence, Interactions, and Treatment. Springer, New York, NY, USA, pp 279–295.

Coates JD, Bruce RA, Patrick JA, Achenbach LA. 1999a. Hydrocarbon bioremediative potential of (per)chlorate-reducing bacteria. Bioremediation J 3:323–334.

Coates JD, Chakraborty R, Lack JG, O'Connor SM, Cole KA, Bender KS, Achenbach LA. 2001. Anaerobic benzene oxidation coupled to nitrate reduction in pure culture by two strains of *Dechloromonas*. Nature 411:1039–1043.

Coates JD, Michaelidou U, Bruce RA, O'Connor SM, Crespi JN, Achenbach LA. 1999b. The ubiquity and diversity of dissimilatory (per)chlorate-reducing bacteria. Appl Environ Microbiol 65:5234–5241.

Coates JD, Michaelidou U, O'Connor SM, Bruce RA, Achenbach LA. 2000. The diverse microbiology of (per)chlorate reduction. In Urbansky ED, ed, Perchlorate in the Environment, 1st ed. Kluwer Academic/Plenum, New York, NY, USA, pp 257–270.

Coleman ML, Ader M, Chaudhuri S, Coates JD. 2003. Microbial isotopic fractionation of perchlorate chlorine. Appl Environ Microbiol 69:4997–5000.

Danielsson-Thorell H, Stenklo K, Karlsson J, Nilsson T. 2003. A gene cluster for chlorate metabolism in *Ideonella dechloratans*. Appl Environ Microbiol 69:5585–5592.

Dasgupta PK, Martinelango PK, Jackson WA, Anderson TA, Tian K, Tock RW, Rajagopalan S. 2005. The origin of naturally occurring perchlorate: the role of atmospheric processes. Environ Sci Technol 39:1569–1575.

de Groot GN, Stouthamer AH. 1969. Regulation of reductase formation in *Proteus mirabilis*. I. Formation of reductases and enzymes of the formic hydrogenlyase complex in the wild type and in chlorate resistant mutants. Arch Microbiol 66:220–233.

Dudley M, Salamone A, Nerenberg R. 2008. Kinetics of a chlorate-accumulating, perchlorate-reducing bacterium. Water Res (Epub, advanced publications).

Earley JES, Tofan DC, Amadei GA. 2000. Reduction of perchlorate ion by titanous ions in ethanolic solution. In Urbansky ET, ed, Perchlorate in the Environment, 1st ed. Kluwer/Plenum, New York, NY, USA, pp 89–98.

Ehrlich HL. 1990. Geomicrobiology. Marcel Dekker, Inc., New York, NY, USA.

Espenson JH. 1999. Atom-transfer reactions catalyzed by methyltrioxorhenium(VII)—mechanisms and applications. Chem Commun 6:479–488.

Giblin TL, Herman DC, Frankenberger WT. 2000. Removal of perchlorate from ground water by hydrogen-utilizing bacteria. J Environ Qual 29:1057–1062.

Gu B, Brown GM. 2006. Recent advances in ion exchange for perchlorate treatment, recovery and destruction. In Gu B, Coates JC, eds, Perchlorate Environmental Occurrence, Interactions and Treatment. Springer, New York, NY, USA, pp 209–249.

Gu B, Dong W, Brown GM, Cole DR. 2003. Complete degradation of perchlorate in ferric chloride and hydrochloric acid under controlled temperature and pressure. Environ Sci Technol 37:2291–2295.

Gurol MD, Kim K. 2000. Investigation of perchlorate removal in drinking water sources by chemical methods. In Urbansky ET, ed, Percholate in the Environment, 1st ed. Kluwer/Plenum, New York, NY, USA, pp 99–107.

Hackenthal E. 1965. The reduction of perchlorate by bacteria – II. The identity of the nitrate reductase and the perchlorate-reducing enzyme of *B. cereus*. Biochem Pharm 14:1313–1324.

Hackenthal EW, Mannheim W, Hackenthal R, Becher R. 1964. The reduction of perchlorate by bacteria - I. Studies with whole cells. Biochem Pharm 13:195–206.

Halvorson HO, Ziegler NR. 1933. Applications of statistics to problems in bacteriology. I. A means of determining bacterial populations by the dilution method. J Bacteriol 25:101–121.

Herman DC, Frankenberger Jr WT. 1999. Bacterial reduction of perchlorate and nitrate in water. J Environ Qual 28:1018–1024.

Holmes DE, Finneran KT, Lovely DR. 2002. Enrichment of *Geobacteraceae* associated with stimulation of dissimilatory metal reduction in uranium-contaminated aquifer sediments. Appl Environ Microbiol 68:2300–2306.

Jendrzejewski N, Eggenkamp HGM, Coleman ML. 2001. Characterisation of chlorinated hydrocarbons from chlorine and carbon isotopic compositions: scope of application to environmental problems. Appl Geochem 16:1021–1031.

Ku TCW, Walter LM, Coleman ML, Blake RE, Martini AM. 1999. Coupling between sulfur recycling and syndepositional carbonate dissolution: Evidence from oxygen and sulfur isotope composition of pore water sulfate, South Florida Platform, U.S.A. Geochim Cosmochim Acta 63:2529–2546.

Logan BE, Wu J, Unz RF. 2001. Biological perchlorate reduction in high-salinity solutions. Water Res 35:3034–3038.

Lovley DR, Chapelle FH. 1995. Deep subsurface microbial processes. Rev Geophysic 33:365–381.

Lyngkilde J, Christensen TH. 1992. Redox zones of a landfill leachate pollution plume (Vejen, Denmark). J Contam Hydrol 10:273–289.

Malmqvist A, Welander T, Moore E, Ternstrom A, Molin G, Stenstrom I. 1994. *Ideonella dechloratans* gen. nov., sp. nov., a new bacterium capable of growing anaerobically with chlorate as an electron acceptor. Syst Appl Microbiol 17:58–64.

Michaelidou U. 2005. An investigation of the environmental significance of microbial (per)chlorate reduction. M.Sc. Thesis. Southern Illinois University, Carbondale, IL, USA.

Michaelidou U, Achenbach LA, Coates JD. 2000. Isolation and characterization of two novel (per)chlorate-reducing bacteria from swine waste lagoons. In Urbansky ED, ed, Perchlorate in the Environment, 1st ed. Kluwer Academic/Plenum, New York, NY, USA, pp 271–283.

Motzer WE. 2001. Perchlorate: problems, detection, and solutions. Environ Forensics 2:301-311.

Nissenbaum A, Presley BJ, Kaplan IR. 1972. Early diagenesis in a reducing fjord, Saanich Inlet, British Columbia, I, Chemical and isotopic changes in major components of interstitial water. Geochim Cosmochim Acta 36:1007–1027.

O'Connor SM, Coates JD. 2002. A universal immuno-probe for (per)chlorate-reducing bacteria. Appl Environ Microbiol 68:3108–3113.

Okeke BC, Frankenberger WT. 2003. Molecular analysis of a perchlorate reductase from a perchlorate-respiring bacterium *perclace*. Microbiol Res 158:337–344.

Okeke BC, Giblin T, Frankenberger WT. 2002. Reduction of perchlorate and nitrate by salt tolerant bacteria. Environ Pollut 118:357–363.

Parette R, Cannon FS. 2006. Perchlorate removal by modified activated carbon. In Gu B, Coates JD, eds, Perchlorate Environmental Occurrence, Interactions and Treatment. Springer, New York, NY, USA, pp 343–372.

Pollock JL. 2005. Potential for *in situ* bioremediation of perchlorate contaminated environments. M.Sc. Thesis, Southern Illinois University, Carbondale, IL, USA.

Rajagopalan S, Anderson T, Rainwater K, Ridley M, Fahlquist L, Jackson A. 2006. Widespread Occurrence of Naturally Occurring Perchlorate in High Plains of Texas and New Mexico. Environ Sci Technol 40:3156–3162.

Rao B, Anderson TA, Orris GJ, Rainwater KA, Rajagopalan S, Sandvig RM, Scanlon BR, Stonestrom DA, Walvoord MA, Jackson WA. 2007. Widespread natural perchlorate in unsaturated zones of dry regions. Environ Sci Technol 41:4487–4832.

Rikken G, Kroon A, van Ginkel C. 1996. Transformation of (per)chlorate into chloride by a newly isolated bacterium: reduction and dismutation. Appl Microbiol Biotechnol 45:420–426.

Roldan MD, Reyes F, Moreno-Vivian C, Castillo F. 1994. Chlorate and nitrate reduction in the phototrophic bacteria *Rhodobacter capsulatus and Rhodobacter sphaeroides.* Curr Microbiol 29:241–245.

Romanenko VI, Korenkov VN, Kuznetsov SI. 1976. Bacterial decomposition of ammonium perchlorate. Mikrobiologiya 45:204–209.

Shrout JD, Parkin GF. 2006. Influence of electron donor, oxygen, and redox potential on bacterial perchlorate biodegradation. Water Res 40:1191–1199.

Song Y, Logan BE. 2004. Effect of O2 exposure on perchlorate reduction by *Dechlorosoma* sp. KJ. Water Res 38:1626–1632.

Stepanyuk V. 1992. New species of the Acinetobacter genus *Acinetobacter thermotoleranticus* sp. nov. Mikrobiologiya 61:347–356.

Sturchio NC, Böhlke JK, Beloso Jr. AD, Streger SH, Heraty LJ, Hatzinger PB. 2007. Oxygen and chlorine isotopic fractionation during perchlorate biodegradation: laboratory results and implications for forensics and natural attenuation studies. Environ Sci and Technol 41:2796–2802.

Sturchio NC, Hatzinger PB, Arkins M, Suh C, Heraty L. 2003. Chlorine isotope fractionation during microbial reduction of perchlorate. Environ Sci Technol 37:3859–3863.

Tan K, Anderson TA, Jackson WA. 2003. Degradation kinetics of perchlorate in sediments and soils. Water Air Soil Pollut 151:245–259.

Tan K, Anderson TA, Jackson WA. 2005. Temporal and spatial variation of perchlorate in streambed sediments: Results from in-situ dialysis samplers. Environ Pollut 136:283–291.

Tan K, Jackson WA, Anderson TA, Pardue JH. 2004. Treatment of perchlorate-contaminated water in an upflow wetland treatment system. Water Res 38:4173–4185.

Thrash JC, Van Trump JI, Weber KA, Miller E, Achenbach LA, Coates JD. 2007. Electrochemical stimulation of microbial perchlorate reduction. Environ Sci Technol 41:1740–1746.

Wallace W, Ward T, Breen A, Attaway H. 1996. Identification of an anaerobic bacterium which reduces perchlorate and chlorate as *Wolinella succinogenes*. J Ind Microbiol 16:68–72.

Waller AS, Cox EE, Edwards EA. 2004. Perchlorate-reducing microorganisms isolated from contaminated sites. Environ Microbiol 6:517–527.

Wolterink AFWM. 2003. Characterization of the chlorate reductase from *Pseudomonas chloritidismutans*. J Bacteriol 185:3210–3213.

Wolterink AFWM, Jonker AB, Kengen SWM, Stams AJM. 2002. *Pseudomonas chloritidismutans* sp. nov., a non-denitrifying, chlorate-reducing bacterium. Int J Syst Evol Microbiol 52:2183–2190.

Zhang HS, Bruns MA, Logan BE. 2002. Chemolithoautotrophic perchlorate reduction by a novel hydrogen-oxidizing bacterium. Environ Microbiol 4:570–576.

APPENDIX 3.1 MEDIUM FOR FRESHWATER PERCHLORATE-REDUCING MICROORGANISMS

Basal Medium

H_2O	1.0 L
$NaClO_4$	0.97 g
NH_4Cl	0.25 g
NaH_2PO_4	0.60 g
$CH_3COONa \cdot 3H_2O$	1.36 g
$NaHCO_3$	2.5 g (primary buffer with CO_2 below)
KCl	0.1 g
Vitamin solution.	10 mL
Mineral solution.	10 mL

Split medium into tubes before sparging with 80% N_2 and 20% CO_2 (at least 6 minutes per tube, the last minute with the stopper in place).

Final pH should be 6.8 - 7.0

Autoclave for 15 minutes at $121°C$

Vitamin and Mineral Solutions

1. Vitamin Mix:

	mg/L
Biotin	2
Folic acid	2
Pyridoxine HCl	10
Riboflavin	5
Thiamin	5
Nicotinic acid	5
Pantothenic acid	5
B-12	0.1
p-aminobenzoic acid	5
Thioctic acid	5

2. Mineral Mix:

	g/L
NTA	1.5
$MgSO_4$	3.0
$MnSO_4 \cdot H_2O$	0.5
NaCl	1.0
$FeSO_4 \cdot 7H_2O$	0.1
$CaCl_2 \cdot 2H_2O$	0.1
$CoCl_2 \cdot 6H_2O$	0.1
ZnCl	0.13
$CuSO_4$	0.01
$AlK(SO_4)_2 \cdot 12H_2O$	0.01
H_3BO_2	0.01
Na_2MoO_4	0.025
$NiCl_2 \cdot 6H_2O$	0.024
$Na_2WO_4 \cdot 2H_2O$	0.025

CHAPTER 4

PERCHLORATE SOURCES, SOURCE IDENTIFICATION AND ANALYTICAL METHODS

Carol E. Aziz[1] and Paul B. Hatzinger[2]

[1]Geosyntec Consultants, Guelph, ON, Canada; [2]Shaw Environmental, Inc., Lawrenceville, NJ 08648

4.1 INTRODUCTION

Given the wide range of anthropogenic and natural perchlorate sources, the remediation practitioner should be aware of effective source identification and analytical methods to determine the origin and extent of perchlorate contamination. Characterization issues discussed in this chapter include: (1) anthropogenic and natural sources of perchlorate and their associated co-contaminants; (2) isotopic techniques to distinguish between natural and anthropogenic sources of perchlorate; (3) analytical methods for perchlorate recommended by the U.S. Department of Defense (DoD) and the U.S. Environmental Protection Agency (USEPA) and their limitations; and (4) chemical and geochemical parameters that should be measured during the characterization and treatment of a perchlorate-contaminated site.

4.2 SOURCES OF PERCHLORATE

The use and disposal of rocket propellant in the defense and aerospace industries is the most widely cited source of perchlorate contamination in the environment (Section 4.2.1.1). However, through monitoring activities mandated by the Unregulated Contaminant Monitoring Rule (USEPA, 1999a), perchlorate has now been detected at low levels (typically less than 50 micrograms per liter [μg/L]) in a significant number of areas without apparent military sources (Brandhuber and Clark, 2005).

Widespread, low concentration perchlorate contamination of groundwater can result from a variety of non-military sources, including the use and manufacture of road flares (Section 4.2.1.2), fireworks displays (Section 4.2.1.3), blasting agents used in mining and construction (Section 4.2.1.4), sodium chlorate (Section 4.2.1.5), sodium hypochlorite (bleach) (Section 4.2.1.6), and perchloric acid (Section 4.2.1.7). In addition to these anthropogenic sources, naturally-occurring perchlorate is likely to account for the low levels of contamination found in some regions of the United States (Section 4.2.2).

4.2.1 Anthropogenic Sources

4.2.1.1 Rocket Propellant

Approximately 90% of perchlorate compounds, primarily ammonium perchlorate, are used in defense activities and the aerospace industry. The widespread manufacture of perchlorate in the United States began in the mid 1940s and, by the 1950s, ammonium perchlorate began replacing potassium perchlorate as the preferred oxidizer for solid propellants in large rocket motors. In the 1960s, solid propellant mixtures of ammonium perchlorate and powdered aluminum replaced liquid propellant systems in intercontinental ballistic missile systems. Other examples of solid rocket motors that use ammonium perchlorate include the space shuttle and commercial satellite vehicles (ITRC, 2005). In the past, munitions manufacturing facilities conducted hydraulic wash out (often referred to as hog-out) of equipment used in solid propellant and munitions production. In some instances, these operations have resulted in the creation of groundwater plumes.

4.2.1.2 Road Flares

Safety flares (or fusées) used in emergency situations for road-side accidents and rail and marine emergencies contain high levels of potassium perchlorate. Although accurate estimates of the number of flares consumed annually are difficult to obtain, it is estimated that between 20 and 40 million flares are produced annually in the United States (Geosyntec, 2005).

A flare generally consists of a waxed cardboard tube casing filled with a burn mixture and a cap at the end to ignite the flare. Based on Material Safety Data Sheets (MSDSs), the burn mixture contains primarily strontium nitrate (75% by weight), potassium perchlorate (<10% by weight), and sulfur (<10% by weight) (Silva, 2003a). Flares from various manufacturers have been found to have perchlorate levels of 5–7% by weight (Geosyntec, 2006). High levels of strontium, nitrate, and possibly sulfur in association with perchlorate in groundwater can be indicative of a road flare source.

Although road flares have high levels of perchlorate, the burning of the flare significantly reduces the potential for perchlorate releases. Silva (2003b) compared perchlorate leaching from unburned flares that had been damaged (i.e., sliced open) to completely burned flares and showed that completely burned flares leached 2,000 times less perchlorate than damaged unburned road flares (i.e., 1.95 milligrams (mg) vs. 3,645 mg perchlorate per flare). An average of 99.8% of perchlorate is consumed upon the complete burning of a flare (Geosyntec, 2006; 2007).

4.2.1.3 Fireworks

Potassium perchlorate is a significant component of fireworks and is used primarily as an oxidizing agent. Because oxidizers must be low in hygroscopicity, potassium salts have been preferred over sodium salts. Potassium perchlorate can be used to produce colored flames, noise, and light when formulated with mixtures of barium (green), strontium (red), copper (blue), aluminum, and magnesium powders (Conkling, 1985). Ammonium perchlorate is also used in some fireworks formulations. Another potential source of perchlorate is the potassium nitrate in the black powder used in the lift charge if the potassium nitrate is of Chilean origin

(Section 4.2.2.2). Large quantities of fireworks are handled and discharged annually in the United States. For example, 220 million pounds (lb) of fireworks were consumed in 2003 (APA, 2004).

Raw perchlorate from fireworks manufacturing facilities and perchlorate residue from launched fireworks have the potential to contaminate surface water and groundwater. For example, perchlorate was detected at a concentration of 270 µg/L in an inactive well near a defunct fireworks site in Rialto, California (CDHS, 2007). Perchlorate derived from fireworks manufacturing also has been detected at a concentration of 122 µg/L in a well near Brookhaven, New York (Groocock, 2002).

Perchlorate contamination linked to fireworks displays was examined by the Massachusetts Department of Environmental Protection (MADEP) at the University of Massachusetts at Dartmouth (UMD). Prior to the 2004 display, soil samples had no detectable levels of perchlorate (MADEP, 2005). Results of soil sampling immediately after the display indicted a maximum perchlorate concentration of 560 micrograms per kilogram (µg/kg). Groundwater concentrations were not substantially different after the display than before (MADEP, 2005). Soil sampling conducted following the 2006 display at UMD indicated a maximum perchlorate concentration of 5 milligrams per kilogram (mg/kg) (Geosyntec, 2007). Perchlorate was also reported to increase appreciably in a municipal lake following a fireworks display in 2006, with concentrations increasing from a mean value of 0.043 µg/L just before the display to as high as 44 µg/L after the display (Wilkin et al., 2007). These values decreased to near background within 80 days after the display. Another study to assess the impacts of firework displays at Columbia Lake on the University of Waterloo's campus has recently been completed (http://www.p2pays.org/ref/22/21726.pdf).

The presence of elevated levels of potassium and magnesium may indicate the potential for perchlorate contamination from fireworks (Geosyntec, 2007). These metals were generally found at elevated concentrations at both the UMD and Columbia Lake sites following firework displays. As previously noted, other metals commonly associated with fireworks include strontium, copper, cobalt, barium, and aluminum. At many sites, natural background levels of these elements may be too high to distinguish contributions from fireworks.

4.2.1.4 Blasting Agents and Explosives

Sodium and ammonium perchlorate salts are components of some blasting agents and explosives. Approximately 2.7 million tons of blasting agents are used in coal mining, quarrying, metal and non-metal mining and construction annually (Kramer, 2003), but the percentage containing perchlorate is unknown.

Unlike explosives, blasting agents require a booster, in addition to a detonator, to initiate. Most water gels and emulsions are classified as blasting agents, as opposed to high explosives, because they are comparatively insensitive materials (i.e., difficult to detonate). This property enhances their ease of handling and safety. However, for certain difficult blasting applications, such as water-saturated construction sites where the explosive is subjected to high static or dynamic pressures, it is desirable to increase the sensitivity by using perchlorate-containing products (IME, 2007; ITRC, 2005). Some water gels and emulsions can contain up to 30% perchlorate (Table 4.1). The inclusion of sodium nitrate of Chilean origin may also introduce perchlorate as discussed in Section 4.2.2.2. Certain seismic explosives can contain 55–72% perchlorate and some non-electric detonators may contain up to 10% perchlorate.

The most common and simplest blasting agent is ammonium nitrate fuel oil (ANFO), which consists of ammonium nitrate (AN) prills soaked with fuel oil (about 5 to 6 weight %). Another popular blasting product consists of a blend of prilled ANFO or AN with AN emulsion in various ratios. Blends containing less than 50% emulsion are sometimes referred to as "heavy ANFO." Their benefits include reduced mining costs, increased water resistance and increased density/strength (ISEE, 1998). MSDSs for some heavy ANFOs list "inorganic oxidizers". Further testing is required to determine if these products contain perchlorate.

Table 4.1. Blasting Agents and Explosives Containing Perchlorate (% Composition)*

Type Product	Blasting Agent (1.5) or Explosive (1.1)	NH_4NO_3	$NaNO_3$	$NaClO_4$
Gel bulk or packaged	Blasting agent	55–85	–	0–4
Packaged gel	Blasting agent	33–40	10–15	–
Package emulsion	Explosive	60–70	0–5	0-15
Package emulsion	Explosive	60–80	0–12	–
Packaged gel	Explosive	<65	<20	<7
Water gel	Blasting agent	<80	–	<5
Water gel	Blasting agent	<75	<5	<5
Water gel	Explosive	<65	<20	<7
Water gel	Explosive	<65	<20	<7
Water gel, presplit	Explosive	<65	<20	<7
Water gel	Blasting agent	10–20	10–20	20–30**

* Data compiled from Material Safety Data Sheets.
** Ammonium perchlorate.

Perchlorate concentrations as high as several hundred µg/L have been measured in groundwater near blasting sites (MADEP, 2005). In response to perchlorate contamination in the Boxborough, Massachusetts area, a ban has been issued by the fire department on the use of perchlorate-based agents for all blasting activities in this area. In addition, the State of Massachusetts is prohibiting its own contractors from using blasting agents that contain perchlorate (Hughes, 2004).

It is theorized that misfires and/or "bad housekeeping" associated with the use of blasting agents are the primary mechanisms that result in groundwater impacts. The DoD Strategic Environmental Research and Development Program (SERDP) is currently funding studies that will attempt to quantify the amount of perchlorate originating from the detonation of blasting agents containing perchlorate (http://www.p2pays.org/ref/22/21726.pdf).

4.2.1.5 Sodium Chlorate

Sodium chlorate, widely used in the pulp and paper industry, often contains perchlorate as an impurity. The total annual consumption of sodium chlorate is approximately 1.2 million tons (USDOC, 2003). The pulp and paper industry uses approximately 94% of all sodium chlorate consumed in the United States to produce chlorine dioxide to bleach pulp fibers (OMRI, 2000). In addition, sodium chlorate is used as a non-selective contact herbicide and a defoliant for cotton, sunflowers, sudan grass, safflower, rice, and chili peppers (OMRI, 2000). As a defoliant, approximately 99% of sodium chlorate application is used on cotton plants in California and Arizona (PAN Pesticides Database, 2002).

Sodium chlorate is produced electrochemically by the electrolysis of aqueous sodium chloride according to the following overall equation (Betts and Dluzniewski, 1997):

$$NaCl + 3H_2O \rightarrow NaClO_3 + 3H_2 \qquad (Rx.\ 4.1)$$

The formation of perchlorate stems from anodic oxidation of chlorate during the electrochemical reaction in accordance with the following reaction (Betts and Dluzniewski, 1997):

$$ClO_3^- + H_2O \rightarrow ClO_4^- + 2H^+ + 2e^- \qquad (Rx.\ 4.2)$$

Recent analyses of several sodium chlorate feedstocks being used for large-scale perchlorate manufacturing suggest that perchlorate is present in industrial-grade chlorate products at concentrations ranging from 50 to 230 mg/kg. Twelve samples of laboratory-grade sodium chlorate were procured and found to contain perchlorate at concentrations ranging from 1.5 to 117 mg/kg, with mean and median concentrations of 42 mg/kg and 26 mg/kg, respectively (Geosyntec, 2007).

4.2.1.6 Bleach (Hypochlorite)

Bleach or sodium hypochlorite may contain perchlorate as an impurity. Hypochlorite is widely used as a household bleach and industrial disinfectant and is also routinely used to disinfect groundwater wells. The most common type of hypochlorite/bleach solution is sodium hypochlorite, NaOCl, a greenish-yellow liquid solution. Calcium hypochlorite, a white powder, is often used for swimming pool chlorination.

Bleach is generally produced by the electrolysis of a weak brine (i.e., NaCl) solution at a pH of 10-12 via the following overall reaction:

$$NaCl + H_2O \rightarrow NaOCl + H_2 \qquad (Rx.\ 4.3)$$

Sodium hypochlorite solutions are not stable, and decomposition is a well-known industry problem and concern. The most prominent degradation pathway results in the production of chlorate:

$$3OCl^- \rightarrow ClO_3^- + 2\ Cl^- \qquad (Rx.\ 4.4)$$

This reaction is minimized during production by maintaining basic pH and keeping the temperature low.

In a recent study of bleach under various storage conditions, perchlorate concentrations increased over time in all six bleach brands tested, from a starting average of 19 µg/L to an average of 154 µg/L after six weeks of storage in the dark. When bleach samples were stored in sealed glass vessels while exposed to sunlight, the perchlorate concentrations were much higher, averaging 3,500 µg/L after 6 weeks. These results show that storage conditions, including light exposure and storage duration, significantly influence perchlorate concentrations in bleach (Geosyntec, 2007). Bleach should be stored in the dark and care should be taken to avoid oxygen and sunlight particularly if bleach is to be used to disinfect wells or irrigation equipment.

4.2.1.7 Perchloric Acid

Perchloric acid or hydrogen perchlorate is used in a wide variety of analyses including acid digestions, Kjeldahl digestions, as an oxidizing agent, as a solvent for extracting sulfide ores, and as a dehydrating agent (Geosyntec, 2005). Perchloric acid or hydrogen perchlorate is sold principally as a 72% acid solution. At room temperature, this solution is not an oxidizing agent and can be safely transported and stored. It becomes a powerful oxidizing agent when heated and used in a concentrated form, allowing for chemical reactions and production processes that can be carefully designed and controlled. This property makes perchloric acid unique among the strong acids.

Perchloric acid discharge was implicated in perchlorate detections in the Merrimack River in Massachusetts during 2004–2005. Investigations undertaken by the Town of Billerica eventually identified the source of perchlorate discharge to the municipal sewerage system: a processor of surgical and medical materials that was using approximately 833 L/month (220 gallons/month) of perchloric acid. Although only a small portion of this acid was discharged as rinse water to the sewer system, it equated to an average of 4.5 kg/day (10 lb/day) of perchlorate (MADEP, 2005).

4.2.2 Natural Sources of Perchlorate

Natural sources of perchlorate include its occurrence in Chilean nitrate and in other mineral deposits. Natural perchlorate is believed to be primarily of atmospheric origin, although other reactions may also contribute to its formation.

4.2.2.1 Atmospheric Origin of Perchlorate

A current theory regarding the origin of naturally occurring perchlorate in the environment is that it is generated via atmospheric processes (Bao and Gu, 2004). While the exact mechanism for natural perchlorate formation is unknown, it has been suggested that chloride, possibly in the form of sodium chloride from the sea or land-based chloride compounds, reacts with atmospheric ozone to create perchlorate. This process probably occurs over much of the earth and is analogous to nitrate formation in the atmosphere (Walvoord et al., 2003). In addition, lightning may play a role in the synthesis of some atmospherically-produced perchlorate (Dasgupta et al., 2005).

Following atmospheric formation, perchlorate returns to the earth's surface dissolved in precipitation. Dasgupta et al. (2005) analyzed precipitation samples and found perchlorate present in 70% of the samples using preconcentration-preelution ion chromatography/mass spectrometry (IC/MS), with concentrations ranging from below the detection limit to 1.6 µg/L. In arid environments, where the rate of deposition exceeds the rate of dissolution by ongoing precipitation, perchlorate can be incorporated into geologic formations as discussed further in the next two subsections. Recent isotopic studies have suggested that nitrate and perchlorate in the Atacama Desert were formed atmospherically (Böhlke et al., 1997; Michalski et al., 2004). Moreover, perchlorate derived from Atacama nitrate ore has been shown to have significant excess in the ^{17}O isotope, an indication of atmospheric production involving ozone (Bao and Gu, 2004). Isotopic analyses to distinguish natural sources from man-made sources are discussed in more detail in Section 4.3.

4.2.2.2 Chilean Nitrate

The presence of perchlorate in the caliche deposits of the Atacama Desert region of Chile, one of the most arid regions of the world, has been documented for over 100 years (Michalski et al., 2004; Schilt, 1979). Perchlorate was first discovered in the caliche deposits in 1886. This discovery was followed in 1896 by the confirmation of perchlorate in "Chilean saltpeter" (sodium nitrate) over the widely varying concentration range of 0 to 6.79% (Schilt, 1979). Since the mid 1800s, Chilean nitrate ore has been imported into the United States for use as fertilizer, for saltpeter used in gunpowder, and as a feedstock for making nitric acid, explosives, fireworks, and additional end products (ITRC, 2005). Historical agronomic literature indicates that Chilean nitrate fertilizers were widely used in citrus, cotton, and tobacco farming in the early to mid 1900s (Howard, 1931; Goldenwieser, 1919; Mehring, 1943).

Little attention was paid to the natural occurrence of perchlorate in Chilean nitrate until the emergence of perchlorate as a chemical of concern at military sites. In 2000, a study of perchlorate in agricultural fertilizers conducted by the USEPA concluded that the occurrence of perchlorate in fertilizer was restricted to fertilizer products derived from Chilean nitrate produced by SQM Corporation and that all fertilizers derived partially or completely from Chilean nitrates contained appreciable perchlorate (Urbansky et al., 2001a, 2001b).

Chilean nitrate fertilizer is still produced by SQM Corporation and makes up 0.14% of the total annual fertilizer application in the United States (Urbansky et al., 2001a). It is sold commercially as "Bulldog Soda" and is primarily used in a few niche markets and specialty products. Currently, world production is 900,000 tons/year, of which 75,000 tons are sold to American farmers for use on cotton, tobacco, and fruit crops (Urbansky et al., 2001a; Renner, 1999). SQM reports that the perchlorate concentration in Chilean nitrate fertilizer has been reduced to 0.01% through changes in the refinement processes since 2002 (Urbansky et al., 2001b).

4.2.2.3 Other Natural Mineral Sources

A study of perchlorate in geologic materials from environments with similar characteristics to the Atacama Desert was recently conducted by the U.S. Geological Survey (USGS). Evaporite and evaporite-related minerals and surface crusts of various ages from North and South America were collected and analyzed by Orris et al. (2003). Samples were originally analyzed by IC, which is non-specific for perchlorate, and perchlorate was found in approximately 50% of the mineral samples tested, including several samples of potash ores and playa crusts (Orris et al., 2003). Reanalysis of these original samples plus several new samples by IC-MS-MS has shown that all evaporite samples containing potassium and/or magnesium contain perchlorate, typically in the few to tens of µg/L (Gu and Coates, 2006). The arid environments in which the samples were collected served to concentrate perchlorate and prevent anaerobic conditions that would promote perchlorate biodegradation.

This USGS study highlights the widespread occurrence of natural perchlorate and the importance of evaporative concentration in producing environmentally significant concentrations of perchlorate. This point is illustrated by several recent reports showing the presence of perchlorate in soils and groundwater over more than 104,000 square kilometers (40,000 square miles) of the arid high plains region of northwest Texas (Jackson et al., 2005, 2006; Rajagopalan et al., 2006), in unsaturated soils located throughout the southwestern United States (Rao et al., 2007) and in ancient groundwater from the Rio Grande Basin of New Mexico (Plummer et al., 2006).

4.3 DISTINGUISHING SYNTHETIC FROM NATURAL PERCHLORATE USING STABLE ISOTOPE ANALYSIS

4.3.1 Stable Isotope Analysis

Isotopes of an element have the same number of protons and electrons but a different numbers of neutrons. Stable isotopes (as opposed to radioactive isotopes) are not subject to nuclear decay. The difference in atomic mass among stable isotopes causes these atoms to exhibit slightly different physical and chemical traits. These differences are particularly notable for light elements, including many of geochemical interest such as H, C, N, O, Cl and S. The differing masses of stable isotopes (and the resulting differences in their charge to mass ratio) result in isotopic "fractionation" whereby various physical, chemical, and biological processes alter isotopic ratios. These fractionation processes often provide unique isotopic signatures, which are indicative of the origin and/or geochemical behavior of a compound in the environment (Sharp, 2007; Clark and Fritz, 1997; Kendall and Caldwell, 1998).

Stable isotopes are generally quantified via isotope ratio mass spectrometry (IRMS). This technique utilizes a mass spectrometer that is designed specifically to measure isotopic proportions of a given element, rather than to determine exact molecular quantities. In general, an element must be present as a pure gas (e.g., O_2, CO_2 CO, N_2O) prior to IRMS analysis. A number of different techniques, including combustion, catalytic oxidation and enzymatic conversion, have been developed to convert liquids, solids and gaseous samples of interest into pure gases suitable for IRMS. An overview of IRMS, including sample preparation techniques, is provided by Sharp (2007).

The stable isotope ratios of light elements gained from IRMS are generally reported relative to those of established reference materials as "delta" (δ) values and measured in parts-per-thousand (denoted "‰" = per mil). As an example, the expression used to report relative abundances of Cl isotopes ($^{37}Cl/^{35}Cl$) is provided below (Eq. 4.1 and Eq. 4.2).

$$\delta \text{ (in ‰)} = \left[(R_x - R_s)/R_s \right] \times (1000) \qquad \text{(Eq. 4.1)}$$

Where: R = ratio heavy/light isotope (e.g., $^{37}Cl/^{35}Cl$)
R_x = sample (e.g., $^{37}Cl/^{35}Cl$ in environmental sample)
R_s = standard (e.g., $^{37}Cl/^{35}Cl$ in "standard mean ocean chloride")

Thus for Cl isotopes:

$$\delta^{37}Cl_{sample} \text{ (in ‰)} = \left[\frac{\left(^{37}Cl/^{35}Cl\right)_{sample} - \left(^{37}Cl/^{35}Cl\right)_{standard}}{\left(^{37}Cl/^{35}Cl\right)_{standard}} \right] \times (1000) \qquad \text{(Eq. 4.2)}$$

The ratio of the heavy to the light isotope is used by convention, and for the case of Cl, the established international reference material is standard mean ocean chloride (SMOC). A positive delta value indicates that the sample is enriched in the heavy isotope relative to the standard, while a negative delta value shows that the sample contains less of the heavy isotope. For example, if $\delta^{37}Cl$ is reported as +15‰, this means that the ratio of $^{37}Cl/^{35}Cl$ is 15 parts-per-thousand (or 1.5%) higher in the sample of interest than in SMOC (for which $\delta^{37}Cl$ is 0.00 ‰).

Stable isotope ratio analysis has been used for several decades by earth scientists to better understand natural geological, geochemical and hydrogeological processes (e.g., Sharp, 2007; Clark and Fritz, 1997). More recently, stable isotope ratio analysis has been applied as an analytical tool to assess the origin and disposition of common industrial and military pollutants. For example, advances in the measurement and application of the stable isotope ratios of carbon and chlorine in chlorinated solvents (Holt et al., 1997; Holt et al., 2001; Drenzek et al., 2002; Jendrzejewski et al., 1997) have led to new approaches for characterizing the behavior of these compounds in contaminated groundwater aquifers (Sturchio et al., 1998; Dayan et al., 1999; Song et al., 2002; Hunkeler et al., 1999; Hunkeler et al., 2005). Similar evaluations also have been performed with nitrogen isotopes to track the fate of explosives such as cyclotrimethlyenetrinitramine (also termed **R**oyal **D**emolition e**X**plosive or RDX) and 2,4,5-trinitrotoluene (TNT) in the environment (Dignazio et al., 1998). Moreover, the development of combined gas chromatography-isotope ratio mass spectrometry (GCIRMS) now provides a technique to gain isotopic ratios of individual chemicals from complex mixtures (Philip, 2002). This approach has been used to determine the origin of various hydrocarbons, including crude oils (Mansuy et al., 1997), gasoline components (Kelly et al., 1997), polycyclic aromatic hydrocarbons (Hammer et al., 1998) and gasoline oxygenates (Smallwood et al., 2001).

4.3.2 Stable Isotope Methods for Perchlorate

Both of the atoms composing a perchlorate molecule (Cl and O) have multiple isotopes. Chlorine has three naturally occurring isotopes, one of which is a long-lived radioactive species (^{36}Cl) and two of which are stable (^{35}Cl and ^{37}Cl, occurring at abundances of 75.77% and 24.23% of naturally occurring chlorine, respectively) (USGS, 2006a). Oxygen has three stable isotopes, ^{16}O, ^{17}O and ^{18}O. These occur in the following percentages in nature: ^{16}O (99.63%), ^{17}O (0.0375%) and ^{18}O (0.1995%) (USGS, 2006b). Techniques to quantify the stable isotope ratio of chlorine ($^{37}Cl/^{35}Cl$) in the perchlorate molecule were reported by Ader et al. (2001) and Sturchio et al. (2003). Subsequently, methods for analysis of $^{18}O/^{16}O$ and $^{17}O/^{16}O$ in perchlorate were described (Bao and Gu, 2004; Böhlke et al., 2005).

4.3.2.1 Sample Preparation and Analysis

For determination of $\delta^{37}Cl$ from pure perchlorate salts, the sample is combusted to produce Cl^-, which is then dissolved and re-precipitated as AgCl. The AgCl is subsequently reacted with CH_3I to produce CH_3Cl, which is further purified and analyzed by IRMS for determination of $\delta^{37}Cl$ (Sturchio et al., 2003; Böhlke et al., 2005). The average precision of this technique is reported as approximately 0.03 ‰. For determination of $\delta^{18}O$, the perchlorate salt is initially reacted with glassy carbon at high temperature to produce CO, which is then purified by GC and analyzed by IRMS (Böhlke et al., 2005). Measurement of ^{17}O (which is generally reported as $\Delta^{17}O$; see Section 4.3.2.2) is performed by combusting the perchlorate salt to produce O_2, which is subsequently analyzed by IRMS. The average variability for measurements of $\delta^{18}O$ and $\Delta^{17}O$, are 0.2 ‰ and 0.1 ‰, respectively (Böhlke et al., 2005).

In order to analyze perchlorate from environmental samples, the anion must first be collected in sufficient quantity (~10 mg), then extracted and purified. For soil samples, the perchlorate is initially extracted with water, and then the extract is passed through small columns of perchlorate-specific anion exchange resin to remove it from solution (Bao and Gu, 2004). A similar approach has been used to collect dilute perchlorate from groundwater. In this case however, the groundwater is pumped from a well (or collected in a secondary container), and then passed through the ion exchange column in the volume required for ~10 mg of perchlorate to be trapped on the resin (Sturchio et al., 2006; Böhlke et al., 2005). Once sufficient perchlorate has been collected on a resin column (from water or extracts), an aqueous solution containing tetrachloroferrate is passed through the column. This ion is preferentially bound and displaces the perchlorate ions from the resin (Gu et al., 2001). The perchlorate-bearing solution is subsequently subjected to a series of purification steps which ultimately result in a pure precipitate of either $KClO_4$ or $CsClO_4$, both of which are relatively insoluble. After verification of the purity of this material, the salts can be prepared and analyzed by IRMS for determination of $\delta^{37}Cl$, $\delta^{18}O$, and/or $\Delta^{17}O$, as described previously. It is critical to ensure that pure perchlorate salts are analyzed as small quantities of oxygen or chlorine-containing impurities can alter the isotopic ratios for these elements.

4.3.2.2 Isotopic Results to Date

Isotopic ratios of Cl and O were recently reported for a variety of different perchlorate salts of laboratory, commercial and military origin as well as for several natural perchlorate samples and fertilizers derived from the Atacama Desert of Chile (Sturchio et al., 2006; Böhlke et al., 2005). Additional samples obtained from road flares, fireworks, chlorate herbicides, bleach, propellants, and other materials are presently being collected and analyzed as part of a DoD Environmental Security Technology Certification Program (ESTCP) project (Hatzinger et al., 2008). Current data from isotopic analyses reveal that the $^{37}Cl/^{35}Cl$ isotope ratio in naturally occurring perchlorate is consistently and significantly lower than that of man-made perchlorate (Figure 4.1). Based on analyses to date, the mean $\delta^{37}Cl$ value (±standard deviation) for synthetic perchlorate is 0.6 ± 1.0 ‰ (n = 18) compared to -12.6 ± 2.0 ‰ for natural perchlorate (n = 6) (Sturchio et al., 2006; unpublished data from Hatzinger et al., 2008). The values for synthetic perchlorate which range from -3.1 to $\pm$ 1.6 ‰, are reasonably close to that of standard ocean chloride (0.00 ‰), reflecting the fact that synthetic perchlorate is synthesized electrochemically from NaCl by a process that is efficient and yields little isotopic fractionation (Sturchio et al., 2006). By comparison, the consistently low $\delta^{37}Cl$ values for Chilean-derived natural perchlorate confirm a different mode of formation of this material, which, based on the corresponding $\Delta^{17}O$ values (see Figure 4.3 and supporting text) suggests oxidation of volatile chlorine by ozone (which is known to have elevated ^{17}O values) in the upper atmosphere (Bao and Gu, 2004; Dasgupta et al., 2005).

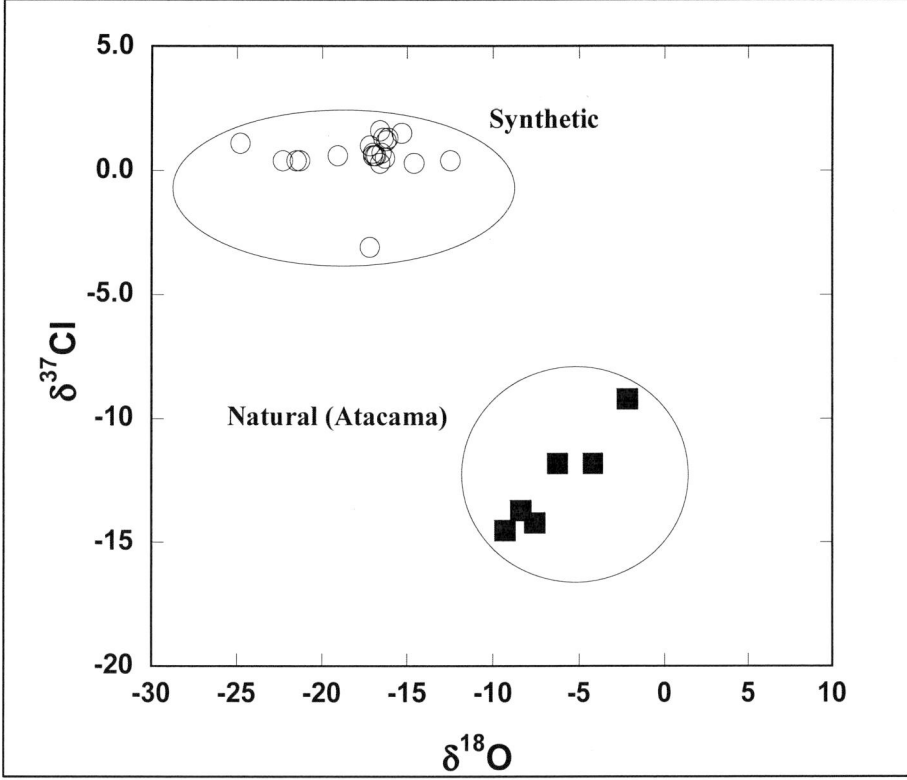

Figure 4.1. Comparison of $\delta^{37}Cl$ and $\delta^{18}O$ for various sources of synthetic perchlorate and for natural perchlorate derived from the Atacama Desert of Chile (from Sturchio et al., 2006 and unpublished data from Hatzinger et al., 2008)

In contrast to the $^{37}Cl/^{35}Cl$ data, the $^{18}O/^{16}O$ isotope ratio in natural perchlorate is appreciably higher than in the synthetic materials. The average $\delta^{18}O$ for natural perchlorate in six samples analyzed to date is - 6.3 ± 2.7 ‰ compared to -17.8 ± 3.2 ‰ for 18 samples of synthetic perchlorate. The $\delta^{18}O$ values in the synthetic samples vary from -12.5 to -24.8 ‰, which is a significantly broader range than that for $\delta^{37}Cl$. Interestingly, however, the $\delta^{18}O$ values from different samples produced by the same manufacturer (e.g., $KClO_4$ and $NaClO_4$ from the same facility) group very tightly together (Figure 4.2). The oxygen in the perchlorate molecule is derived from H_2O during the electrochemical formation of perchlorate. This process is less efficient than for chloride, and a 7 ‰ isotopic enrichment of oxygen in the perchlorate molecule compared to the source water has been reported (Sturchio et al., 2006). However, the limited information gathered so far (i.e., data presented in Figure 4.2) suggest that $\delta^{18}O$ data may be useful for distinguishing synthetic perchlorate sources.

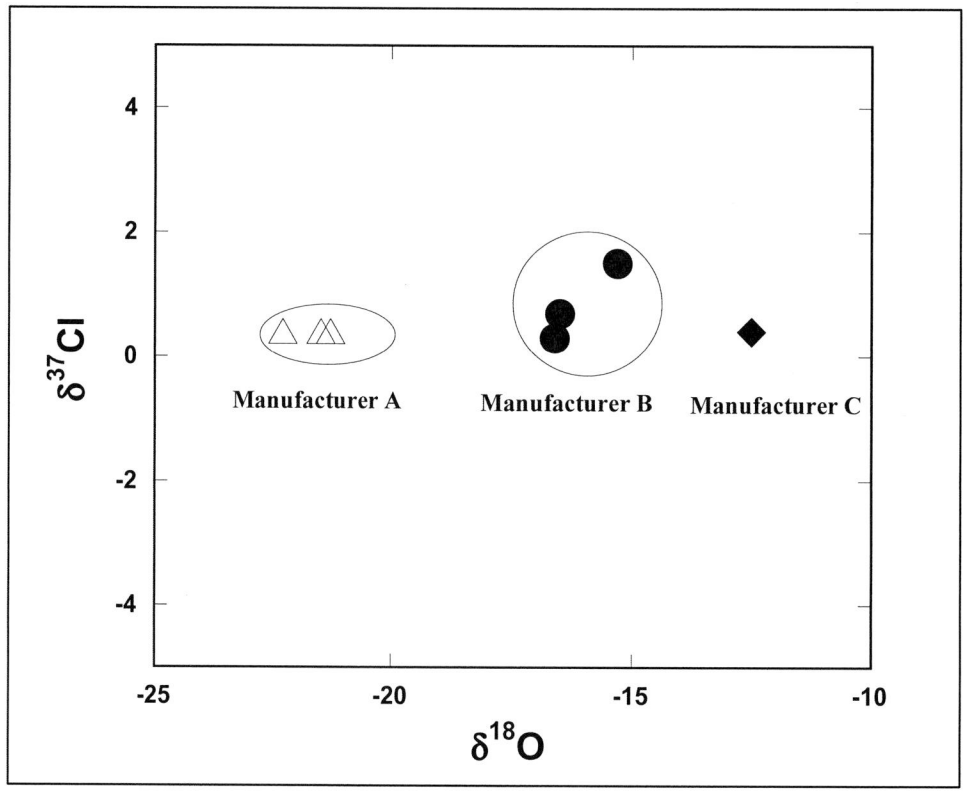

Figure 4.2. Comparison of δ³⁷Cl and δ¹⁸O for three sources of synthetic perchlorate. Analytical error is approximately ±0.3 per mil (modified from Sturchio et al., 2006)

Perhaps the most important isotopic difference between natural perchlorate derived from Chile and synthetic perchlorate comes from analysis of ¹⁷O. There appears to be a consistent and reproducible excess of ¹⁷O in natural perchlorate, relative to the abundance that would be consistent with simple mass-dependent isotopic fractionation processes (Bao and Gu, 2004; Böhlke et al., 2005; Sturchio et al., 2006). A similar enrichment is not seen in synthetic perchlorate. More simply, there is an expected ratio of ¹⁷O to ¹⁸O for terrestrial materials (δ¹⁷O = 0.525 δ¹⁸O) and natural perchlorate (but not synthetic perchlorate) shows a significant deviation from this ratio. The excess ¹⁷O in natural perchlorate is shown in Figure 4.3 as Δ¹⁷O, which represents the deviation in ¹⁷O from the expected value. The equation used to derive Δ¹⁷O is as follows:

$$\Delta^{17}O\,(‰) = \left[\left(\frac{(1+\delta^{17}O/1000)}{(1+\delta^{18}O/1000)^{0.525}}\right) - 1\right] \times 1000 \qquad (Eq.\ 4.3)$$

As shown in Figure 4.3, synthetic samples analyzed to date have a Δ¹⁷O value in the vicinity of 0 (0.02 ± 0.05 ‰), which is as expected for man-made materials. In contrast, the mean Δ¹⁷O of natural samples averages + 9.6 ‰. As previously noted, the elevated Δ¹⁷O in the Chilean perchlorate is consistent with atmospheric formation (Bao and Gu, 2004; Dasgupta et al., 2005).

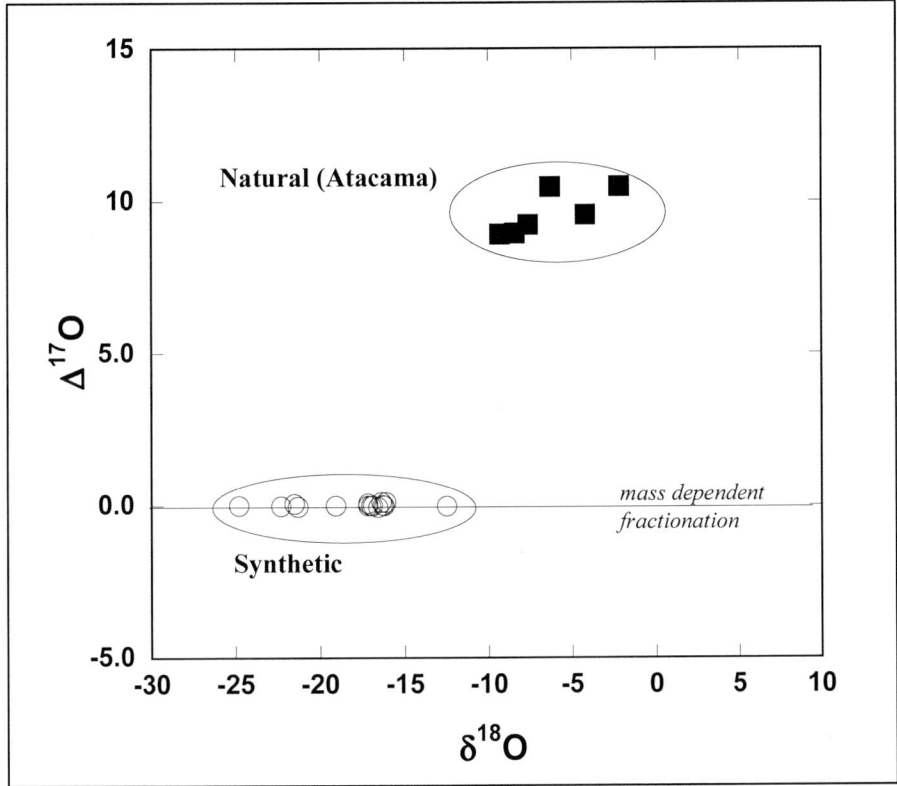

Figure 4.3. Comparison of Δ^{17}O and δ^{18}O for synthetic perchlorate and natural perchlorate derived from the Atacama Desert of Chile (from Sturchio et al., 2006 and unpublished data from Hatzinger et al., 2008)

Current data suggest that stable isotope analysis of Cl and O represents a practical forensic tool to distinguish natural perchlorate of Chilean origin from synthetic perchlorate. A recent study also shows that isotopic fractionation due to biodegradation of perchlorate is unlikely to cause isotopic signatures of synthetic and Chilean-derived perchlorate to overlap (Sturchio et al., 2007). The potential utility of stable isotopes to distinguish perchlorate from different synthetic sources is less clear. It appears that δ^{37}Cl values are too similar among different samples to be of use, but that δ^{18}O may differ enough in some instances to distinguish source materials. Additional studies are required to quantify the variability in δ^{18}O values between batches of perchlorate from a single manufacturing plant, and to further quantify differences in δ^{18}O among different production plants in the United States and abroad. Natural perchlorate also has been detected in evaporites and mineral deposits in the United States and Canada (see Section 4.2.2) as well as in surface soils throughout the southwestern United States (Rajagopalan et al., 2006; Rao et al., 2007). The isotopic values for Cl and O in natural perchlorate collected from evaporites in the Mojave Desert and other locations are presently under investigation, but results are not yet available (Hatzinger et al., 2008). Thus, it is currently unknown whether perchlorate derived from these materials is isotopically distinct from either synthetic material or Chilean-derived perchlorate.

4.4 ANALYTICAL METHODS FOR PERCHLORATE ANALYSIS

A variety of methods exist for the analysis of perchlorate in groundwater. The DoD Perchlorate Handbook (DoD EDQW, 2007) discusses perchlorate sampling and analysis. The following sections discuss the DoD-approved analytical methods for perchlorate, other widely used USEPA methods, and their respective detection limits and limitations.

4.4.1 DoD-Approved Analytical Methods

According to recent DoD perchlorate policy, only methods employing MS are to be used for environmental restoration/cleanup or range assessment projects. Methods employing IC with conductivity detection alone (e.g., USEPA Methods 314.0 and 314.1) are not appropriate for these purposes. A summary of each of the DoD-recommended perchlorate methods, their applicability, limitations and target reporting limits are provided in Table 4.2 (DoD EDQW, 2007). Of the methods listed in Table 4.2, only USEPA Methods 6850 and 6860 have been approved by the DoD for groundwater analysis. The DoD has approved methods 331.0 and 332.0 for analysis of drinking water. These methods are briefly described as follows:

4.4.1.1 USEPA Methods 6850 (HPLC/ESI/MS) and 6860 (IC/ESI/MS)

The USEPA Office of Solid Waste (OSW) has developed and validated two new methods for the determination of perchlorate in various environmental media, including soil, sludge, wastewater and high salt water. Method 6850 uses high performance liquid chromatography/electrospray ionization/mass spectrometry (HPLC/ESI/MS) and Method 6860 uses ion chromatography/electrospray ionization/mass spectrometry (IC/ESI/MS). The Methods 6850 and 6860 were published in January 2007 and are available on the USEPA OSW Methods Web Site (http://www.epa.gov/epaoswer/hazwaste/test/new-meth.htm#6850).

4.4.1.2 USEPA Method 331.0—Liquid Chromatography Electrospray Ionization Mass Spectrometry

Method 331.0 is a liquid chromatography/electrospray ionization/mass spectrometry (LC/ESI/MS) method for the determination of perchlorate in raw and finished drinking water (USEPA, 2005a). In this method, water samples are collected in the field using a sterile filtration technique. Prior to analysis, isotopically enriched perchlorate is added to the sample as an internal standard. The sample is injected without cleanup or concentration onto a chromatographic column, which separates perchlorate from other anions and background interferences. Perchlorate is subsequently detected by negative electrospray ionization mass spectrometry and is quantified using the internal standard technique. The reporting limit in water is 0.02 µg/L (USEPA, 2005a).

4.4.1.3 USEPA Method 332.0—Ion Chromatography with Suppressed Conductivity and Electrospray Ionization Mass Spectrometry

This IC/MS method is an ion chromatography method with an MS and electrospray interface (USEPA, 2005b). The method requires the use of a suppressor to avoid inorganic salt buildup and uses a conductivity meter to check its efficiency. It uses m/z 99 and 101 ions for peak identification of perchlorate. The advantages of IC/MS are increased sensitivity and increased specificity. One should be aware that high sulfate content (~1000 mg/L) will elevate the baseline at m/z 99 because it elutes prior to perchlorate. However, even with a sulfate concentration of 1000 mg/L, 0.1 μg/L perchlorate can still be detected. If the baseline is elevated, there is a mandatory cleanup step to remove the sulfate prior to sample injection. The quantitation limit in water is reported to be 0.1 μg/L (USEPA, 2005b; ITRC, 2005).

Table 4.2. DoD Recommended Methods for Perchlorate Analysis*

Method (Technique)	Applicability	Limitations	Target Reporting Limits
USEPA Method 6850 (LC/ESI/MS)	• Environmental Restoration • Operational Ranges • Wastewater • Aqueous samples including those with high TDS • Soil and sludge samples	• Requires a proprietary column	• Drinking Water and Groundwater: 0.2 μg/L • Soil: 2 μg/kg • Wastewater: <1 μg/L
USEPA Method 6860 (IC/ESI/MS) (IC/ESI/MS/MS)	• Environmental Restoration • Operational Ranges • Wastewater • Aqueous samples including those with high TDS • Soil and sludge samples	• Pretreatment recommended for samples with high concentrations of sulfate	• Drinking Water and Groundwater: 0.2 μg/L • Soil: 2 μg/kg • Wastewater: <1 μg/L
USEPA 331.0 (LC/MS) (LC/MS/MS)	• DoD-Owned Drinking Water Systems (proposed for UCMR 2) • Applicable to drinking water samples, including those with high TDS	• Pretreatment recommended for samples with high concentrations of sulfate (proposed for UCMR 2) • Validated for drinking water samples only	• Drinking Water: 0.1 μg/L (LC/MS) 0.02 μg/L (LC/MS/MS)
USEPA 332.0 (IC/MS) (IC/MS/MS)	• DoD-Owned Drinking Water Systems (proposed for UCMR 2) • Applicable to drinking water samples, including those with high TDS	• Pretreatment recommended for samples with high concentrations of sulfate (proposed for UCMR 2) • Validated for drinking water samples only	• Drinking Water: 0.1 μg/L (IC/MS) 0.02 μg/L (IC/MS/MS)

TDS = Total dissolved solids UCMR = Unregulated Contaminant Monitoring Rule

* Adapted from DoD Environmental Data Quality Workgroup *DoD Perchlorate Handbook,* August 2007 (DoD EDQW, 2007).

4.4.2 Other Analytical Methods for Perchlorate

Table 4.3 provides information about other widely used perchlorate methods that are not recommended for groundwater analysis in accordance with the DoD Perchlorate Policy (DoD EDQW, 2007). These methods include USEPA Method 314.0 (IC), USEPA Method 314.1, and USEPA Method 9058 (IC). The methods are discussed briefly below.

Table 4.3. Other Available Methods for Perchlorate Analysis*

Method (Technique)	Applicability	Limitations	Target Reporting Limits
USEPA 314.0 (IC)	• Mandatory for Drinking Water samples reported under UCMR 1 • Aqueous samples with low dissolved solids (conductivity <1 milliSiemens per centimeter [mS/cm] TDS) and chloride, sulfate, and carbonate concentrations < 100 mg/L each • Not proposed for UCMR 2	• Subject to false positives due to lack of specificity of the conductivity detector • Validated for drinking water samples only • Inappropriate for use in samples with high TDS	Drinking Water 4 µg/L
USEPA 314.1 (IC)	• Drinking Water samples • Proposed option for UCMR 2	• Reduces but does not eliminate the potential for false positives • Validated for drinking water samples only • Long analytical run time • Limited commercial availability. • Requires confirmation of perchlorate results above reporting limit when used for UCMR 2	Drinking Water 0.13 µg/L
USEPA Method 9058 (IC)	• Aqueous samples with low dissolved solids (conductivity <1 mS/cm TDS) and chloride, sulfate, and carbonate concentrations <100 mg/L each	• Subject to false positives due to lack of specificity of the conductivity detector • Inadequate quality control criteria • Method is expected to undergo significant revision prior to publication	Low TDS Groundwater 4 µg/L

*Reproduced from DoD Environmental Data Quality Workgroup *DoD Perchlorate Handbook*, August 2007 (DoD EDQW, 2007).

4.4.2.1 USEPA Method 314.0—Ion Chromatography

USEPA Method 314.0 (USEPA, 1999b), an ion chromatography method, has been the most-widely used method to date. Aqueous samples are introduced into an ion chromatograph and the perchlorate ion is separated from other ions based on its affinity for the chromatographic column. A conductivity detector is used to differentiate the perchlorate ion based solely on retention times.

The use of Method 314.0 involves many sources of uncertainty including (1) non-specificity for perchlorate, (2) possible interferences, (3) a relatively high method reporting limit (MRL) of 4 µg/L and (4) absence of systematic validation in matrices other than potable water (DoD EDQW, 2007). It should be emphasized that USEPA Method 314.0 was developed for perchlorate detection in drinking water, not other matrices. In most instances this method provides reliable results when applied to drinking water, and it is by far the least expensive and most widely available commercial analytical method for this purpose. However, because no other method for perchlorate detection was available until a few years ago, USEPA Method 314.0 has been applied to many different matrices for which it was not designed, including soil extracts, saline waters and industrial and residential wastewaters. Even in these cases, the method has generally provided dependable results, although as noted below, specific interferences have been observed. Most of these interferences would not typically be associated with drinking water, but rather are an artifact of application of the method to other types of samples.

Sample matrices with high concentrations of common anions, such as chloride, sulfate and carbonate, can destabilize the baseline in the retention time window for perchlorate and also increase or suppress the response of the detector to perchlorate. The concentration of these anions can be indirectly assessed by monitoring the conductivity of the matrix. The laboratory must determine its instrument-specific matrix conductivity threshold (MCT), and all sample matrices must be monitored for conductivity prior to analysis. When the MCT is exceeded, sample dilution and/or pretreatment must be performed. However, sample dilution leads to elevated reporting limits, and pretreatment to remove potential interfering ions at low concentrations has the potential to reduce the actual perchlorate content of the sample (USEPA, 1999b).

There is evidence of cases where Method 314.0 has resulted in the reporting of false positives, falsely elevated concentrations and false negatives when applied to non-drinking water matrices. For example, investigation of polluted groundwater at an industrial site in the Henderson, Nevada, area showed that the compound p-chlorobenzenesulfonate (p-CBS) co-elutes with perchlorate during analysis using USEPA Method 314.0, causing falsely elevated perchlorate concentrations (Johnson et al., 2003). In addition, data from a study comparing perchlorate analytical methods for analysis of groundwater at 13 locations from the United States and Canada using USEPA 314.0 compared to analysis by IC/MS (Geosyntec, 2007). Six samples were polluted groundwater collected near residential or public septic systems, two were from a plume near a dichlorodiphenyltrichloroethane (DDT) manufacturing facility, one was from a landfill and one was from a site where groundwater was known to contain surfactants. The interfering compounds were not positively identified in any of the cases of false positives in groundwater near septic systems, but this study suggests that Method 314.0 is not appropriate for analysis of sewage or surfactant impacted water. p-CBS was identified by the MS analyses in four of the other samples that had falsely high detects. p-CBS is a by-product of DDT manufacturing and is used as a solvent in some paints (Johnson et al., 2003).

4.4.2.2 USEPA Method 314.1—Inline Column Concentration/Matrix Elimination Ion Chromatography with Suppressed Conductivity Detection

USEPA Method 314.1 is intended to increase sensitivity, tolerance of TDS and selectivity through the use of a confirmation column and in-line concentration (USEPA, 2005c). Water samples are collected in the field using a sterile filtration technique. The sample, without cleanup, is concentrated onto the concentrator/trap column, which is placed in the sample loop position and binds perchlorate more strongly than other matrix anions. The sample matrix anions are rinsed from the concentrator column with 1 mL of 10 millimolar (mM) NaOH. This weak rinse solution allows the concentrator to retain the perchlorate while eluting the majority of the matrix anions, which are directed to waste. The concentrator column is switched in-line and the perchlorate is eluted from the concentrator column with a 0.50 mM NaOH solution. Following elution from the concentrator, the perchlorate is refocused onto the front of the guard column. The eluent strength is then increased to 65 mM NaOH, which elutes the perchlorate from the guard column and onto the analytical column where perchlorate is separated from other anions and remaining background interferences. Perchlorate is subsequently detected using suppressed conductivity and is quantified using an external standard technique. Confirmation of any perchlorate concentration reported at or above the MRL on the primary column is accomplished with a second analytical column that has a dissimilar separation mechanism (USEPA, 2005c).

4.4.2.3 USEPA Method 9058—Ion Chromatography with Chemical Suppression Conductivity Detection

USEPA Method 9058 is the USEPA's OSW IC method and is essentially the same as Method 314.0, with the exception of the MCT requirement. The method is stated to perform adequately on water samples with conductivities up to 1000 microsiemens per centimeter (µS/cm) and is potentially applicable to surface water, mixed domestic water, and industrial wastewaters. The limitations described above for Method 314.0 apply similarly to Method 9058. OSW is in the process of revising the November 2000 version of Method 9058 given the known interferences and the high probability of false positive and false negative results. Optimization of the method may include an extraction procedure for solids, making the method applicable for high TDS aqueous samples, lowering the detection limit to sub-µg/L levels, better separation, and minimization of false positive and negative results. After the revised method is drafted, an interlaboratory validation study will be conducted (USEPA, 2000; ITRC, 2005).

4.5 SITE CHARACTERIZATION FOR PERCHLORATE TREATMENT

In addition to characterizing the concentration and distribution of perchlorate, several other chemical and geochemical parameters should be assessed during evaluation of a perchlorate-contaminated site as they may play a role in the effectiveness of perchlorate treatment. Table 4.4 presents a list of chemical and geochemical indicators and the importance of their measurement for either bioremediation or ion exchange, the two most widely used *ex situ* treatment technologies for perchlorate-contaminated water. A more detailed discussion of groundwater characterization and monitoring during *in situ* bioremediation is provided in Chapter 3, Principles of Perchlorate Treatment.

Table 4.4. Additional Chemical and Geochemical Parameters to Measure during Site Characterization

Parameter	Rationale
Oxidation-Reduction Potential (ORP)	Low ORP or anaerobic groundwater required for bioremediation. Addition of sufficient electron donor to stimulate bioremediation without achieving highly reduced, sulfate reducing conditions.
pH	pH outside range of 5 to 8.5 is inhibitory to perchlorate-degrading bacteria.
Specific Conductance	High values indicative of high TDS, which can interfere with effectiveness of ion exchange.
Ferrous Iron and Manganese	Measured during bioremediation. Minimize mobilization through control of ORP and amount of electron donor added.
Nitrate	High nitrate levels will require additional electron donor for reduction during bioremediation and may interfere with removal of perchlorate via ion exchange.
Sulfate, Chloride	High sulfate and chloride levels may compete with perchlorate treatment via ion exchange.
Total Dissolved Solids (TDS)	Gross measure of anions such as sulfate, chloride etc. High levels may compete with perchlorate treatment via ion exchange.
Bicarbonate, Carbonate	High bicarbonate and carbonate levels may compete with perchlorate treatment via ion exchange.
Bromide	High bromide levels may compete with perchlorate treatment via ion exchange.

The most common co-contaminant found at perchlorate-contaminated sites is nitrate. Nitrate concentrations are generally far greater than those of perchlorate; however, nitrate is commonly removed along with perchlorate during *in situ* or *ex situ* bioremediation because most perchlorate-reducing bacteria are denitrifiers as well (Logan, 2001; Coates and Achenbach, 2004). Other anions, such as sulfate and carbonate, generally do not adversely impact perchlorate biodegradation, as perchlorate is generally reduced before sulfate. The anions sulfate, nitrate, bicarbonate, carbonate, and bromide compete with perchlorate during the ion exchange process (ITRC, 2005). Ion exchange resins have differing affinities for each of these anions, so levels of each must be considered when selecting resins and determining system operating parameters. Additional information on ion exchange for perchlorate treatment is provided in Boodoo (2003) and Gu and Coates (2006).

4.6 SUMMARY

Over the past several years, various sources of perchlorate in groundwater have been identified. These include various natural sources (e.g., Chilean nitrate, other evaporite deposits, and precipitation) as well as a host of anthropogenic sources such as fireworks, road flares, sodium chlorate, bleach, and perchloric acid. SERDP Project ER-1429 is attempting to identify key co-contaminants to aid in source identification. In addition, current data suggest that stable isotope analysis of Cl and O represents a practical forensic tool to distinguish natural perchlorate of Chilean origin from synthetic perchlorate. The potential utility of stable isotopes to distinguish perchlorate from different synthetic sources is less clear as it appears that $\delta^{37}Cl$ values are too similar among different samples to be of use. However, $\delta^{18}O$ may differ enough in some instances to distinguish source materials.

Several analytical methods are available to analyze perchlorate in environmental media. However, the DoD Environmental Data Quality Workgroup specifically recommends the use of IC/MS or LC/MS methods (e.g., USEPA Methods 6850 and 6860) for contaminated groundwater because of the potential for false positives with ion chromatography methods that were initially developed for drinking water. For example, the presence of sulfate, carbonate, chloride, and p-CBS have the potential to interfere with IC methods. Depending on the proposed groundwater treatment technology, other anions, such as sulfate, nitrate, bicarbonate, carbonate, and bromide, also should be measured, as the presence of these anions can potentially adversely impact treatment performance.

REFERENCES

Ader M, Coleman ML, Doyle SP, Stroud M, Wakelin D. 2001. Methods for the stable isotopic analysis of chlorine in chlorate and perchlorate compounds. Anal Chem 73:4946–4950.

APA (American Pyrotechnics Association). 2004. APA Anticipates Robust Year for Fireworks Retailers but Tougher Times for Professional Display Industry. APA Press Release. Bethesda, MD, USA. June 23. http://www.americanpyro.com/pdf/ 0406_pr_release.pdf. Accessed February 24, 2008.

Bao H, Gu B. 2004. Natural perchlorate has a unique oxygen isotope signature. Environ Sci Technol 38:5073–5077.

Betts JA, Dluzniewski TJ. 1997. Impurity Removal for Sodium Chlorate. U.S. Patent No. 5,681,446, October 28, 1997.

Böhlke JK, Ericksen GE, Revesz K. 1997. Stable isotope evidence for an atmospheric origin or desert nitrate deposits in northern Chile and southern California, USA. Chem Geol 136:135–152.

Böhlke JK, Sturchio NC, Gu B, Horita J, Brown GM, Jackson WA, Batista J, Hatzinger PB. 2005. Perchlorate isotope forensics. Anal Chem 77:7838–7842.

Boodoo F. 2003. POU/POE removal of perchlorate. Water Cond Purif 45:52–55.

Brandhuber P, Clark S. 2005. Perchlorate Occurrence Mapping. Submitted to American Water Works Association, Washington, DC, USA. http://www.awwa.org/files/Advocacy/ PerchlorateOccurrenceReportFinalb02092005.pdf. Accessed February 24, 2008.

CDHS (California Department of Health Services). 2007. History of Perchlorate in California Drinking Water. http://www.cdph.ca.gov/certlic/drinkingwater/Pages/Perchloratehistory.aspx. Accessed February 21, 2008.

Clark ID, Fritz P. 1997. Environmental Isotopes in Hydrogeology. CRC Press LLC, Boca Raton, FL, USA. 328p.

Coates JD, Achenbach LA. 2004. Microbial perchlorate reduction: Rocket-fuelled metabolism. Nature Rev Microbiol 2:579–580.

Conkling JA. 1985. Chemistry of Pyrotechnics. Basic Principles and Theory. Marcel Dekker, Inc., New York, NY, USA.

Dasgupta PK, Martinelango PK, Jackson WA, Anderson TA, Tian K, Tock RW, Rajagopalan S. 2005. The origin of naturally occurring perchlorate: The role of atmospheric processes. Environ Sci Technol 39:1569–1575.

Dayan H, Abrajano T, Sturchio NC, Winsor L. 1999. Carbon isotopic fractionation during reductive dechlorination of chlorinated solvents by metallic iron. Org Geochem 30:755–763.

Dignazio FJ, Krothe NC, Baedke SJ, Spalding RF. 1998. $\delta^{15}N$ of nitrate derived from explosive sources in a karst aquifer beneath the ammunition burning ground, Crane Naval Surface Warfare Center, IN, USA. J Hydrol 206(3-4):164–175.

DoD EDQW (Department of Defense Environmental Data Quality Workgroup). 2007. DoD Perchlorate Handbook. August 2007. http://www.fedcenter.gov/_kd/Items/actions.cfm?action=Show&item_id=8172&destination=ShowItem. Accessed February 24, 2008.

Drenzek N, Tarr C, Eglinton T, Heraty L, Sturchio NC, Shiner V, Reddy C. 2002. Stable chlorine and carbon isotopic compositions of selected semi-volatile organochlorine compounds. Org Geochem 33:437–444.

Geosyntec. 2005. Alternative Causes of Wide-Spread, Low Concentration Perchlorate Impacts to Groundwater. White Paper submitted to DoD SERDP, Arlington, VA, USA.

Geosyntec. 2006. Evaluation of Alternative Causes of Wide-Spread, Low Concentration Perchlorate Impacts to Groundwater. SERDP Project ER-1429 Annual Report. Submitted to DoD SERDP, Arlington, VA, USA.

Geosyntec. 2007. Evaluation of Alternative Causes of Wide-Spread, Low Concentration Perchlorate Impacts to Groundwater. SERDP Project ER-1429 Annual Report. Submitted to DoD SERDP, Arlington, VA, USA.

Goldenwieser EA. 1919. A Survey of the Fertilizer Industry. U.S. Department of Agriculture (USDOA), Bulletin No. 798. Washington, D.C., USA, October 20.

Groocock G. 2002. New problem: Perchlorate. The Suffolk Times Online.

Gu B, Coates JD (eds). 2006. Perchlorate Environmental Occurrence, Interactions and Treatment. Springer, New York, NY, USA.

Gu B, Brown GM, Maya L, Moyer BA. 2001. Regeneration of perchlorate (ClO_4^-)-loaded anion exchange resins by novel tetrachloroferrate ($FeCl_4^-$) displacement technique. Environ Sci Technol 35:3363–3368.

Hammer BT, Kelly CA, Coffin RB, Cifuentes LA, Mueller J. 1998. $\delta^{13}C$ values of polycyclic aromatic hydrocarbons collected from two creosote contaminated sites. Org Geochem 152:43–58.

Hatzinger PB, Böhlke JK, Sturchio NC, Gu B. 2008. Validation of Chlorine and Oxygen Isotope Ratio Analysis to Differentiate Perchlorate Sources and to Document Perchlorate Biodegradation. Project ER-0509 Fact Sheet. ESTCP, Arlington, VA, USA. http://www.estcp.org/Technology/ER-0509-FS.cfm. Accessed February 24, 2008.

Holt BD, Sturchio NC, Abrajano TA, Heraty LJ. 1997. Conversion of chlorinated volatile organic compounds to carbon dioxide and methyl chloride for isotopic analysis of carbon and chlorine. Anal Chem 69:2727–2733.

Holt BD, Heraty LJ, Sturchio NC. 2001. Extraction of chlorinated aliphatic hydrocarbons from groundwater at micromolar concentrations for isotopic analysis of chlorine. Environ Pollut 113:263–269.

Howard PE. 1931. Survey of the Fertilizer Industry. USDOA, Circular No. 129. Washington, DC, USA.

Hughes V. 2004. Contamination Might be Traced to Lowell Treatment Plant. The Lowell Sun, September 2. http://www.lowellsun.com/. Accessed February 21, 2008.

Hunkeler DR, Aravena, Butler BJ. 1999. Monitoring microbial dechlorination of tetrachloroethene (PCE) in groundwater using compound-specific stable carbon isotope ratios: Microcosm and field studies. Environ Sci Technol 33:2733–2738.

Hunkeler DR, Aravena R, Spark KB, Cox E. 2005. Assessment of degradation pathways in an aquifer with mixed chlorinated hydrocarbon contamination using stable isotope analysis. Environ Sci Technol 39:5975–5981.

IME (Institute of Makers of Explosives). 2007. Perchlorate. http://www.ime.org/tmp_downloads/Perclhorate%20statement07.pdf. Accessed February 21, 2008.

ISEE (International Society of Explosives Engineers). 1998. Blasters' Handbook, 17th edition. ISEE, Cleveland, OH, USA.

ITRC (Interstate Technology & Regulatory Council). 2005. Perchlorate: Overview of Issues, Status, and Remedial Options. ITRC Perchlorate Team, September. http://www.itrcweb.org/Documents/PERC-1.pdf. Accessed February 21, 2008.

Jackson WA, Anandam S, Anderson TA, Lehman T, Rainwater KA, Rajagopalan K, Ridley M, Tock WR. 2005. Perchlorate occurrence in the Texas southern high plains aquifer system. Groundwater Monit Remediat 25:137–149.

Jackson WA, Anderson TA, Harvey G, Orris G, Rajagopalan S, Namgoo K. 2006. Occurrence and formation of non-anthropogenic perchlorate. In Gu B, Coates JD, eds, Perchlorate: Environmental Occurrence, Interactions, and Treatment. Springer, New York, NY, USA, pp 49–66.

Jendrzejewski N, Eggenkamp HGM, Coleman ML. 1997. Sequential determination of chlorine and carbon isotopic composition in single microliter samples of chlorinated solvent. Anal Chem 69:4259–4266.

Johnson J, Grimshaw D, Richman K. 2003. Analysis for perchlorate by ion chromatography: significant recent findings. American Pacific Corporation, Las Vegas, NV, USA.

Kelly CA, Hammar BT, Coffin RB. 1997. Concentrations and stable isotope values of BTEX in gasoline contaminated groundwater. Environ Sci Technol 31:2469–2472.

Kendall C, Caldwell EA. 1998. Fundamentals of isotope geochemistry. In Kendall C, McDonnell JJ, eds, Isotope Tracers in Catchment Hydrology. Elsevier Science, Amsterdam, Netherlands, pp 51–86.

Kramer DA. 2003. Explosives. U.S. Geological Service. http://minerals.usgs.gov/minerals/pubs/commodity/explosives/explomyb03.pdf. Accessed February 24, 2008.

Logan BE. 2001. Assessing the outlook for perchlorate remediation. Environ Sci Technol 35:482A–487A.

MADEP (Massachusetts Department of Environmental Protection). 2005. The Occurrence and Sources of Perchlorate in Massachusetts, Draft Report. MADEP, Boston, MA, USA, August. http://www.mass.gov/dep/cleanup/sites/percsour.pdf. Accessed September 21, 2008.

Mansuy L, Philip RP, Allen J. 1997. Source identification of oil spills based on the isotopic composition of individual components in weathered oil samples. Environ Sci Technol 31:3417–3425.

Mehring AL. 1943. Fertilizer Consumption in 1941 and Trends in Usage. USDOA, Circular No. 689. Washington, DC, USA.

Michalski G, Böhlke JK, Thiemens M. 2004. Long term atmospheric deposition as the source of nitrate and other salts in the Atacama Desert, Chile: new evidence from mass independent oxygen isotopic compositions. Geochim Cosmochim Acta 68:4023–4038.

OMRI (Organic Materials Review Institute). 2000. National Organics Standards Board (NOSB) Technical Advisory Panel (TAP) Review Compiled by OMRI for Sodium Chlorate. http://www.omri.org/sodium_chlorate.pdf. Accessed February 21, 2008.

Orris GH, Harvey GJ, Tsui DT, Eldridge JE. 2003. Preliminary analyses for perchlorate in selected natural materials and their derivative products. USGS Open File Report 03-314. USGS, USA.

PAN Pesticides Database. 2002. Sodium Chlorate California Pesticide Use Statistics for 2002. http://www.pesticideinfo.org/Search_Use.jsp#SearchCAUse.

Philip RP. 2002. Application of stable isotopes and radioisotopes in environmental forensics. In Murphy BL, Morrison RD, eds, Introduction to Environmental Forensics. Elsevier, New York, NY, USA. 560p.

Plummer LN, Böhlke JK, Doughten MW. 2006. Perchlorate in Pleistocene and Holocene groundwater in north-central New Mexico. Environ Sci Technol 40:1757–1763.

Rajagopalan S, Anderson TA, Fahlquist L, Rainwater KA, Ridley M, Jackson WA. 2006. Widespread presence of naturally occurring perchlorate in high plains of Texas and New Mexico. Environ Sci Technol 40:3156–3162.

Rao B, Anderson TA, Orris GJ, Rainwater KA, Rajagopalan S, Sandvig RM, Scanlon BR, Stonestrom DA, Walvoord MA and Jackson WA. 2007. Widespread natural perchlorate in unsaturated zones of the southwest United States. Environ Sci Technol 41:4522–4528.

Renner R. 1999. Study finding perchlorate in fertilizer rattles industry. Environ Sci Technol 33:394A–395B.

Schilt AA. 1979. Perchloric Acid and Perchlorates. GFS Chemicals, Inc., Columbus, OH, USA.

Sharp Z. 2007. Principles of Stable Isotope Geochemistry. Pearson Prentice Hall, Upper Saddle River, NJ, USA. 344p.

Silva MA. 2003a. Perchlorate from Safety Flares: A Threat to Water Quality. Santa Clara Valley Water District Publication. Santa Clara Valley Water District, San Jose, CA, USA.

Silva MA, 2003b. Safety Flares Threaten Water Quality with Perchlorate. Santa Clara Valley Water District Publication. Santa Clara Valley Water District, San Jose, CA, USA.

Smallwood BJ, Philip RP, Burgoyne TW, Allen J. 2001. The use of stable isotopes to differentiate specific source markers for MTBE. J Environ Forensics 2:215–221.

Song DL, Conrad ME, Sorenson KS, Alvarez-Cohen L. 2002. Stable carbon isotope fractionation during enhanced in situ bioremediation of trichloroethene. Environ Sci Technol 36:2262–2268.

Sturchio NC, Clausen JC, Heraty LJ, Huang L, Holt BD, Abrajano T. 1998. Stable chlorine isotope investigation of natural attenuation of trichloroethene in an aerobic aquifer. Environ Sci Technol 32:3037–3042.

Sturchio NC, Hatzinger PB, Arkins MD, Suh C, Heraty LJ. 2003. Chlorine isotope fractionation during microbial reduction of perchlorate. Environ Sci Technol 37:3859–3863.

Sturchio NC, Böhlke JK, Gu B, Horita J, Brown GM, Beloso A, Patterson LJ, Hatzinger PB, Jackson WA, Batista J. 2006. Stable isotopic composition of chlorine and oxygen in synthetic and natural perchlorate. In Gu B, Coates JD, eds, Perchlorate: Environmental Occurrence, Interactions, and Treatment. Springer, New York, NY, USA, pp 93–110.

Sturchio NC, Böhlke JK, Beloso Jr AD, Streger SH, Heraty L, Hatzinger PB. 2007. Oxygen and chlorine isotopic fractionation during perchlorate biodegradation: Laboratory results and implications for forensics and natural attenuation studies. Environ Sci Technol 41:2796–2802.

Urbansky ET, Brown SK, Magnuson ML, Kelly CA. 2001a. Perchlorate levels in samples of sodium nitrate fertilizer derived from Chilean caliche. Environ Pollut 112:299–302.

Urbansky ET, Collette TW, Robarge WP, Hall WL, Skillen JM, Kane PF. 2001b. Survey of Fertilizers and Related Materials for Perchlorate. EPA 600/R/01/047. USEPA, Washington, DC, USA.

USDOC (U.S. Department of Commerce). 2003. Inorganic Chemicals: 2002, Product code 325188A141. USDOC Economics and Statistics Administration, U.S. Census Bureau, Washington DC, USA.

USEPA (U.S. Environmental Protection Agency). 1999a. Revisions to the Unregulated Contaminant Monitoring Regulation for Public Water Systems; Final Rule. Federal Register, September 17, 64:50555–50620.

USEPA. 1999b. Method 314.0. Determination of Perchlorate in Drinking Water Using Ion Chromatography, Rev. 1.0. USEPA National Exposure Research Laboratory, Office of Research and Development, Cincinnati, OH, USA. www.epa.gov/safewater/methods/sourcalt.html. Accessed February 21, 2008.

USEPA. 2000. Method 9058. Determination of Perchlorate Using Ion Chromatography with Chemical Suppression Conductivity Detection, Rev. 0.

USEPA. 2005a. Method 331.0. Determination of Perchlorate in Drinking Water by Liquid Chromatography Electrospray Ionization Mass Spectrometry, Rev. 1.0. EPA 815/R/05/007. Technical Support Center, Office of Groundwater and Drinking Water, USEPA, Cincinnati, OH, USA.

USEPA. 2005b. Method 332.0. Determination of Perchlorate in Drinking Water by Ion Chromatography with Suppressed Conductivity and Electrospray Ionization Mass Spectrometry. EPA/600/R/05/049. USEPA National Exposure Research Laboratory, Office of Research and Development, Cincinnati, OH, USA. www.epa.gov/nerlcwww/ordmeth.htm. Accessed February 21, 2008.

USEPA. 2005c. Method 314.1, Determination of Perchlorate in Drinking Water Using Inline Column Concentration/Matrix Elimination Ion Chromatography with Suppressed Conductivity Detection, Rev 1.0. EPA 815/R/05/009. Technical Support Center, Office of Ground Water and Drinking Water, USEPA, Cincinnati, OH, USA. www.epa.gov/safewater/methods/sourcalt.html. Accessed February 21, 2008.

USGS (U.S. Geological Survey). 2006a. Resources on Isotopes: Periodic Table – Chlorine. http://wwwrcamnl.wr.usgs.gov/isoig/period/cl_iig.html. Accessed February 21, 2008.

USGS. 2006b. Resources on Isotopes: Periodic Table – Oxygen. http://wwwrcamnl.wr.usgs.gov/isoig/period/o_iig.html. Accessed February 21, 2008.

Walvoord MA, Phillips FM, Stonestrom DA, Evans RD, Hartsough PC, Newman BD, Striegl RG. 2003. A reservoir of nitrate beneath desert soils. Sci 302:1021–24.

Wilkin RT, Fine DD, Burnett, NG. 2007. Perchlorate behavior in a municipal lake following fireworks displays. Environ Sci Technol 41:3966–3971.

CHAPTER 5

ALTERNATIVES FOR *IN SITU* BIOREMEDIATION OF PERCHLORATE

Hans F. Stroo[1] and Robert D. Norris[2]

[1]HydroGeoLogic, Inc., Ashland, OR 97520; [2]Brown and Caldwell, Golden, CO 80401

5.1 INTRODUCTION

Perchlorate can be reduced to the innocuous anion chloride under slightly reducing conditions by several microorganisms that are apparently ubiquitous in soils and groundwaters (Coates et al., 1999; Logan, 2001). This reduction can happen naturally, but stimulation of the activity of indigenous perchlorate reducers by addition of an electron donor to the subsurface environment is often needed because natural attenuation alone may not be sufficient to ensure environmental protection. The deliberate enhancement of natural biodegradation (bioremediation) has proven to be a successful technology for treating perchlorate contaminated soils and groundwaters (Xu et al., 2003). Currently there are few other mechanisms available for cost-effectively treating perchlorate *in situ*.

The overall reaction stoichiometry, with acetate as the electron donor, is:

$$CH_3COO^- + ClO_4^- \rightarrow 2HCO_3^- + H^+ + Cl^-, \qquad \text{(Rx. 5.1)}$$

and the reaction apparently proceeds through chlorate and chlorite intermediates:

$$ClO_4^- \rightarrow ClO_3^- \rightarrow ClO_2^- \rightarrow Cl^- + O_2 \qquad \text{(Rx. 5.2)}$$

The fundamental requirements for stimulating perchlorate reduction are relatively simple – slightly reducing conditions are necessary, an electron donor source is needed, and other anions that can inhibit perchlorate reduction should be absent or present in limited concentrations (Chaudhuri et al., 2002). The key anions of concern are sulfate and particularly nitrate, because nitrate and perchlorate reduction occur under similar mildly reducing conditions, and nitrate can compete with perchlorate for the enzymes responsible for perchlorate reduction (Herman and Frankenberger, 1999).

But subsurface conditions vary, as do the remedial objectives for different sites, so the most effective, and cost-effective, strategy for stimulating perchlorate reduction may also vary from site to site. Remediation professionals have therefore devised several different approaches to *in situ* bioremediation and have adapted the basic technology for a wide variety of site conditions (Cox et al., 2000; Hatzinger et al., 2002).

There are essentially three approaches to *in situ* bioremediation of perchlorate contaminated groundwater. These have been termed active, semi-passive and passive remediation (Figure 5.1). Table 5.1 summarizes the characteristics of these approaches.

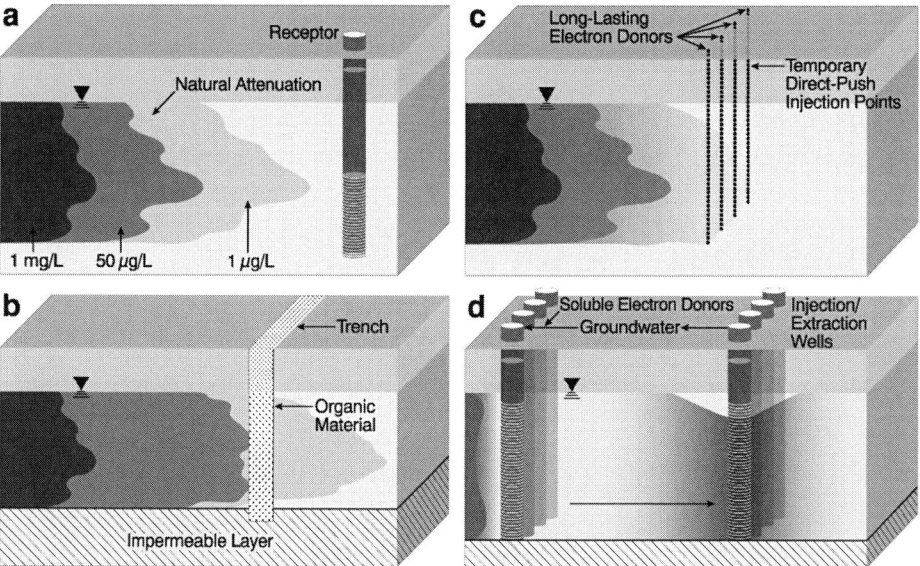

Figure 5.1. Alternatives for *in situ* bioremediation of perchlorate
a) Natural attenuation; b) Passive system using mulch biowall to intercept plume; c) Passive biobarrier using injection of semi-soluble electron donors such as vegetable oil; and d) Active or semi-passive system, using injection of soluble electron donors in recirculated groundwater extracted from downgradient.

The complex set of site and management factors that impact remediation decisions make it difficult to develop simple and straightforward guidance regarding which approach is best for a given site. Such decisions are best made with a sound understanding of the biodegradation mechanism, associated microbiology, site hydrogeology and project goals. This initial guidance is intended to help identify the critical factors influencing the selection process and the relative strengths and limitations of these alternatives.

Table 5.1. Characteristics of Alternative Approaches for Implementing *In Situ* Bioremediation for Perchlorate in Groundwater

Approach	Pumping Frequency	Injection Frequency	Electron Donor
Active	Continuous to Near-Continuous	Continuous to Frequent Intervals	Water-Soluble (e.g., Lactate, Ethanol, Citrate, Benzoate)
Semi-Passive	Intermittent or Pulsed	Weekly to Monthly Intervals	Water-Soluble
Passive – Injected Biobarrier	No Pumping	One-Time to 3- to 6-Year Intervals	Semi-Soluble (HRC®, Vegetable Oil) or Possibly Solid (e.g., Chitin)[1]
Passive – Biowall Trench	Little or None	None or Recharging at 3- to 6-Year Intervals	Solid (e.g., Mulch, Compost) and Semi-Soluble to Recharge (e.g., Vegetable Oil)

[1] Co-injections of soluble and less soluble donors may also be used, especially as initial treatments.

This chapter provides an initial comparison of the available *in situ* bioremediation approaches for perchlorate, some of the options for implementing them and a discussion of the types of situations for which each is best suited. It is intended to assist in screening-level evaluations and technology selection efforts to assist managers in identifying the approaches that best fit the conditions at a particular site. This overview serves as an introduction to the following chapters on Active Bioremediation (Chapter 6), Semi-Passive Bioremediation (Chapter 7), Passive Bioremediation Using Emulsified Edible Oils (Chapter 8), and Passive Bioremediation Using Permeable Biowalls (Chapter 9). These chapters provide more detail on technology selection and design and operation, with illustrative case studies of field-scale applications.

5.2 TECHNOLOGY SELECTION PROCESS

The first step is selecting a technology is to develop an adequate Conceptual Site Model (CSM) and Remedial Action Objectives (RAOs). In essence, these steps involve diagnosis (defining the problem) and setting realistic goals for treatment (USACE, 2003). For *in situ* technologies, whose performance and costs can be highly sensitive to site constraints, the CSM often needs to be relatively detailed or the design needs to be flexible and/or adjustable as needed during operation. RAOs need to be developed with care, especially for *in situ* technologies, because it is often difficult to select an appropriate technology and develop an acceptable and successful design, unless there is consensus on and clear expectations for treatment and for the results that would allow turning off the treatment system.

The CSM is a summary of the environmental conditions at a site that identifies the types and locations of all potential sources of contamination and past and future transport, as well as the factors that control how and where people, plants or animals may be exposed to the contamination. For *in situ* bioremediation, it is generally helpful to have a three-dimensional model of the subsurface and contaminant distribution, hydrogeological conditions (including significant temporal and spatial variations and heterogeneities), any geochemical conditions that can impact biological perchlorate reduction and any surface features that could affect access or reagent delivery (Mayer and Greenberg, 2005). One can then define RAOs for different locations at a site and/or for different times in the remedial process. For example, at perchlorate sites, *in situ* bioremediation often will be useful for plume containment, but may or may not be applicable for source control. After defining the objectives for overall site management, one can then assess whether *in situ* bioremediation is appropriate, and if so, what approach to its implementation will best suit the site conditions and management objectives.

5.2.1 *In Situ* Bioremediation

In situ bioremediation in general has many advantages, notably reduced costs and time for cleanup as compared to pump-and-treat. *In situ* treatment also does not require disruption of ongoing activities at a site in most cases and generally needs little or no aboveground infrastructure. Finally, *in situ* treatment generates few if any waste materials that will require disposal or treatment.

There are several limitations common to all *in situ* bioremediation approaches. Performance is strongly dependent on the ability to deliver the reagents to the target treatment zones and the distribution of the reagents within that zone to meet the spatial reagent demands. Delivery and distribution are strongly dependent on the degree of mixing possible within the subsurface. Therefore it is important to understand the permeability of the aquifer materials, the amount and locations of spatial heterogeneity within the subsurface and the variability in direction and velocity of groundwater flow, as all of these factors can impact performance. However, some of the alternative approaches are more sensitive to distribution limitations than others. In fact, these distribution limitations are often the primary technical determinants of the approach that can be used at any given site.

In situ bioremediation can be infeasible or excessively expensive if contamination extends too deep to allow delivery of remedial agents, if groundwater seepage velocities combined with other electron donor demands are too great or if contaminants are located in inaccessible areas such as within fractured bedrock. *In situ* treatment can also be more difficult, or in some cases impossible, if surface structures impede access or limit the ability to inject materials or install monitoring points. Perchlorate at high concentrations may form a denser-than-water brine solution, and there may therefore be little mixing with the rest of the groundwater, reducing transport and complicating the delivery of remedial agents (Flowers and Hunt, 2007).

There is also an inherent degree of uncertainty involved with any *in situ* treatment. The key uncertainties include the actual extent and distribution of contamination, the time required for treatment to acceptable levels, and the ability to achieve adequate contact between contaminants and reactants. Finally, *in situ* anaerobic bioremediation can cause undesirable changes in groundwater chemistry such as mobilization of some metals (e.g., arsenic or iron), excess acidity or generation of methane or sulfides. These side effects are generally limited in duration and spatial extent, but may cause problems at some sites (Suthersan et al., 2002). Of course, metal reduction may also be a beneficial side effect (e.g., converting chromium from Cr^{6+} to less toxic forms).

All of the *in situ* bioremediation alternatives discussed in this volume have been used for perchlorate remediation at a field scale, but only for a few sites. But, as documented in the recent U.S. Environmental Protection Agency (USEPA) review of the status of various treatment technologies for perchlorate (USEPA, 2005), none of the applications of *in situ* bioremediation have had long enough operating histories to allow definitive conclusions regarding longevity or long-term performance. The active approach was the first to be tested under field conditions for perchlorate remediation, in the late 1990s (Cox et al., 2000), and it has been used full scale at five or more sites. The passive permeable biowall approach was first attempted soon thereafter, and has since been implemented full scale at a similar number of sites. The semi-passive strategy has been successfully demonstrated in field trials, but to date it has had little full-scale use for treating perchlorate sites.

Bioremediation technology for perchlorate largely has been transferred from the successful use of enhanced anaerobic bioremediation of chlorinated solvents, and the use of denitrification to remove nitrates from groundwaters. The technology has been adapted to the particular issues related to perchlorate remediation in groundwater, and it is still evolving as we continue to learn more about the process and gain field experience with its application.

5.2.2 Active Treatment

Active systems use pumping to inject and distribute electron donors. The donors used are soluble compounds, such as lactate, ethanol, citrate, benzoate, and acetate, and injections are generally frequent or even continuous, usually by metering the donor solution into recirculated groundwater (Hatzinger et al., 2006). Injection and extraction wells are generally used to recover water and reinject it to deliver the donor solution, although injection-only systems can also be used.

Typical variations include alternating operations (switching between extraction and injection at individual wells) and/or recirculation within the subsurface by injection and extraction at different depths (Parr et al., 2003). The system may rely on upgradient extraction and reinjection into the target treatment zone, or injection and extraction along a transect at wells located cross-gradient to the direction of groundwater flow. To cost-effectively distribute amendments, cross-gradient injection/recovery systems may be operated intermittently to make use of both induced and natural gradients.

An active treatment system has several advantages (see Table 5.2 for a summary of the pros and cons of each of the alternative approaches). Donor distribution throughout the treatment area is maximized by continuous or near-continuous recirculation. Because of recirculation, donor distribution is not dependent on natural groundwater flow, and effective treatment is possible over a wide range of hydraulic conductivities and flow velocities. Active treatment is also highly flexible. Injection rates and concentrations can be adjusted to respond to changing or uncertain flow conditions or donor demands. For sites with deeper contamination, active treatment is more applicable than many of the more passive barrier systems.

Table 5.2. Advantages and Limitations of Perchlorate *In Situ* Bioremediation Approaches

Approach	Advantages	Limitations
Active	• Greater extent of treatment is possible • Flexible – numerous site adaptations possible • Relatively rapid treatment • Can modify amount and type of electron donor added and donor choice • Greatest distribution of reactants • More flexible around buildings and other infrastructure	• Usually the most costly approach • Possible degradation of water quality • Continuous O&M required • Greatest potential for biofouling problems
Semi-Passive	• Moderate cost due to lower O&M than for active system • Can modify amount and type of electron donor added	• Capital cost similar to active reinjection system • Donor distribution dependent on groundwater flow • Some potential for biofouling • More O&M than semi-passive and passive systems
Passive – Injected Biobarrier	• Low cost • Little disruption of ongoing operations • Little ongoing O&M	• Distribution depends on natural groundwater flow • May require numerous wells for high flow rates or low permeability
Passive – Biowall Trench	• Potentially the lowest cost • Heterogeneity is less of a problem • Little disruption of ongoing operations • Little ongoing O&M	• Limited to depths approximately <10–15 m (35–50 ft) bgs based on costs • Not applicable for bedrock • May not be suitable for high permeability aquifers • Flows >0.3–3.0 m/day (1–10 ft/day) may result in breakthrough • High concentrations of other e^- acceptors may limit performance (e.g., >1–10 mg/L nitrate-N)

The greatest disadvantage of active treatment is the relatively high cost, for both the ongoing operation and maintenance (O&M) and the capital equipment (wells, pipes, pumps, tanks, etc.) needed for recirculation and donor solution metering systems. Also, biofouling can be a particularly difficult problem because of the constant input of rapidly-degradable electron donors near the injection wells, although there are methods available to control such biofouling (Chopra et al., 2005; Cox, 2005). These methods include the intermittent use of biocides such as chlorine dioxide, the use of electron donors such as citric acid that inhibit biofouling near the injection points, and the continuous injection of water, with only intermittent additions of electron donors.

In addition, it can be very difficult to operate an active *in situ* bioremediation system successfully, or cost-effectively, in tight (low-permeability) formations because of limitations to electron donor delivery and distribution. Finally, high-yield aquifers can require a large number of closely-spaced wells to pump unrealistic (or at least uneconomical) amounts of water.

5.2.3 Semi-Passive Treatment

The inclusion of "semi-passive" treatment as a separate category is a reflection of the fact that many applications of *in situ* bioremediation have been hybrid systems, adapted to reduce costs and on-site O&M, while still enhancing the distribution of reactants. Semi-passive treatment was originally proposed as a way to reduce costs by pulsing nutrient additions (Devlin and Barker, 1994) and has been frequently used for *in situ* bioremediation of chlorinated solvents (Devlin and Barker, 1996) as well as for *in situ* denitrification (Gierczak et al., 2007).

Generally, semi-passive treatment is implemented by intermittent pumping of soluble donors, sometimes with mobile equipment that can be transported to different injection wells or sites. Semi-passive treatment takes advantage of the fact that the biomass that flourishes soon after addition of a soluble substrate can serve when decomposed as a longer-term electron donor source that can enhance perchlorate reduction for extended periods (several weeks to months) after each addition.

The primary advantage of the semi-passive strategy is that electron donor distribution can be increased during the active injection period, but costs are generally lower than for the active approach because most of the time no pumping or recirculation is needed. Also, the impacts to secondary water quality are generally reduced compared to other approaches, because less electron donor is added and donor costs are minimized. The number of injection points also may be fewer than when using a completely passive approach, because the injection pressures allow a larger effective radius of influence (ROI) than can be achieved by relying on natural advection and diffusion alone. A larger ROI can be particularly helpful when attempting to remediate wide plumes or sites with surface obstructions.

The primary disadvantage of semi-passive technology is that a permanent injection system must still be installed, and as a result the capital costs generally will be higher than for passive systems (although in some cases passive systems may require so many injection points that the capital costs can be higher). The O&M costs generally will be higher than for passive systems but lower than for continuous operation, since some active maintenance is still required to operate the system. Also, the aboveground equipment needed may be similar to a continuous (active) recirculation system, though it can be smaller in size and mobile (requiring fewer pumps and tanks) for transport between injection points as needed. The distribution of the donor will be largely dependent on natural groundwater flow, hence active systems may be more appropriate for lower permeability or higher flow rate aquifers.

5.2.4 Passive Treatment

Passive treatment systems rely on slow-release electron donor sources that can be placed in wells or trenches, or directly injected into the subsurface. These sources can include edible oils such as EOS^{TM} (AFCEE, 2007; Borden, 2007), organic wastes such as mulch (Ahmad et al., 2007; Aziz et al., 2001) or slow-release electron donor sources such as HRC® (USEPA, 2005). For perchlorate remediation, these materials generally have been used to create permeable reactive barriers (or "biobarriers"), to cut off the contaminated plume (Cowan, 2000). Because there are important differences between biobarriers created by injecting donor sources into the subsurface (using rows of injection wells or injection points placed perpendicular to groundwater flow) and "biowalls" created by filling trenches with organic materials (Parsons, 2004), these technologies will be discussed separately at greater depth in later sections of this chapter.

The greatest advantage of a passive system is the relatively low cost, particularly for O&M and capital. The cost of the donor may be relatively high for some of the commercial products, but some sources, such as mulch and compost, are very inexpensive. However, it is critical to focus on the total costs for treatment and not simply the unit costs for the substrate used. Passive treatment also avoids serious biofouling concerns, and the slow release of the donor limits the need for ongoing site O&M and interferences in site activities.

The limitations of passive systems include the reliance on natural groundwater flow to distribute injected reactants and the depths that can be treated by trenching or injections. Sites with rapid groundwater flow velocities can be difficult to treat effectively with passive systems, particularly where the sulfate or nitrate levels are high, as these will determine the electron donor demand. Installing any barrier is generally limited to depths of less than 15 meters (m) (approximately 50 feet [ft]) below ground surface (bgs) (Parsons, 2004), and there should be a competent material at the base of the treatment zone to prevent underflow. Short-circuiting, flow around the barrier and overflow due to water table fluctuations may also occur in some situations (ITRC, 2005).

Some electron donors are introduced as emulsions of edible oils or similar compounds. These electron donor sources tend to sorb to the soil matrix and not move through the formation. Slow degradation of the oils releases mobile electron donors. This type of electron donor can be augmented with a more soluble electron donor, such as lactate, which will accelerate the achievement of low redox conditions as well as more quickly provide electron donors down gradient of the injection points. Mobile electron donors are rapidly consumed so they are not appropriate for passive treatment when used alone. The use of edible oils may lower the groundwater pH to below desirable levels. Addition of sodium lactate can act as a mild buffer and partially offset the pH drift. Also, buffering materials may be added to prevent such pH changes.

Since the longevity of the various donors is not adequately known, reinjection or recharging of the materials may be needed, at roughly 3- to 6-year intervals (Craig et al., 2006). Close spacing of injection points may be needed. Since there is little flexibility in passive remediation designs, the understanding of site conditions must be adequate for a treatment scheme based on infrequent applications.

5.3 DECISION GUIDELINES

The remainder of this chapter briefly summarizes the factors that will influence the selection of the system best suited to a given set of site conditions. Although it is not realistic or desirable to develop prescriptive guidance, this section will focus on comparing the different approaches, identifying problematic conditions for each, and evaluating their abilities to meet typical management objectives for perchlorate site remediation.

5.3.1 Ability to Meet Management Objectives

In many cases, minimizing cost is the primary decision driver. However, other factors are also important and in many cases are more so than costs. And even though cost efficiency can seem to be a clear goal, some responsible parties may be primarily concerned with minimizing initial costs, while others are more concerned with the total life-cycle costs and still others are concerned primarily about the rate of spending and/or the predictability of the costs. It is important to be clear about the economic objectives, because minimizing the initial costs for implementing a technology may not lead to the same decisions as minimizing overall life-cycle costs for plume restoration or controlling the rate of spending. The use of spread sheets, such as the one discussed in the cost chapter (Chapter 10), allows for a comparison of capital as well as life cycle costs and rate of spending over the life time of the project.

The key non-economic objectives that often influence the decisions regarding which type of *in situ* bioremediation approach to use include reducing the time of restoration, minimizing downtime and other reliability issues, minimizing disruption of on-site activities, and maximizing flexibility to optimize operations over time. The ability of the different approaches to meet these goals is discussed below, and summarized in Table 5.3. It is important to realize that there may be considerable range within some of these relative rankings, depending on specific conditions, but these are useful screening-level comparisons designed to focus on the key advantages and constraints of each approach.

Table 5.3. Comparisons of Ability of Perchlorate *In Situ* Bioremediation Approaches to Meet Typical Management Objectives

Technology	Performance Relative to Management Objectives					
	Cost (Initial)	Cost (Life Cycle)	Speed	Reliability	Site Disruption	Flexibility
Active	High	High	High	Moderate	High	High
Semi-Passive	High to Moderate	Moderate	Moderate	High	Moderate	High
Passive – Injected Biobarrier	Moderate	Low	Moderate to Low	Moderate	Low	Low
Passive – Biowall Trench	Low	Low	Low	Moderate	Low	Low

5.3.1.1 Costs

If reducing initial implementation cost is the primary goal, biobarriers are generally the least costly option, if they are viable based on other criteria. Trenching may represent a significant cost, but the electron donor materials are generally inexpensive, and the costs are usually less than the costs for permanent injection wells and the pumps, piping and tanks needed for more active approaches. However, for contamination deeper than roughly 10 to 15 m bgs, or approximately 35 to 50 ft bgs, the trenching costs may become noncompetitive relative to direct injection. Direct injection of slow-release electron donors may have to be done at relatively close spacing, raising the initial costs. The benefits of biobarriers are more obvious when considering the life-cycle costs, because of the minimal O&M needed. Semi-passive treatment is often (but not always) less costly initially than active treatment because less aboveground equipment may be needed.

The rankings for long-term costs are similar. Active treatment is the most costly because of the need for continuous recirculation. The passive systems generally require little O&M, although if the electron donor demand is high (due to high concentrations of perchlorate and/or other electron acceptors, such as sulfate or nitrate, or rapid groundwater flux), the need for relatively frequent recharging may raise the life cycle costs considerably. A major consideration in life cycle costs is the cost of long-term monitoring after treatment, which will be greater for the least rapid or most passive systems. The monitoring plan should include only the essential analyses needed to adequately understand system performance and should be designed to include fewer wells and less frequent sampling over time.

5.3.1.2 Speed

In most cases, biological treatment of perchlorate in groundwater will be used for plume control. In such cases, the restoration time frame will be controlled by the groundwater flow rates, sorption to subsurface solids (which is generally minimal for perchlorate) and any source depletion efforts undertaken. However, all of the approaches except biowalls could be used for source treatment as well, so the rankings reflect the possibility for this deployment and the time that could be required for plume-wide restoration using these technologies.

As might be expected, active treatment should result in the fastest treatment although semi-passive may not be much different in many cases. The passive treatment systems, because they rely on diffusion and natural advective flow to distribute the electron donors, will generally result in slower overall biodegradation rates and longer overall treatment times. Specific design features, such as multiple barriers, might also impact the time to achieve closure.

5.3.1.3 Reliability

Reliability refers to minimizing the need for optimization, maintenance or modifications due to performance difficulties. The major threat to reliable performance is biofouling, which can be most problematic for active treatment because of the continuous injection of readily-degradable materials.

Short-circuiting (i.e., preferential flow through the barrier with little contact between contaminants and the remediation reagents) usually poses the greatest threat to passive systems. Passive systems also can be susceptible to changes in the groundwater flow direction and/or the depth to groundwater, and it can be difficult and costly to modify the system to respond to such hydrogeological changes. Mounding of groundwater upgradient of the passive barrier can occur if the permeability of the barrier decreases over time to below the natural gradient flow due to biofouling and/or precipitation within the barrier. In such cases, groundwater may flow around or over the barrier, reducing its effectiveness.

5.3.1.4 Disruption of Site Activities

Disruption of site activities is often a critical concern for operating facilities. Installation of passive systems may cause some problems, particularly when trench lines cross existing subsurface infrastructures (e.g., utility lines), but the disruptions are temporary. Active treatment can disrupt activities both for the initial installation of injection wells and for activities required for the ongoing operations of the system. With additional cost it is possible to design an active system to function beneath operating facilities, for example by plumbing all of the injection points to a more remote location. In some cases horizontal wells can be used. Semi-passive systems are more flexible and can be designed to work around the facility's operations to minimize ongoing disruption.

5.3.1.5 Flexibility

The flexibility to modify and optimize operations can be an important feature in process selection, particularly when the site conditions are relatively uncertain. Passive systems generally offer one chance to "get it right". If the contamination extends beneath the trench, for example, or if water levels, flow rates, or flow directions change, there is little opportunity to modify the system. However, increasing the horizontal extent of the system (e.g., adding more linear feet of biowall) is not difficult.

Active treatment offers significant flexibility and the opportunity for continual tweaking and adjusting of injection rates and concentrations. This flexibility often is the greatest strength of active treatment, but if increasing the horizontal extent of the system is required or if greater well depth is required, the fix can be expensive.

5.3.2 Problematic Site Conditions

For *in situ* biological treatment of perchlorate in general, the most common site-specific problems include very high donor demands (due to rapid flow rates and/or high perchlorate, nitrate, sulfate, or dissolved oxygen concentrations), inhibition due to high salinity and infrastructure restrictions that limit access. Nitrate in particular can pose serious problems for cost-effective biological treatment of perchlorate because nitrate and sulfate reducers may compete for the electron donors, and nitrate may directly inhibit perchlorate reducing enzymes (Herman and Frankenberger, 1999; Tan et al., 2004).

Contamination that extends too deep can prevent the use of some passive systems, such as biowalls, because cost-efficient trenching equipment is generally limited to depths of roughly 10 to 15 m (35 to 50 ft) bgs. Another potential limitation is highly acidic or alkaline conditions that can inhibit the activity of perchlorate reducers. These problems usually can be more easily dealt with when using active systems although, as noted earlier, buffered slow-release donors are available. In addition, some plumes are simply too wide for cost-effective passive biological treatment, particularly when using biobarriers, because of the large number of injection points that will be required.

Low-permeability aquifers also may be very difficult to treat *in situ* using bioremediation technology. Use of biowalls may be a good choice where reactant distribution is difficult due to low permeability materials because biowalls can better intercept plumes and retain contaminants within the treatment zone.

5.4 SUMMARY

In situ bioremediation of perchlorate has progressed rapidly over the last decade, from basic microbiological research to full-scale applications. Practitioners have adapted anaerobic bioremediation technologies to stimulate native perchlorate reducing bacteria, and have developed several variations of the technology to address different site conditions and remedial objectives.

In situ bioremediation for perchlorate is still a rapidly evolving technology, and to date it has been used full-scale at only a few sites. However, it has proven successful and cost-effective in several controlled field demonstrations and has the potential to greatly reduce the costs of managing perchlorate contaminated groundwaters. In many cases *in situ* bioremediation is the only feasible option for treatment other than the limited and costly *ex situ* options that are available. Given the large number of perchlorate sites and the high costs for their remediation, it is expected that *in situ* bioremediation will continue to evolve and play a major role in managing the risks of perchlorate in the environment.

REFERENCES

AFCEE (Air Force Center for Environmental Excellence). 2007. Final Protocol for *In Situ* Bioremediation of Chlorinated Solvents Using Edible Oil. Prepared by Solutions IES, Inc.; Terra Systems, Inc.; and Parsons Corporation. AFCEE, Brooks City-Base, San Antonio, TX, USA. http://www.afcee.af.mil/shared/media/document/AFD-071203-094.pdf. Accessed September 21, 2008.

Ahmad F, Schnitker SP, Newell CJ. 2007. Remediation of RDX- and HMX-contaminated groundwater using organic mulch biowalls. J Contam Hydrol 90:1–20.

Aziz CE, Hampton MM, Schipper M, Haas P. 2001. Organic mulch biowall treatment of chlorinated solvent-impacted groundwater. In Leeson A, Alleman BC, Alvarez PJ, Magar VS, eds, Bioaugmentation, Biobarriers and Biogeochemistry. Battelle Press, Columbus, OH, USA, pp 73–112.

Borden RC. 2007. Concurrent bioremediation of perchlorate and 1,1,1-trichloroethane in an emulsified oil barrier. J Contam Hydrol 94:13–33.

Chaudhuri SK, O'Connor SM, Gustavson RL, Achenbach LA, Coates JD. 2002. Environmental factors that control microbial perchlorate reduction. Appl Environ Microbiol 68:4425–4430.

Chopra G, Dutta L, Nuttall E, Anderson W, Hatzinger P, Goltz MN. 2005. Investigation of methods to control biofouling during in situ bioremediation. Proceedings, Eighth International In Situ and On-Site Bioremediation Symposium, Baltimore, MD, USA, June 6–9, 2005, paper A-22.

Coates JD, Michaelidou U, Bruce RA, O'Connor SM, Crespi JN, Achenbach LA. 1999. The ubiquity and diversity of dissimilatory (per)chlorate-reducing bacteria. Appl Environ Microbiol 65:5234–5241.

Cowan D. 2000. Innovative abatement and remediation of perchlorate at McGregor, Texas weapons plant site. Soil Sediment Groundw 5:25–26.

Cox EE. 2005. A Review of Biofouling Controls for Enhanced Bioremediation of Groundwater. Prepared for the DoD Environmental Security Technology Certification Program (ESTCP), Arlington, VA, USA. http://www.estcp.org/Technology/upload/ER-0429-WhtPaper.pdf. Accessed February 4, 2008.

Cox EE, Edwards E, Neville S. 2000. In situ bioremediation of perchlorate in groundwater. In Urbansky ET, ed, Perchlorate in the Environment. Kluwer Academic/Plenum Publishers, New York, NY, USA, pp 231–240.

Craig M, Jacobs A, Britto R. 2006. Biological PRB application expanded to accelerate perchlorate degradation in groundwater. EPA 542-N-06-004. U.S. Environmental Protection Agency, Solid Waste and Emergency Response, Cincinnati, OH, USA.

Devlin JF, Barker JF. 1994. A semipassive nutrient injection scheme for enhanced in situ bioremediation. Groundw 32:374–380.

Devlin JF, Barker JF. 1996. Field investigation of nutrient pulse mixing in an in situ biostimulation experiment. Water Resour Res 32:2869–2877.

Flowers TC, Hunt JR. 2007. Viscous and gravitational contributions to mixing during vertical brine transport in water-saturated porous media. Water Resour Res 43, WO1407, doi:10.1029/2005WR004773.

Gierczak R, Devlin JF, Rudolph D. 2007. Field test of a nutrient injection wall for stimulating *in situ* denitrification near a municipal water supply well. J Contam Hydrol 89:48–70.

Hatzinger PB, Diebold J, Yates CA, Cramer RJ. 2006. Field demonstration of *in situ* perchlorate bioremediation in groundwater. In Gu B, Coates JC, eds, Perchlorate: Environmental Occurrence, Interactions, and Treatment. Springer, New York, NY, USA, pp 311–341.

Hatzinger PB, Whittier CM, Arkins MD, Bryan CW, Guarini WJ. 2002. In-situ and ex-situ bioremediation options for treating perchlorate in groundwater. Remediat 12:69–86.

Herman DC, Frankenberger WTJ. 1999. Bacterial reduction of perchlorate and nitrate in water. J Environ Qual 28:1018–1024.

ITRC (Interstate Technology & Regulatory Council). 2005. Permeable reactive barriers: Lessons learned/new directions. PRB-4. ITRC, Permeable Reactive Barriers Team. Washington, DC, USA.

Logan BE. 2001. Assessing the outlook for perchlorate remediation. Environ Sci Technol 35:482A–487A.

Mayer HJ, Greenberg, MR. 2005. Using integrated geospatial mapping and conceptual site models to guide risk-based environmental clean-up decisions. Risk Anal 25:429–446.

Parr JF, Goltz MN, Huang JQ, Hatzinger PB, Farhan YH. 2003. Modeling in situ bioremediation of perchlorate-contaminated groundwater. Proceedings, Seventh International In Situ and On Site Bioremediation Symposium. Orlando, FL, USA, June 2–5, 2003, paper C-07.

Parsons. 2004. Principles and Practices of Enhnaced Anaerobic Bioremediation of Chlorinated Solvents. Prepared for the AFCEE, Brooks City-Base, TX, USA; Naval Facilities Engineering Service Center, Port Hueneme, CA, USA; DoD ESTCP, Arlington, VA, USA. http://www.afcee.af.mil/shared/media/document/AFD-071130-020.pdf. Accessed September 21, 2008.

Suthersan SS, Lutes CC, Palmer PL, Lenzo F, Payne FC, Liles DS, Burdick J. 2002. Technical Protocol for Using Soluble Carbohydrates to Enhance Reductive Dechlorination of Chlorinated Aliphatic Hydrocarbons. Prepared for the AFCEE, Brooks City-Base, San Antonio TX, USA, and ESTCP, Arlington, VA, USA. http://www.estcp.org/Technology/upload/CU-9920-PR-01.pdf. Accessed February 4, 2008.

Tan K, Anderson TA, Jackson WA. 2004. Degradation kinetics of perchlorate in sediments and soils. Water Air Soil Pollut 151:245–249.

USACE (U.S. Army Corps of Engineers). 2003. Conceptual Site Models for Ordnance and Explosives (OE) and Hazardous, Toxic, and Radioactive Waste (HTRW) Projects. Pub. No. EM 1110-1-1200. USACE, Washington, DC, USA. http://www.usace.army.mil/publications/eng-manuals/em1110-1-1200/toc.htm. Accessed February 4, 2008.

USEPA (U.S. Environmental Protection Agency). 2005. Perchlorate Treatment Technology Update: Federal Facilities Forum Issue Paper. EPA 542-R-05-015. USEPA, Solid Waste and Emergency Response, Washington, DC, USA. http://www.epa.gov/tio/download/remed/542-r-05-015.pdf. Accessed February 4, 2008.

Xu J, Song Y, Min B, Steinberg L, Logan BE. 2003. Microbial degradation of perchlorate: Principles and applications. Environ Eng Sci 20:405–422.

CHAPTER 6

ACTIVE BIOREMEDIATION

Paul B. Hatzinger,[1] Charles E. Schaefer[1] and Evan E. Cox[2]

[1]Shaw Environmental, Inc., Lawrenceville, NJ 08648; [2]Geosyntec Consultants, Inc., Guelph, ON, Canada

6.1   BACKGROUND AND GENERAL APPROACH

The primary engineering approaches that have been tested for *in situ* perchlorate treatment are as follows: (1) "active systems" that meter and mix soluble electron donors into groundwater during continuous active pumping; (2) "semi-passive systems" that mix soluble electron donors into groundwater during intermittent pumping; and (3) "passive systems" that apply slow-release electron donors in trenches, wells, or using direct-push methods and rely upon natural groundwater flow to mix electron donor with contaminated water. This chapter focuses on the application of active treatment systems for perchlorate, whereas Chapters 7, 8 and 9 address the alternate approaches.

Active treatment systems can be implemented using a variety of different designs based on site conditions, but generally function either through subsurface groundwater recirculation or extraction and reinjection of groundwater. A soluble electron donor is metered and mixed into groundwater during pumping, usually at a concentration that is determined based on the levels of perchlorate and other co-contaminants requiring treatment. A wide variety of different electron donors are suitable for this approach (see Section 6.5.3 for more detail). To date, ethanol, lactate, acetic acid, citric acid and benzoate have been utilized in the field as electron donors for active pumping systems. The donor selection is based on four factors: (1) site-specific effectiveness for biotreatment of perchlorate and co-contaminants; (2) cost; (3) regulatory considerations; and (4) tendency to cause biofouling of injection wells. These factors are discussed further in Section 6.5.3.

The primary advantages of an active treatment system design are:

- It is applicable and potentially cost-effective in deep as well as shallow aquifers.

- It can be utilized as a groundwater capture and treatment system to prevent plume migration.

- The effective treatment zone does not rely upon natural groundwater flow and is applicable over a range of hydraulic conductivities.

- Active mixing and metering of electron donor based on stoichiometry of electron acceptors (perchlorate and co-contaminants) reduces cost and provides means for control of subsurface oxidation-reduction potential (ORP).

- It minimizes the potential for secondary impacts on groundwater geochemistry, including the production of methane and sulfide, pH changes and the reduction and mobilization of manganese, iron and arsenic.

- The systems are flexible, so modifications in pumping rates, electron donor type and electron donor quantity are possible at any time in order to respond to changing flow and transport conditions.

As with any treatment approach, there are also disadvantages to active *in situ* treatment. The primary disadvantages are as follows:

- Biofouling of injection wells is a frequent problem which must be actively controlled.
- System infrastructure and operation and maintenance (O&M) requirements are often greater than for comparable passive or semi-passive systems.
- Re-injection of groundwater containing contaminants is prohibited or only allowed under permit in some states.

6.2 WHEN TO CONSIDER AN ACTIVE TREATMENT SYSTEM

The choice of an active versus a passive system for *in situ* perchlorate treatment will depend upon many different factors including site conditions, remedial objectives, regulatory considerations, and cost. The electron donors used in passive systems are generally complex, slowly biodegradable substrates such as vegetable oils, chitin and/or polylactate ester. The advantage of these substrates is that they are long-lived, but the disadvantage is that they do not distribute well with groundwater and must be injected via closely-spaced wells or injection points. For this reason, passive approaches tend to be most practical and economical where perchlorate contamination is shallow (< ~15 meters [m] or ~50 feet [ft]) and can be reached by direct-push methods, and where plume width is limited (e.g., less than a few hundred meters).

Passive treatment also depends on natural groundwater flow to bring the contaminant in contact with the injected electron donor. Thus, hydraulic gradients must be taken into consideration. In contrast, active systems rely on pumping wells to capture perchlorate-contaminated groundwater, and they meter and mix soluble electron donor with this water. These systems can be implemented in both deep and shallow aquifers regardless of plume width, with site hydrogeological conditions determining the number of wells necessary for plume capture and treatment. While active systems often require more infrastructure than passive approaches and are subject to higher O&M costs, an active system can be a more economical and effective treatment option than a passive approach, especially for deep contamination and/or wide plumes.

A second general difference between active and passive systems is the potential impact of these approaches on groundwater geochemistry. Because passive systems are designed to provide a long-term remedial solution with infrequent (e.g., yearly) addition of electron donor, the amount of total organic carbon (TOC) added to an aquifer (as vegetable oil, polylactate ester, molasses, etc.) is orders of magnitude greater than that minimally required for biological reduction of perchlorate. As a consequence, passive systems generally result in the production of sulfide (from sulfate reduction), methane (from methanogenesis) and the mobilization of redox-sensitive metals such as iron, manganese and sometimes arsenic. Because active treatment systems are designed to optimize mixing, they will generally create less secondary groundwater impacts than passive systems.

This is not to say that active systems will not produce sulfide or mobilize various metals—secondary groundwater impacts are common, but the extent of such impacts is dependent on the concentration of electron donor added and how well this donor is mixed with groundwater. Because electron donors can be metered into the groundwater as needed to ensure perchlorate biodegradation, the extent of geochemical change tends to be less significant with active systems than with passive approaches. All of the compounds produced or mobilized in the bioactive zones of either a passive or an active system will generally re-oxidize at some distance downgradient of the zone of substrate injection, and geochemistry will return to the conditions similar to those in the upgradient groundwater. However, the scope of geochemical change and the distance that the impact zones extend downgradient will generally differ between active and passive systems.

The potential for biofouling to occur within injection wells is another issue to consider when deciding between an active or passive *in situ* bioremediation approach. Biofouling is one of the more significant O&M issues with active treatment systems. Issues related to biofouling and approaches to control the detrimental impacts of this process are discussed in Section 6.5.5.2.

6.3 TREATMENT SYSTEM CONFIGURATIONS

The active approaches tested to date for perchlorate remediation have been primarily groundwater extraction and reinjection (ER) designs. Among these systems, both plume cutoff (active biobarriers) and source area treatment (recirculation loop) technologies have been tested, and one full-scale active biobarrier is presently in operation in Nevada at a flow rate of ~950 liters per minute (L/min) (~250 gallons per minute [gpm]) (see Section 6.7). One subsurface recirculation design utilizing paired horizontal flow treatment wells (HFTWs) has also undergone field testing. The basic design of each of these systems is provided below, and a detailed case study of an active ER biobarrier system is provided in Section 6.6.

6.3.1 Groundwater Extraction and Reinjection (ER)

Groundwater ER systems can be designed in a variety of ways based on site hydrology and remedial objectives. As the name implies, a basic groundwater ER design relies upon one or more extraction wells to pump contaminated groundwater from a subsurface aquifer, an engineered system to add and mix a soluble electron donor (and possibly inorganic nutrients or buffer) to the extracted water, and one or more injection wells to return the amended water to the ground, subsequently creating a bioactive zone. Within the bioactive zone, the added electron donor is oxidized and perchlorate is reduced to chloride and water, generally by indigenous bacteria. Other common electron acceptors, particularly nitrate and oxygen, as well as co-contaminants subject to reductive biodegradation processes, such as chlorinated solvents and nitramine explosives, can also be treated in this zone. A schematic of a basic ER system is provided in Figure 6.1.

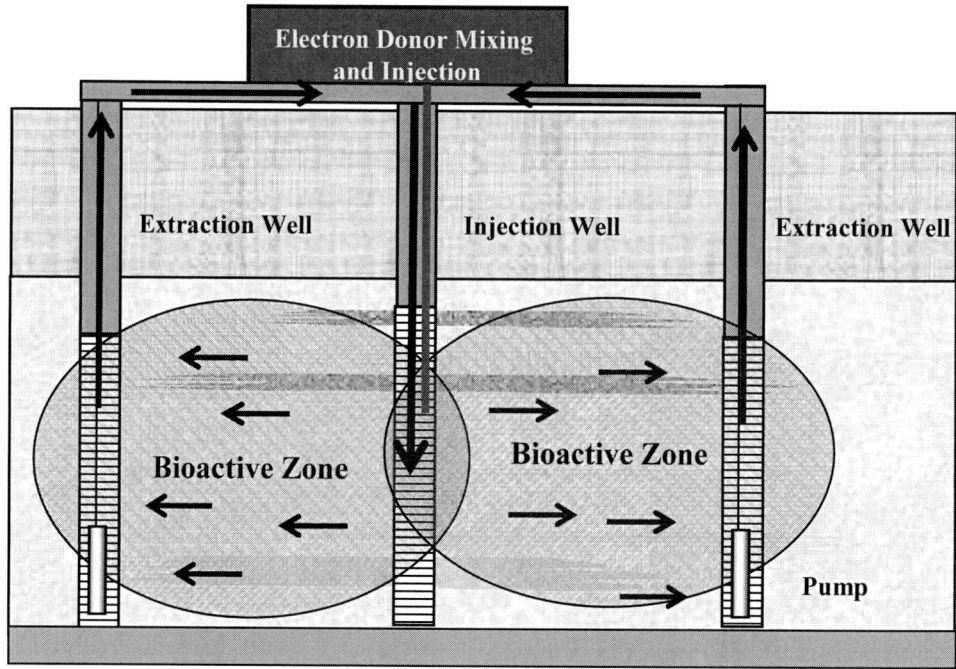

Figure 6.1. Schematic of an active ER system consisting of two extraction wells and a single injection well. Water is pumped to the surface from each extraction well, amended with electron donor, and then re-injected into the formation through the injection well.

6.3.2 Horizontal Flow Treatment Wells (HFTWs)

Unlike an ER system in which water is pumped to the surface, amended and then reinjected, HFTWs recirculate groundwater in the subsurface. This type of system has been tested at the pilot scale for aerobic treatment of trichloroethene (TCE) (McCarty et al., 1998; Gandhi et al., 2002a; Gandhi et al., 2002b) and more recently for perchlorate (Parr et al., 2003; Knarr, 2003; Secody, 2007). In each of the previous pilot tests, a single pair of HFTWs was installed. Each of the treatment wells has two screened intervals in the subsurface, one serving as a zone for groundwater extraction (extraction screen) and one as a zone for groundwater injection (injection screen) (Figure 6.2). One HFTW (upflow well) pumps water upward, drawing groundwater from the formation through the lower screened interval in the well, and injecting it back into the formation through the upper screen. The second HFTW (downflow well) operates in reverse, pulling water from the upper screened interval and injecting it back into the formation through the lower screen. The screened intervals in each well are separated by a blank section of casing fitted with an inflatable packer.

The net effect of this pumping system design is the circulation of water between the two HFTWs. Due to hydraulic conductivity anisotropy, such as is typically seen in aquifers (Fetter, 1999), groundwater flow between the injection and extraction screens of a well pair is primarily horizontal. This is in contrast to conventional groundwater circulation wells that depend on vertical flow between the injection and extraction screens of a single well.

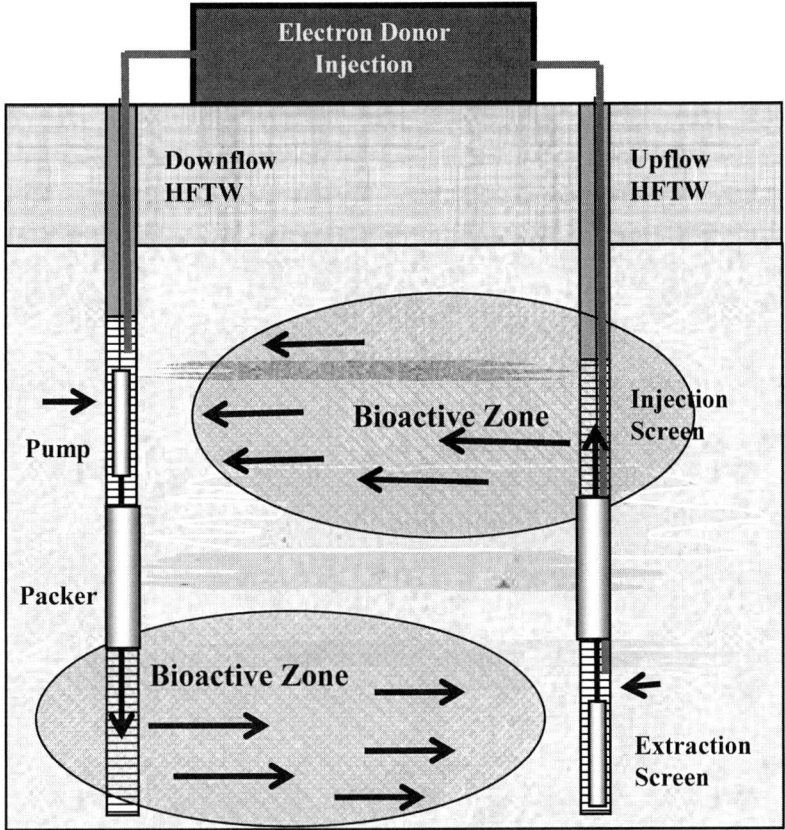

Figure 6.2. Schematic of an active HFTW system consisting of two treatment wells. In the upflow HFTW, water is amended with electron donor and pumped from the lower screened interval (extraction screen) through the packer to the upper screened interval (injection screen), where it is released into the formation. The downflow HFTW operates in reverse.

Soluble electron donor is supplied to each HFTW through piping that terminates near the pump (Figure 6.2). The pump then pushes the water and electron donor through the packer and releases it in the zone of the injection screen for each well. Due to the circulation between treatment wells, the contaminated water is treated multiple times, so that perchlorate removal efficiencies (comparing concentrations upgradient and downgradient of the treatment wells) can be greatly increased over the removal achieved by a single pass of perchlorate-contaminated water.

6.4 SYSTEM APPLICATIONS

6.4.1 Biobarriers

Because perchlorate is a highly mobile contaminant, some of the more significant releases at Department of Defense (DoD) facilities, manufacturing plants and aerospace companies have led to extensive groundwater plumes. For example, a plume emanating from a former flare manufacturing facility in Morgan Hill, California, is reported to be approximately 16 kilometers (km) (10 miles [mi]) in length (Woods, 2006). A cutoff barrier design is often desirable to prevent further migration of large plumes either onsite or, more significantly, to offsite locations.

Hydraulic barriers coupled to *ex situ* treatment systems (either fluidized bed reactors or ion exchange systems) have been utilized at several locations to prevent plume migration. With this coupled design, perchlorate-contaminated groundwater is captured by a series of extraction wells, pumped through a central treatment plant for removal of perchlorate and other contaminants, then discharged to the surface or used for aquifer recharge (Hatzinger, 2005). For example, hydraulic control systems for perchlorate are presently in operation at the Aerojet Site in Rancho Cordova, California (Aerojet), the aforementioned site in Morgan Hill, California, and at the former Kerr-McGee Site in Henderson, Nevada.

The same hydraulic barrier approach described above potentially can be coupled directly to an *in situ* treatment regimen to create an active biobarrier. In this case, perchlorate-contaminated groundwater is captured by a series of extraction wells, amended with a soluble electron donor and then recharged into the aquifer through one or more injection wells. This approach has now been applied at full-scale to treat a groundwater plume at the former PEPCON perchlorate manufacturing site near Henderson, Nevada (see Section 6.7 and ITRC, 2008). The *in situ* system consists of 9 extraction wells (operating at ~950 L/min or ~250 gpm total flow) and 6 injection wells where water amended with sodium benzoate is recharged to the aquifer. Alternatively, rows of paired HFTWs could be used to mix electron donor into the groundwater at the front of a plume, subsequently preventing further migration. The advantage of this *in situ* approach is the potential reduction in infrastructure and equipment, and subsequently in cost, compared to conventional *ex situ* treatment.

For ER systems, various well patterns can be used to create a biobarrier. Some of the possible designs are presented in Figure 6.3. These designs include the following: (1) upgradient extraction and downgradient injection wells (Figure 6.3a); (2) upgradient injection and downgradient extraction wells (Figure 6.3b); and (3) alternating injection and extraction wells in a line perpendicular to groundwater flow (Figure 6.3c).

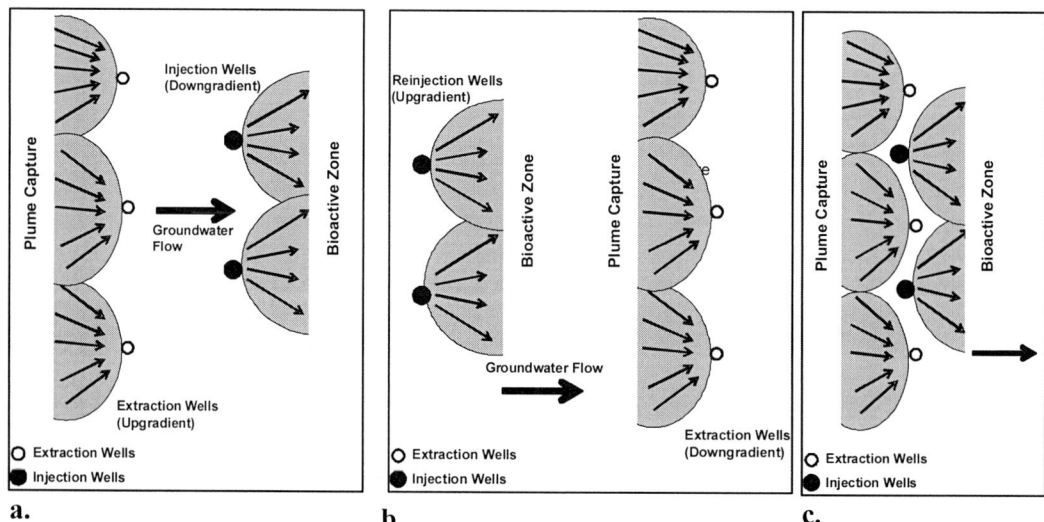

Figure 6.3. Schematic showing different configurations of active biobarrier systems: (a) upgradient extraction and downgradient injection wells; (b) upgradient injection and downgradient extraction wells; (c) alternating injection and extraction wells in a line perpendicular to groundwater flow.

The well pattern in Figure 6.3c could also be implemented using a series of paired upflow and downflow HFTWs. Variations of each of these designs are possible based on site hydrogeologic conditions, contaminant distribution, contaminant and electron donor degradation kinetics, accessibility and remedial goals. The design shown in Figure 6.3a represents a single pass approach. After amendment with electron donor, the groundwater is reinjected downgradient and will not be recaptured by the extraction wells. The advantage of this design is that the potential for biofouling of the extraction wells is minimized. With each of the latter two designs (Figures 6.3b and c), recirculation of groundwater between the extraction and injection wells will occur to some extent (based on pumping rates and well spacing). The resulting recirculation can be used to provide additional residence time of electron donor in the groundwater and to further enhance mixing. However, the potential for fouling of the extraction wells (biofouling or metals precipitation) is also increased in recirculation designs (see Section 6.5.5.2).

Field pilot investigations have been conducted to evaluate the performance of various biobarrier designs including both ER and HFTW approaches. One of the first field demonstrations of an active treatment approach was conducted at Aerojet (Cox et al., 2001). During this pilot study, groundwater was extracted from the subsurface via two downgradient wells screened at depths of approximately 28 to 33 m (92 to 108 ft) below ground surface (bgs), amended with ethanol as an electron donor and chlorine dioxide for biofouling control, and then reinjected into the same aquifer via a single injection well. The wells were in line and perpendicular to groundwater flow as illustrated in Figure 6.3c. The data from this demonstration showed that perchlorate concentrations of nearly 8 milligrams per liter (mg/L) could be reduced to below 4 micrograms per liter (µg/L) within ~5 m (~16 ft) of the injection well, with a perchlorate degradation half-life of less than one day. The details of this field trial, including the site characteristics, site investigation work, system design, analytical results and conclusions, are provided as a case study in Section 6.6.

A pilot demonstration of a HFTW system was also conducted at Aerojet, approximately 0.3 km (0.2 mi) upgradient of the aforementioned ER demonstration. This project utilized a single pair of HFTWs installed approximately 10 m (33 ft) apart and perpendicular to groundwater flow (Figure 6.4). The screened interval of each HFTW was 14 to 19 m (46 to 61 ft) bgs for the upper screen and 24 to 30 m (80 to 100 ft) bgs for the lower screen.

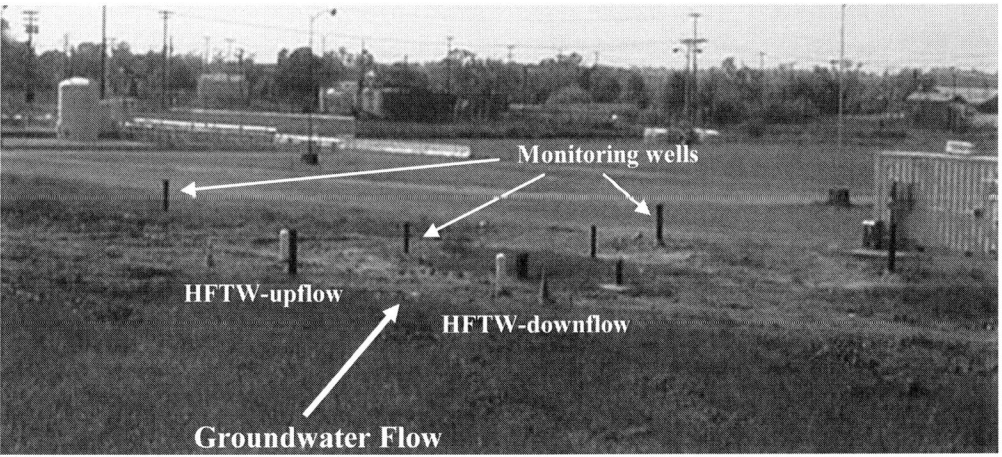

Figure 6.4. Photograph of the HFTW system layout at Aerojet during construction.

The system design was based on extensive hydrogeological characterization and modeling. Citric acid was added in pulses as the electron donor in this study, and chlorine dioxide was used to prevent biofouling within the HFTWs (see Section 6.5.5.2). After 5 months of citric acid addition at levels ranging from 2 to 4 times the calculated stoichiometric requirement (based on levels of dissolved oxygen [DO], perchlorate and nitrate), perchlorate levels in seven shallow monitoring wells (14–19 m or 46–61 ft bgs) declined by 95% to a mean concentration of 120 µg/L from the starting average of 2230 µg/L preceding citric acid addition. The perchlorate levels in 5 deep downgradient wells (24–30 m or 80–100 ft bgs) averaged 220 µg/L after 5 months, compared to 3340 µg/L prior to electron donor addition (a decline of 93%).

In addition, mobilization of the secondary groundwater contaminants iron and manganese was minimal during this study (generally less than 300 µg/L for each). The pilot data showed that a HFTW system can provide good mixing of electron donor into groundwater and that significant reductions in perchlorate can be achieved through this approach. It should be noted, however, that perchlorate levels of <4 µg/L (the practical quantitation limit [PQL] for U.S. Environmental Protection Agency Method 314.0) were not consistently achieved in monitoring wells during the initial system test. Downgradient mixing of treated groundwater with surrounding groundwater still containing perchlorate most likely accounted for the low levels of residual perchlorate.

6.4.2 Source Area Treatment

In addition to a biobarrier approach, active systems are also applicable for source area treatment. One common source of perchlorate at military and aerospace facilities is open burn/open detonation (OB/OD) areas. Before strict environmental regulations were implemented, propellants that did not meet correct particle size standards or that were not mixed to specification were often burned at these locations. Residual material from OB/OD areas contributes to plumes at several sites including the Massachusetts Military Reservation (Cape Cod, Massachusetts), Aerojet, Whittaker-Bermite (Santa Clarita, California) and ATK Thiokol (Elkton, Maryland). Another significant source of environmental perchlorate contamination is past discharges of wastewater from hog-out operations. During the hog-out process, propellant is removed from missiles and casings, often using a high pressure water stream. Currently, this water is collected and treated, but in past decades wastewater from this process was discharged to ground surface at some locations. Other areas, including sumps, landfills and wastewater lagoons, also serve as significant sources of perchlorate at manufacturing, military and aerospace facilities.

Active systems can be designed to treat dissolved perchlorate in groundwater underlying or just downgradient of source areas. A groundwater treatment system may need to be combined with soil and/or vadose zone treatment in some instances to be effective, with the main goal being removal of as much of the source perchlorate as possible. One design that is applicable for this purpose is a small recirculation cell installed cross-gradient to groundwater flow. This design can be coupled with an infiltration gallery or other flushing system to remove perchlorate from soils and/or unsaturated aquifer solids to the groundwater, where biological removal occurs. Pilot tests of vadose zone treatment designs, including a coupled vadose zone-groundwater treatment system, are presently being validated through the DoD Environmental Security Technology Certification Program (ESTCP). Information on these projects is available at http://www.estcp.org/technology/er-perchlorate.cfm (Projects ER-0435 and ER-0511).

Active Bioremediation 99

A cross-gradient recirculation system was tested successfully on a concentrated plume originating from a hog-out facility at the Indian Head Division, Naval Surface Warfare Center (IHDIV) in Indian Head, Maryland (Hatzinger, 2005; Hatzinger et al., 2006). An initial site investigation revealed a small plume of concentrated perchlorate behind the IHDIV hog-out facility (Building 1419), with dissolved perchlorate levels ranging from 8 to 430 mg/L in perched groundwater. In addition, the pH of site groundwater was generally below 5.0, which was found to be too low for *in situ* perchlorate biodegradation in laboratory studies. A field-pilot demonstration employing a recirculation cell design was undertaken based on site geochemical and hydrogeologic data. Two plots were installed (Test Plot and Control Plot). Each plot consisted of two extraction wells, two injection/recharge wells and 9 monitoring wells (Figure 6.5). The extraction and injection wells in each plot were spaced 3.7 m (12 ft) apart, and the two plots were installed 6 m (20 ft) apart, with the Control Plot located southwest of the Test Plot.

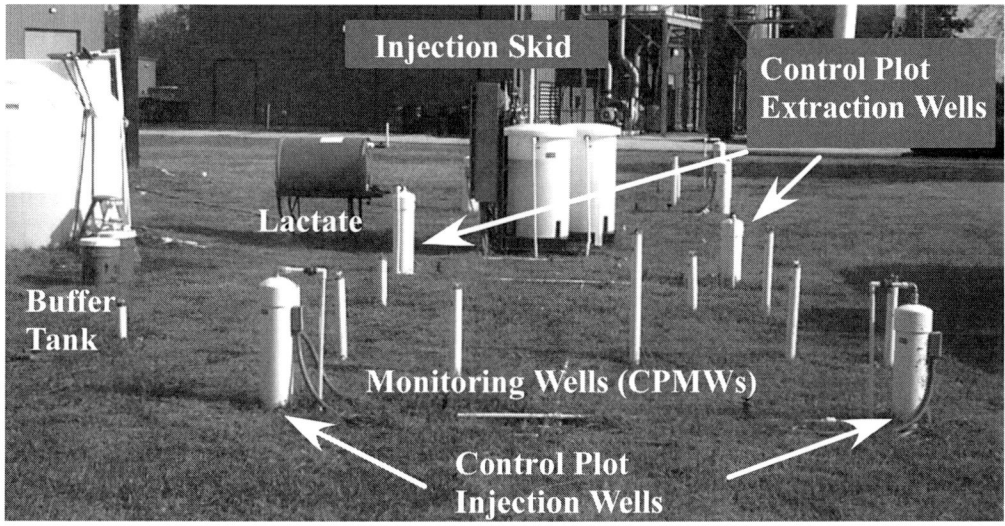

Figure 6.5. Photograph of the field pilot system installed to treat perchlorate in a source area at the Indian Head Division, Naval Surface Warfare Center (IHDIV), Indian Head, MD. Reprinted from Hatzinger et al., 2006.

In the Test Plot, groundwater was pumped to the surface, amended with electron donor (lactate) and buffer (carbonate/bicarbonate mixture), and then reinjected into the aquifer. In the Control Plot, groundwater was extracted and reinjected without substrate or buffer amendment. During the 5-month study, ~80,000 L (~21,000 gallons [gal]) of groundwater was re-circulated through each plot. Groundwater pH was elevated to at least 5.7 standard units (SU) in all Test Plot wells during the demonstration, and lactate was measured throughout the Test Plot within 3 weeks of system operation. Perchlorate levels were reduced by greater than 95% in 8 of 9 monitoring wells within the Test Plot during the demonstration, with 5 wells reaching below 1 mg/L, and 2 below the 5 µg/L PQL for this project (Figure 6.6a). Conversely, there was no significant increase in pH or reduction in perchlorate levels within the Control Plot (Figure 6.6b). The data from this demonstration indicate that active *in situ* biostimulation is a viable remediation option for treating perchlorate source areas.

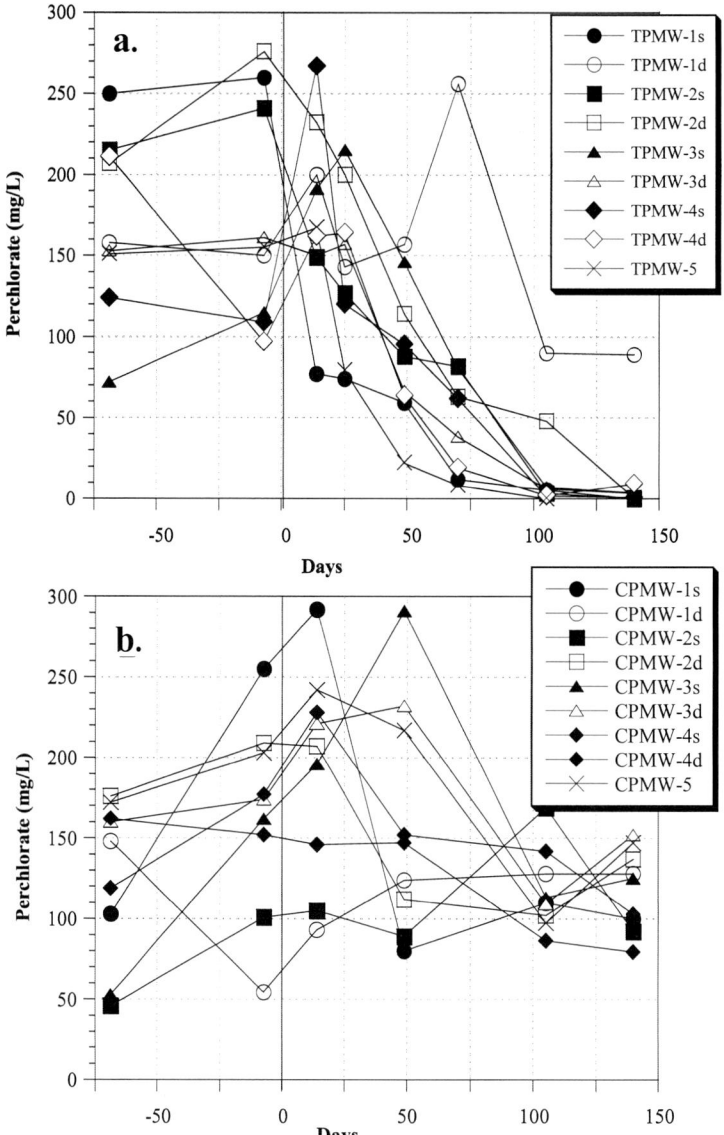

Figure 6.6. Perchlorate levels in groundwater during the IHDIV pilot test: (a) test plot monitoring wells receiving electron donor and buffer; (b) control plot monitoring wells receiving no amendments. Reprinted from Hatzinger et al., 2006.

6.5 SYSTEM DESIGN, OPERATION AND MONITORING

This section summarizes the basic requirements for designing and operating an active *in situ* bioremediation system for perchlorate treatment.

6.5.1 Site Assessment Needs

A basic site assessment, including a determination of the extent of perchlorate contamination (depth, concentrations, plume length and width), presence of co-contaminants and characterization of groundwater geochemistry (pH, ORP, DO, anions, cations, total dissolved solids and other parameters as needed on a site specific basis), is necessary prior to designing any treatment system for perchlorate. Understanding site geology and hydrogeology is also crucial for properly designing and assessing an *in situ* bioremediation system, particularly for active systems. Groundwater velocity and flow direction must be accurately determined in order to properly orient the system (particularly if a cutoff barrier approach is employed) and to estimate the volume of groundwater that needs to be re-circulated in order to effectively capture and treat the dissolved plume. Groundwater velocity will also determine the residence time of the groundwater through the treatment zone. In addition, vertical hydraulic gradients and flow can impact mixing efficiency and overall treatment effectiveness. All of these hydrologic factors will impact system design parameters such as well spacing, groundwater flow rates and the number of injection and extraction wells required.

Of the parameters requiring assessment prior to treatment system design, hydraulic conductivity is among the most important. Both horizontal and vertical hydraulic conductivity information is required to determine aquifer response to injection and extraction groundwater flows. These responses are typically simulated in the design phase using hydrogeologic models, as discussed in Section 6.5.2. Hydraulic conductivities can be determined by performing slug tests and/or pump tests (Weight and Sondregger, 2001). Pump tests are typically more useful than slug tests, in that the tests measure hydraulic conductivity over a much larger subsurface volume, and data can be collected at multiple observation points and at multiple depth intervals (which facilitate calculation of vertical hydraulic conductivities). Pump testing also provides an additional means to verify that design flow rates can be attained for the treatment system.

A thorough understanding of site geology is important for designing an active treatment system. In particular, identification of regions or layers of either atypically low or high hydraulic conductivity is required for the development of an appropriate site conceptual model and for proper system design (Section 6.5.2). The presence of such layers or regions can have a substantial impact on the subsurface flow field. Several methods are available for investigating subsurface lithology and stratigraphy including (but not limited to) collection of soil cores via direct push methods (for shallow sites), split spoons or rotosonic drilling, cone penetrometer testing, ground penetrating radar and use of electrical resistance/gamma logs (Weight and Sondregger, 2001).

Low hydraulic conductivity layers may act as confining layers (as determined by pump testing), potentially limiting amendment distribution and mixing *in situ*. Delivery of amendments within these low hydraulic conductivity regions also poses a significant challenge. If improperly designed, injected amendments will simply by-pass these regions and distribute within high-flow (path of least resistance) zones. System design, in particular well screen locations, must take such geology into consideration.

6.5.2 Modeling

6.5.2.1 Modeling Overview

System design typically requires developing a site numerical model to simulate the coupled groundwater flow, solute transport and biodegradation processes that are involved in the application of an active bioremediation system. Specifically, these models are used to ensure that the treatment system will:

- **Completely intercept or capture the contaminant plume in the targeted zone.** Hydraulic capture of the contaminant plume is evaluated by assessing the hydraulic radius of influence of the simulated extraction wells and/or by evaluating particle flow paths.

- **Provide sufficient mixing of injected amendments with groundwater.** Simulated amendment concentrations in the treatment zone are evaluated as a function of depth and distance from the injection well(s) to determine the injection/extraction well flow rates, spacing and screen intervals needed to ensure proper mixing.

- **Biologically degrade perchlorate within the treatment zone, thereby preventing downgradient contaminant migration.** Simulated contaminant biodegradation kinetics typically are based on laboratory microcosms or column studies (using site groundwater and solids), or literature-based values. The model is then used to verify that the system design provides sufficient residence time such that perchlorate and co-contaminant concentrations decrease to target levels within the effective influence of the treatment system.

- **Limit excess delivery of electron donor.** Using electron donor biological decay rates measured in laboratory microcosm or column experiments, the fate and transport of injected electron donor can be simulated in the model. Thus, electron donor delivery can be optimized to limit downgradient migration (and subsequent secondary impacts such as metals mobilization) while still providing a sufficiently large biological treatment zone.

- **Provide a monitoring well network sufficient to evaluate system performance.** Models are used to determine locations and screen intervals for monitoring wells so that system performance can be readily assessed. The models can also be used to develop an initial estimate of sampling schedule and frequency.

Ultimately, the model will be used to determine well spacing, number of injection and extraction wells or HFTWs, groundwater flow rates, screened intervals, amendment delivery rates, the optimum locations for monitoring wells and the time required to reduce contaminants to target levels. Commercial models that specifically couple perchlorate fate with groundwater transport are not presently available. One approach that has been used is to substitute perchlorate for an alternate electron acceptor (e.g., sulfate, nitrate) in commercially available fate and transport models that simulate the sequential decay of multiple electron acceptors (e.g., SEAM3D; see Waddill and Widdowson, 1998). In addition, a perchlorate-specific biodegradation model recently has been developed and coupled to a groundwater flow model for HFTW systems (Parr et al., 2003; Secody, 2007).

Model development is associated closely with the site assessment, as many of the site hydrogeologic parameters (e.g., stratigraphy, hydraulic conductivities and hydraulic gradients) need to be incorporated in the model. MODFLOW (McDonald and Harbaugh, 1988) is a three-dimensional numerical grid-based hydrogeologic model that has been widely used to simulate groundwater flow during groundwater extraction and reinjection activities. MODFLOW was also the basis for simulating groundwater flow in a model developed to predict the impact of a HFTW system on groundwater movement (Parr, 2002; Parr et al., 2003; Secody, 2007) and for the Aerojet ER pilot test described in Section 6.6. There are several commercially available solute fate and transport models that interface with the MODFLOW platform. Examples include MT3DMS (Zheng and Wang, 1999), RT3D (Clement, 1997) and SEAM3D. Selection of the most appropriate fate and transport model should be based, at least in part, on the model's ability to simulate the necessary biological kinetics with respect to degradation of perchlorate and alternate electron acceptors. As previously noted, none of these models specifically incorporate perchlorate as a target contaminant.

A mathematical model was recently described to simulate perchlorate biodegradation in groundwater containing oxygen and nitrate as alternate electron acceptors (Parr, 2002). This biodegradation model (which includes cell growth and decay, electron donor and acceptor utilization parameters and an inhibition term to account for the effect of competing electron acceptors like nitrate on perchlorate biodegradation) was coupled to a MODFLOW-based fate and transport model to assist in the design and operation of the HFTW pilot system at Aerojet (see Section 6.4.1) (Parr et al., 2003; Secody, 2007). More detail on this coupled model is provided in Section 6.5.2.2.

Typically, use of a model for system conceptual design involves a trial-and-error process, as several different well configurations, amendment injection schemes and amendment dosages may be considered. As such, the model serves as a very useful tool for developing an optimum system design and increases the likelihood of project success. Model development is generally performed in two stages. The initial phase, as described above, is useful for system design. Injection and extraction well installations, pump sizing and monitoring well installations are typically based on this initial modeling phase. Initial system testing, which might include evaluation of hydraulic gradients during pumping and tracer testing (discussed in Section 6.5.4), can then be used to refine the conceptual model. If needed, minor adjustments (e.g., flow rates, amendment delivery rates and monitoring schedule) to the system design can be made to further optimize performance.

6.5.2.2 Example of Model Application: HFTWs

Numerous design and engineering decisions must be made when implementing a HFTW system. Key parameters include treatment and monitoring well spacing, groundwater pumping rates, and electron donor concentration. Due to the complexity of groundwater flow with this type of recirculation design and the dependence of system performance on complex biological, hydrogeological and geochemical variables, fate and transport modeling is an important component of system design. A site-specific fate and transport model was developed to simulate the operation of the HFTW system installed at Aerojet for perchlorate treatment. Details of this model are described in Parr (2002), Parr et al. (2003), and Secody (2007). Many of the design parameters for the field demonstration, including well spacing, pumping rates and electron donor delivery schedule, were selected based on model simulations.

A conceptual geologic layering pattern was utilized to develop the model. This layering was based on the lithography observed in rotosonic cores obtained from the aquifer. Flow modeling (using MODFLOW) and optimization techniques were used to estimate layer hydraulic conductivities in the model. The conductivities selected for the layers were those that provided the best fit between model-simulated draw-downs and measured draw-down data recorded in various monitoring wells during a stepped pump test conducted at the site. Layer depths for the site model used during the field test are shown in Figure 6.7a along with the screen intervals of the HFTWs. Each layer represents a specific conductivity value. Using calibrated conductivities, the model was successfully validated by comparing model-simulated and measured draw-downs at a monitoring well (Well 3633) that was not used for calibration. Figure 6.7b shows the goodness-of-fit of the model simulation to the draw-down data at Well 3633.

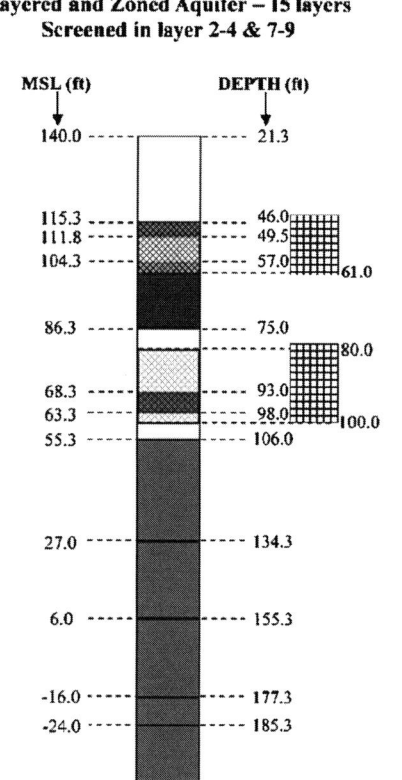
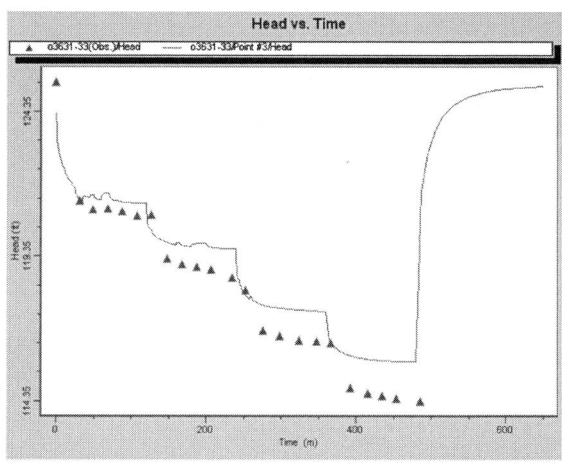

Figure 6.7. (a) Schematic showing the conceptual geologic layering pattern used in the fate and transport model developed for the Aerojet HFTW system. Screened intervals are shown on the right. (b) Comparison of model-simulated and measured water elevations in Well 3633 during pump testing at Aerojet.

After completing the model calibration and validation, the multi-layer flow model was used to simulate the flow regime created by the pair of HFTWs. The model simulations were performed assuming screened intervals for the HFTWs of 14 to 19 m bgs (46 to 61 ft bgs) for the shallow screen and 24 to 30 m bgs (80 to 100 ft bgs) for the deep screen (Figure 6.7a). Several model simulations were run to assess the impact of varying the spacing between the HFTWs and the pumping rate on the interflow ratios (i.e., the vertical water flow short-circuiting between the screens of the same HFTW compared to the water flowing horizontally between the HFTWs). The modeling simulations showed that increasing the spacing between the HFTWs decreased horizontal flow (i.e., interflow between wells) but had very little impact on vertical short-circuiting. Moreover, interflow versus pumping rate simulations indicated that increasing the pumping rate of each well above 38 L/min (10 gpm) had only a marginal impact on the ratio between vertical and horizontal flows.

Draw-down observations in model cells near the downflow HFTW indicated the potential for dewatering within layer 3 from 15.0 to 17.4 m (49.5 to 57 ft). Based on the observed draw-down and flow ratios associated with different spacing and pumping scenarios, a spacing between wells of 10 m (33 ft) and an initial pumping rate of 26.5 L/min (7 gpm) in each well were selected. Modeling was also used to simulate two-dimensional streamlines in each

Active Bioremediation

horizontal layer, as well as to simulate advective-dispersive transport of a conservative tracer for the proposed design. Based on the streamline and tracer simulations, plume capture and treatment widths of up to 70 m (230 ft) within the upper treatment zone and 44 m (144 ft) within the lower treatment zone were expected. A plan view of simulated streamlines in layer 3 of the site model (within the upper treatment zone, from 15.0 to 17.4 m [49.5 to 57 ft] bgs) shows the modeled capture zone when the HFTWs operate at 26.5 L/min (7 gpm) (Figure 6.8).

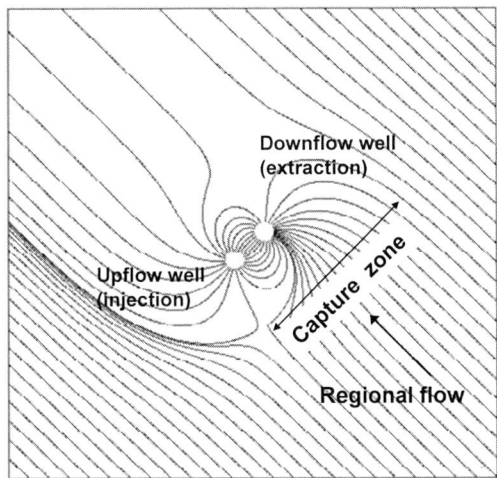

Figure 6.8. Plan view of streamlines showing the simulated capture of groundwater in layer 3 of the conceptual site model at a pumping rate of 26.5 L/min and 10 m HFTW spacing.

6.5.3 Electron Donor

Laboratory studies performed during the past several years reveal that perchlorate-reducing bacteria are naturally occurring in most environments including soils, sludges, surface waters and groundwater aquifers (Waller et al., 2004; Coates and Achenbach, 2004; Coates et al., 1999; Wu et al., 2001). Therefore, unlike remediation of chlorinated solvents where augmentation of an aquifer with exogenous organisms (primarily *Dehalococcoides* spp.) is often beneficial, bioaugmentation with perchlorate reducing bacteria is rarely necessary. Moreover, these indigenous organisms can often be stimulated to degrade perchlorate using a variety of different electron donors including various fatty acids, sugars and alcohols (Wu et al., 2001; Hatzinger, 2005; Waller et al., 2004; Xu et al., 2003). It should be noted that the range of effective electron donors tends to vary somewhat by site, probably reflecting the make-up of the indigenous microflora. Thus, simple microcosm studies are recommended to test different possible electron donors prior to field application (see next section). However, as noted, a number of different electron donors will generally be effective for promoting perchlorate reduction at nearly every site.

6.5.3.1 Microcosm Testing

Laboratory microcosm tests can be used to select an effective electron donor for *in situ* perchlorate treatment. The microcosm data can also be utilized to provide kinetic parameters for perchlorate biodegradation in fate and transport modeling, to ensure that perchlorate reduction to desired levels is achievable and to evaluate whether addition of inorganic nutrients or other amendments (e.g., buffer) is necessary. Laboratory microcosm tests should be planned in conjunction with site assessment work, as the aquifer solids required for microcosms can be obtained during a geologic investigation or well installation. Microcosm testing should, if possible, always be performed using both groundwater and aquifer solids collected from the targeted treatment area.

In a typical microcosm test, homogenized subsamples of the aquifer sediment and site groundwater are added to sterile serum bottles under an anoxic headspace, the bottles are amended with one of several electron donors, sealed with stoppers and incubated at typical site groundwater temperatures (often 15 degrees Celsius [°C]). An unamended sample (i.e., without electron donor) and a killed control containing an electron donor and a microbial inhibitor can also be prepared to account for any abiotic losses of perchlorate or losses due to biodegradation with naturally-occurring electron donors. Treatments with nitrogen and phosphorus added as inorganic nutrients are also recommended if nutrient limitation is suspected. Groundwater in the bottles is then sampled periodically under a nitrogen headspace and analyzed for concentrations of the added electron donor, perchlorate and any other electron acceptors or co-contaminants of interest.

6.5.3.2 Example of a Microcosm Test

The design of a typical microcosm test in support of a field demonstration of active perchlorate treatment is provided in this section. In this study, aquifer sediments and groundwater were collected from the demonstration location and added to sterile 160-milliliter (mL) serum bottles. In each of the 21 bottles prepared, 30 grams (g) of homogenized aquifer solids and 120 mL of site groundwater were added. Triplicate bottles then received the following:

- Treatment 1 – ethanol
- Treatment 2 – lactate
- Treatment 3 – citrate
- Treatment 4 – acetate
- Treatment 5 – acetate and inorganic nutrients (i.e., nitrogen and phosphorous)
- Treatment 6 – unamended (a live control)
- Treatment 7 – formaldehyde (a killed control to inhibit biological activity)

Treatments 1 through 5 were amended with electron donor such that a final concentration of 3 millimolar [mM] was obtained in each bottle. This equates to approximately 140, 180, 270 and 570 mg/L for ethanol, acetate, lactate and citrate, respectively. After amendments were added to the microcosms, each bottle was sealed with a sterile Teflon-lined septum and flushed thoroughly with nitrogen gas. The bottles were then incubated at 15°C.

At designated sampling times (Days 0, 5, 8, 12, 16, 20 and 27), 10-mL aqueous subsamples were collected from each bottle under a nitrogen headspace using a syringe. The samples were passed through a 0.22-micrometer [μm] pore size cellulose-acetate filter for preservation and stored at 5°C to limit biological activity prior to analysis. Samples were analyzed for perchlorate, nitrate, nitrite, sulfate, chloride, fatty acids and ethanol. Perchlorate levels from Day 0 through Day 27 are shown in Figure 6.9. Perchlorate concentrations in the samples receiving lactate, citrate, acetate, and ethanol declined from ~250 μg/L to <5 μg/L (PQL) during the first 16 days of incubation; no significant decreases in perchlorate concentrations were observed for the killed or live controls. The addition of inorganic nutrients to samples receiving acetate further enhanced the rate of perchlorate degradation, as concentrations decreased to <5 μg/L within 12 days. In addition, nitrate levels were observed to decrease from 14 mg/L to less than 1 mg/L within 12 days for all the electron-donor amended treatments, and no measurable decreases in sulfate concentrations were observed in any of the treatments, indicating that appreciable hydrogen sulfide was not formed during the test period.

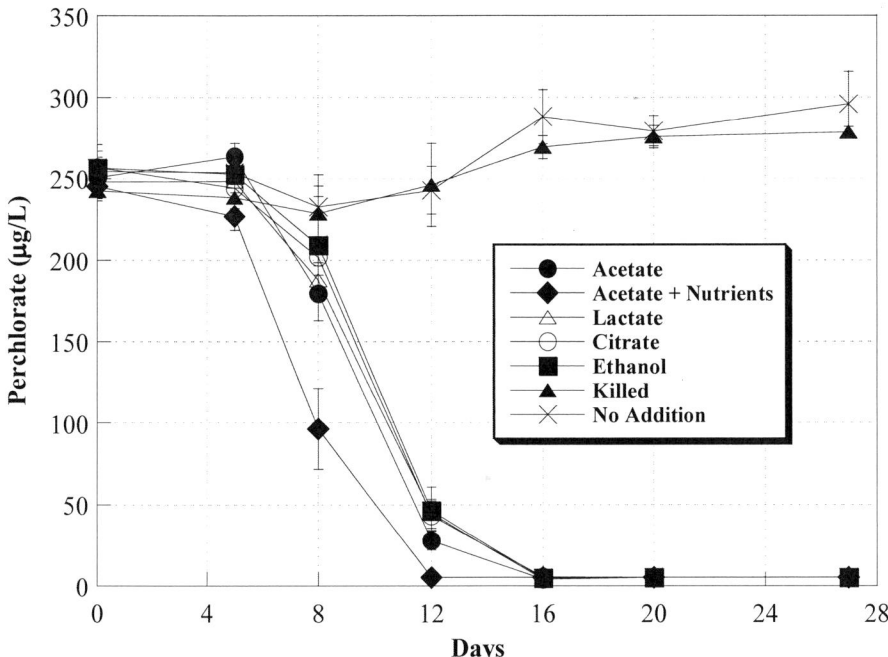

Figure 6.9. Perchlorate biodegradation in laboratory microcosms amended with different electron donors. Data from unamended samples (no addition) and killed controls are also presented.

Overall, the data from this treatability study show the following: (1) perchlorate-reducing bacteria are present in the aquifer at this site; (2) these organisms can be stimulated to degrade perchlorate and nitrate using one of several different organic substrates; (3) existing perchlorate concentrations (~250 μg/L) can be reduced to <5 μg/L (PQL) using substrate addition; and (4) small quantities of inorganic nutrients in addition to an organic substrate may enhance degradation rates. From these data, rates of degradation were derived for perchlorate, nitrate, and selected electron donors and incorporated into the modeling effort for this site.

6.5.3.3 Basis for Electron Donor Selection

Soluble electron donors that have been tested in the laboratory for perchlorate reduction include numerous fatty acids (acetic, lactic, citric, benzoic, valeric, butyric, propionic, pyruvic), alcohols (ethanol, methanol) and sugars (glucose, sucrose), as well as mixed substrates such as molasses, casamino acids and soluble components of cheese whey and yeast extract (Wallace et al., 1998; Coates et al., 1999; Hatzinger et al., 2002). Many other possibilities exist as well. Only a small percentage of these donors have been used in the field in active applications including ethanol, acetic acid, lactate, citric acid and benzoate. Others have been tested in passive and/or semi-passive applications. As the number of field trials and full-scale installations increase, the number of electron donors examined is certain to increase as well. However, efficacy for promoting *in situ* biodegradation is only one of many factors that determine which compounds are tested in the field. Other factors include commercial availability, ease and safety of handling and storage, availability as a food-grade product, potential to promote well fouling, regulatory considerations and cost.

A summary comparison of three common electron donors—ethanol, acetic acid and citric acid—is provided in Table 6.1. On a cost basis (calculated based on number of electrons donated per mole), ethanol is by far the cheapest electron donor of the three. It is for this reason that ethanol is used as the donor in the two largest *ex situ* bioreactor systems currently treating perchlorate (located at Aerojet and the former Kerr-McGee site; see Hatzinger, 2005). However, ethanol is also controlled by the U.S. Bureau of Alcohol, Tobacco, and Firearms, and permits are required to use it. Moreover, because ethanol is flammable, explosion-proof storage containers, pumps and other equipment must be used when handling it for bioremediation applications. In addition, some recent laboratory data suggest that ethanol is more likely to promote well fouling than some alternative substrates (Chopra et al., 2004). In comparison to ethanol, citric acid is more expensive. However, special permitting is not required for this material, it is available in food grade (which decreases regulatory concerns for aquifer injection), it reduces local pH and can act as a chelating agent, helping to remove metal precipitates from well screens, and it appears to promote less or slower biofouling than some alternate donors (Chopra et al., 2004). Thus, a number of different factors come into play when selecting an electron donor for *in situ* perchlorate treatment, and all must be considered together when making a choice for field application.

Table 6.1. Cost Comparison of Three Common Soluble Electron Donors Used for Perchlorate Treatment

Electron donor	Formula wt (g/mol)	Mols e⁻/mol	Mols e⁻/kg	Solution[1] (cost basis)	Cost[2] ($/kg)	Cost ($/mol e⁻)
Citric acid	192	18	94	50% (w/v)	5.06	0.054
Ethanol	46	12	261	95% (v/v)	2.17	0.008
Acetic acid	60	8	133	56% (w/v)	2.55	0.019

[1]The costs were based on current 55-gal drum prices from a chemical supplier in New Jersey. The chemical percentages are those supplied by the manufacturer.
[2]Costs were corrected based on chemical percentage and are given on a per kg basis.

6.5.4 Performance Monitoring

Performance monitoring for an active remediation system can generally be divided into two phases: (1) pre-amendment testing (prior to addition of electron donor and any other required amendments), and (2) treatment testing (during delivery of electron donor and amendments). A third phase of monitoring is also desirable in some instances after the electron donor addition period ends. This period can be used to determine the rate and extent of contaminant rebound after active treatment ceases. Table 6.2 summarizes the performance monitoring activities associated with each of these phases.

Table 6.2. Typical Performance Monitoring Activities for an Active Treatment System

Phase	Performance Monitoring	Purpose
Pre-amendment	Hydraulic heads	• Confirm hydraulic radius of influence
	Tracer testing	• Verify solute travel times and distribution • Calculate dispersivity
	Baseline concentrations	• Determine baseline contaminant concentrations after *in situ* mixing
Amendment delivery	Electron donor distribution	• Verify distribution of electron donor throughout targeted zone
	Changes in biogeochemistry	• Confirm biological activity • Identify zone of influence • Identify potential metals mobilization
	Electron donor decay	• Confirm biological activity • Calculate electron donor consumption rates
	Contaminant biodegradation	• Confirm treatment of perchlorate and co-contaminants • Verify production and/or biodegradation of daughter products
	Injection well pressures	• Identify potential issues related to biofouling
Rebound monitoring	Contaminant and biogeochemical parameters	• Assess contaminant rebound and continuing sources • Evaluate return of upgradient aquifer geochemistry

Collection of water table elevation data from all wells is recommended during the pre-amendment phase (with active pumping but prior to electron donor addition) to verify the hydraulic radius of influence of the treatment system. Water table elevations measured at system monitoring wells and piezometers should be compared to the appropriate model simulations. Tracer testing should also be performed to ensure that all monitoring wells are hydraulically connected to the treatment wells (injection wells or HFTWs) and that the groundwater flow rates and travel times are as anticipated based on modeling results.

Because contaminant concentrations may change significantly after pumping begins, it is recommended that two or more rounds of groundwater sampling be performed to clearly establish baseline contaminant levels throughout the plot prior to amendment delivery. These measurements will serve to differentiate concentration changes due to mixing from those caused by biodegradation. All results from the pre-amendment testing should be compared to the relevant model simulations. If needed, modifications to the model parameters and/or system operation (e.g., pumping rates) should be made at this time.

Once active treatment commences, performance monitoring consists primarily of sampling groundwater from monitoring wells (and often treatment wells) and measuring contaminant concentrations and various other chemical and geochemical parameters with time. A typical list of sampling parameters for an active treatment demonstration at a site containing perchlorate and volatile organic compounds (VOCs) as primary contaminants is provided in Table 6.3. Ethanol was the electron donor in this case. Measurement of perchlorate and co-contaminant concentrations (and relevant daughter products) during treatment provides verification of overall treatment effectiveness and allows estimation of *in situ* biodegradation kinetics. Calculation of *in situ* biodegradation rates is particularly useful during pilot demonstrations, as these rates may differ substantially from those measured in the laboratory during microcosm or column testing.

Table 6.3. Typical Monitoring Parameters for an Active Perchlorate Treatment System Utilizing Ethanol as an Electron Donor and with One or More VOCs as Co-Contaminants

Parameter	Method/Procedure	Preservative	Bottle Size
Nitrate	EPA 300.0	4°C	100 mL[1]
Sulfate	EPA 300.0	4°C	100 mL[1]
Chloride	EPA 300.0	4°C	100 mL[1]
Bromide	EPA 300.0	4°C	100 mL[1]
Perchlorate	EPA 314.0	Sterile 0.22 μm syringe filter	50 mL sterile[2] screw-cap tube
Volatile Organic Compounds	EPA 8260	acid	40 mL VOA vial
Volatile Fatty Acids	EPA 300.0m	Sterile 0.22 μm syringe filter	50 mL sterile[2] screw-cap tube
Ethanol	EPA 8015	Sterile 0.22 μm syringe filter	50 mL sterile[2] screw-cap tube
Dissolved Fe, Mn, and As	EPA 200.7	0.45 μm filter	500 mL plastic or glass—nitric acid rinsed
Dissolved Gases; methane, ethene, ethane	EPA 3810, RSK-175	acid	40 mL VOA vial
Redox Potential	Field Meter	–	–
Dissolved Oxygen	Field Meter	–	–
pH	Field Meter	–	–
Conductivity	Field Meter	–	–

[1] The same sample bottle will be used for the analyses noted.
[2] The same sample bottle will be used for all analyses noted.

Electron donor concentrations should be quantified throughout the demonstration period to verify distribution and to ensure that levels are sufficient to treat the target contaminants but not supplied in excess (unless that is the intent). Common degradation products of the chosen donor (e.g., acetate, propionate and other volatile fatty acids from ethanol) should also be measured because often the parent electron donor is quickly oxidized, but degradation intermediates persist as a source of reducing equivalents for bacteria. Moreover, electron donor consumption rates may increase or decrease with time as biomass and electron acceptor levels within the aquifer change. Thus, electron donor levels during active treatment may periodically require adjustment.

Various chemical and biogeochemical parameters including DO, ORP, pH, alternate electron acceptor levels (particularly nitrate and sulfate), methane concentrations and dissolved iron, manganese and arsenic also provide important information on system performance as well as potential secondary impacts of the treatment approach on aquifer geochemistry. These parameters can be used in conjunction with contaminant concentrations to adjust electron donor dosing rates during active treatment.

As discussed in Section 6.5.5.2, biofouling at the injection wells can adversely impact system operation and performance. As such, it is important to monitor injection well (or HFTW injection well screen) pressures during active pumping. Transducers should be installed in each well for this purpose. If pressures within the injection wells or HFTWs increase significantly during operation, expected rates of electron donor addition can be reduced due to backpressure, water leakage from well casings to the surface can occur, and down-well pumps and equipment can be damaged. In addition, once biofouling has occurred to a significant extent within a treatment well, physical redevelopment is often necessary to restore pressures and resume system operation.

After the active phase of system operation is complete, post-treatment monitoring is often desirable for evaluating overall treatment effectiveness. Evaluation of geochemical parameters is useful in order to determine the time required for the aquifer to return to baseline conditions once electron donor addition ceases. This is particularly important if mobilization of iron, manganese and/or arsenic has occurred during the demonstration period. Evaluating contaminant rebound during the post-treatment phase is also useful. Although this evaluation will differ somewhat between source area treatment and biobarrier applications, an increase in perchlorate and/or other contaminant concentrations during this phase indicates that sources still exist in the treatment zone (perhaps diffusing or slowly advecting from low permeability materials), that dissolved contaminants are infiltrating the treatment area from overlying vadose zone soils or that contaminants are migrating from upgradient sources.

6.5.5 Operational Issues

6.5.5.1 Undesirable Geochemical Impacts

One potential limitation with any *in situ* technology where organic carbon is added to an aquifer is that the carbon addition will result in negative impacts on groundwater geochemistry. Undesirable impacts are generally observed after addition of the large quantities of substrates commonly used in passive systems such as vegetable oil, molasses or polylactate ester. The injection of these electron donors at TOC levels in vast excess of that required for perchlorate treatment (ORP similar to denitrification) usually yields significant hydrogen sulfide (from sulfate reduction), methane (from methanogenesis), dissolved iron and manganese (from biological and chemical reduction and solubilization of these species) and sometimes arsenic (as arsenate is biologically reduced to the more mobile, more toxic arsenite species). These endpoints are undesirable, particularly in an aquifer used as a drinking water source. Moreover, regulatory agencies and water purveyors are becoming increasingly concerned about such secondary impacts.

The same geochemical concerns noted above for passive systems are also issues with active treatment systems. However, unlike passive approaches, the quantities of electron donor added using active systems are generally much lower since they are continuously (or very frequently) added, and mixing of the added donor with groundwater is much more extensive. Thus, the undesirable consequences of excess TOC addition, such as sulfate reduction and methanogenesis, tend to be less significant with active treatment approaches. In addition, with an active treatment system, the electron donor can be decreased in concentration or changed completely if undesirable endpoints are observed. This is not true for passive systems. Once a slow-release donor is added to an aquifer in high concentration, the material (or various degradation intermediates) can persist for years after application.

6.5.5.2 Biofouling

Bacteria can grow either in solution (sessile) or attached to solid surfaces. Surface-associated cells are frequently found in a matrix of extracellular polysaccharide, forming a microbial biofilm (e.g., Costerton et al., 1995; Donlan, 2002). As biofilms develop, their microbial diversity increases, and the growing matrix traps numerous other organic and inorganic materials (e.g., diatoms, clay particles and mineral precipitates). Thus, these films become increasingly complex with time. In the context of groundwater wells, biofilms forming on the surface of well screens and casings, filter pack, and in the regional aquifer near the well can lead to significant reductions in water production and/or increases in well pressure. The negative impact of biofilms, generally termed biofouling (or bioclogging), is one of the more significant operational issues affecting many *in situ* bioremediation applications, and this is a particular problem for active treatment systems which rely on continual pumping of groundwater for efficacy.

The overall nature and extent of biofouling in a treatment well depends on a multitude of factors including groundwater geochemistry, type of amendments added, characteristics of the aquifer, well design, pumping rates and others. Some of these general issues are addressed in Cullimore (2000). Often, aquifers contaminated with perchlorate are aerobic, contain a few to perhaps 100 mg/L of nitrate and are low in available carbon and electron donors. When an organic electron donor is added to this environment via an injection well (or HFTW), the main limiting factor for microbial growth (i.e., absence of available carbon and energy) is overcome, and growth of aerobic bacteria and denitrifiers is expected to be rapid. It is generally the local region where the electron donor is added and mixed with groundwater (i.e., the injection well or the injection well screen of a HFTW) where biofilm formation occurs (presumably consisting initially of aerobes and denitrifiers) and biofouling ultimately becomes a significant issue for well performance. This process can occur in a matter of days if a preventive strategy is not in place when electron donor is added. It is much less common for extraction wells in an ER system or for the extraction well screens in a HFTW system to become fouled, but this can occur if electron donor reaches an extraction well in a recirculation application.

There are a number of preventive strategies to control well fouling during *in situ* active treatment (ESTCP, 2005). These approaches include the following: (1) choice of an electron donor that promotes less or slower biofilm formation; (2) application of one or more chemical agents to kill bacteria and remove biofilms as they form; (3) pulsed rather than continuous application of electron donor; (4) physical removal of the biofilm by groundwater pumping/surging within the well; and (5) intermittent rather than continuous pumping (i.e., semi-passive design). There is not currently a "magic bullet" treatment for fouling, but often a combination of the different strategies described above is effective at controlling if not completely preventing the problem.

Among electron donors, specific chemical characteristics can be used to prevent or slow biofouling. For example, organic acids, such as acetic, lactic, and citric acid, can be applied not only as electron donors but also to significantly reduce the local pH in an injection well, slowing or preventing local microbial growth. This pH effect is particularly evident when high concentrations of the organic acids are added in short pulses (rather than adding low concentrations continuously) such that the natural alkalinity and buffering capacity of the groundwater is overcome. Moreover, citric acid acts as a metal chelator and, combined with its low pH, is likely to dissolve and remove some iron and manganese precipitates from an injection well.

A variety of different chemical agents have been tested for prevention of biofilm formation and, thus, to control biofouling. In general, these chemicals must be added in pulses at least on a daily basis to be effective. In some instances, the electron donor and biocide can be added in programmed cycles such that electron donor addition is followed closely by a short pulse of biocide. Chemicals that have been tested to control biofouling include strong oxidants, such as chlorine dioxide, sodium hypochlorite and hydrogen peroxide, glycolic, phosphoric and other acids, sodium azide, enzymes that degrade polysaccharides, liquid carbon dioxide and Tolcide (tetrakis(hydroxymethyl)phosphonium sulfate [THPS]). Laboratory testing of several of these agents is described in Chopra et al. (2004; 2005), and field application of hydrogen peroxide for prevention of well fouling in a HFTW system is reported by McCarty et al. (1998). A more extensive overview of these agents and other biofouling control strategies is provided in an ESTCP report entitled "A Review of Biofouling Controls for Enhanced *In Situ* Bioremediation of Groundwater" (ESTCP, 2005).

In addition to electron donor selection and biocide addition, injection well design and operation can also be utilized to reduce the impacts of well biofouling on system operation. Strategies that have proven successful include: (1) injecting amended groundwater through a pressurized packer to promote movement of water into the formation, even if well pressures increase modestly; (2) placing a pump in the injection well with one or more discharge lines in the vicinity of the well screen and using the turbulence created during water discharge to remove biomass and precipitates; and (3) designing injection wells using continuously-wrapped rather than slotted well screens to maximize open area within the screened interval.

As previously noted, multiple control strategies may be required to prevent or minimize injection well fouling. This is exemplified in a recent demonstration of *in situ* perchlorate treatment using a HFTW system (see Section 6.4.1). Prior to system installation, laboratory column studies were performed to assess both the impact of different electron donors on biofouling (measured via pressure drop across a sand column) and the potential effectiveness of different chemical control agents, including chlorine dioxide, commercial enzymes and THPS, to prevent and/or remove biological growth (Chopra et al., 2004; 2005).

Based on the results from these column studies (combined with laboratory microcosms quantifying contaminant biodegradation), citric acid was selected as the electron donor for the demonstration, and chlorine dioxide was chosen as a biocidal treatment. The two chemicals were applied in sequential pulses on daily cycles to provide electron donor (citric acid) while inhibiting growth in the injection well (chlorine dioxide). The length and timing of these cycles were varied during the demonstration in order to optimize the dosing regimen. Pressures at the HFTW injection well screens were measured in conjunction with perchlorate concentrations in the monitoring well network in the aquifer to assess the effectiveness of each injection strategy. Although biofouling was not completely prevented during the HFTW demonstration, the selection of electron donor and the application of that donor in sequential pulses with a chemical anti-fouling agent appeared to significantly slow the process. The system operated for five months with these biofouling control measures in place, which compares well to previous *in situ* tests at this location in which well fouling caused shut-down of ER systems within a few weeks.

6.6 CASE STUDY: AEROJET AREA 20 GROUNDWATER EXTRACTION – REINJECTION SYSTEM

A field pilot test was conducted by Geosyntec Consultants and Aerojet in the summer of 2001 to evaluate the applicability of a pilot-scale groundwater extraction-reinjection system (active biobarrier) for treatment of perchlorate. The details of this test are provided in the following case study (GeoSyntec Consultants, 2002).

6.6.1 Site Description

Perchlorate is present in soil and groundwater at Aerojet as a result of the production and testing of solid rockets for DoD use. The pilot test area (PTA) selected at the Aerojet facility is located within a perchlorate plume that originates from a former disposal/burn area.

Chlorinated solvents, consisting predominantly of TCE, are also present in the groundwater in this area. The perchlorate groundwater plume is approximately 1,500 m (5,000 ft) long and approximately 900 m (3,000 ft) wide in the vicinity of the PTA. The main impacted aquifer is located at a depth of about 30 m (100 ft) bgs. Figure 6.10 shows the location of the Aerojet facility within California, the location of the PTA at the facility and the locations of existing groundwater monitoring wells in the vicinity of the PTA.

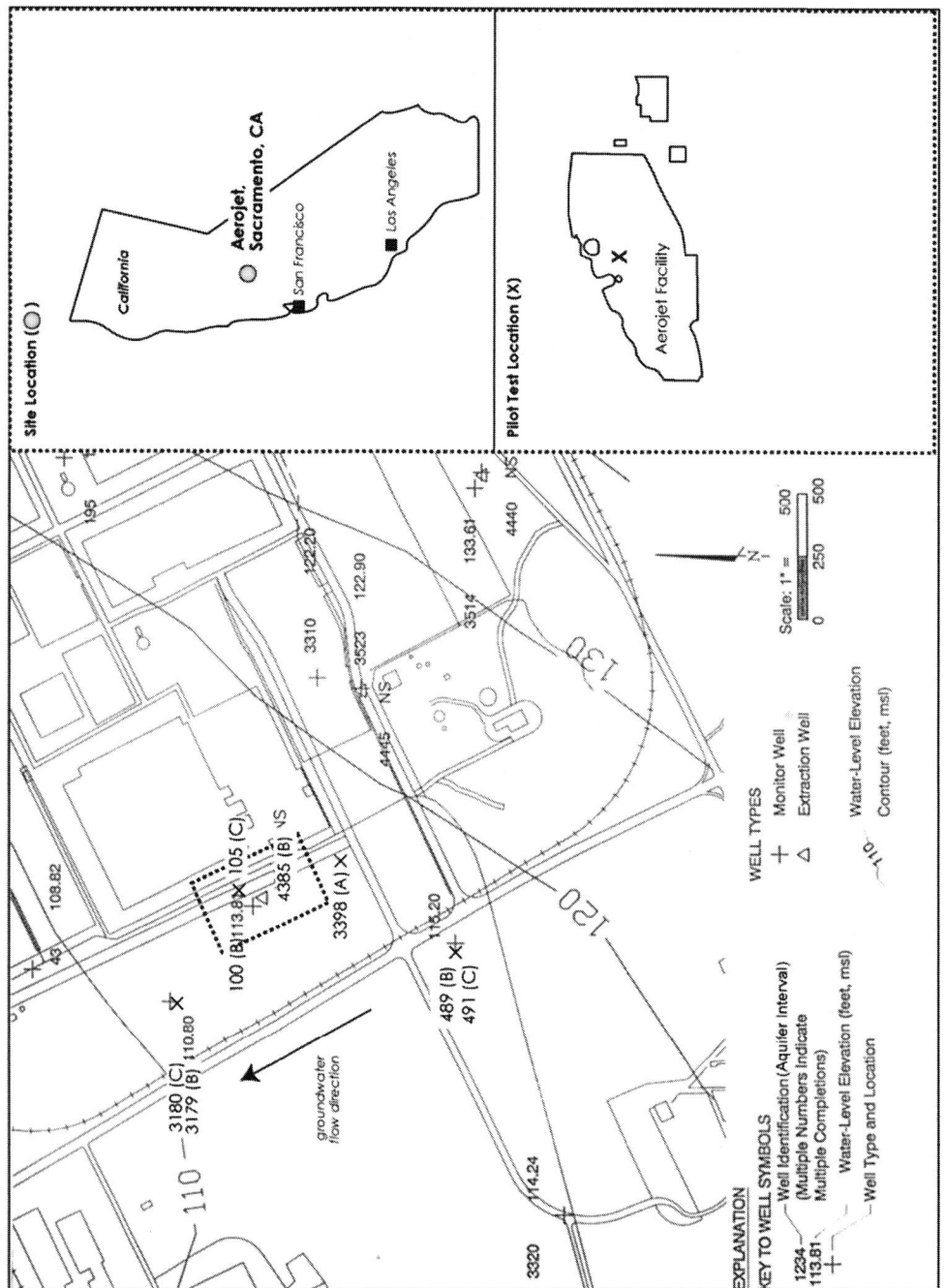

Figure 6.10. Map of the Pilot Test Area (PTA). Insets show the location of the Aerojet site in California and of the PTA on the Aerojet Site, respectively.

6.6.2 Site Geology and Hydrogeology

The Aerojet site is located on fluvial deposits from the ancestral American River. Three distinct river terraces are recognized across the Site, and the PTA is located on the youngest of these terraces. The geologic materials beneath the PTA have been previously divided into four main aquifer units, designated as Aquifers A, B, C and D. Aquifer A is an unconfined aquifer predominantly composed of unconsolidated sand and gravel to depths of about 26 m (85 ft) bgs in the PTA. Aquifer B (26–47 m [85–155 ft] bgs) is also predominantly composed of sand and gravel but contains several low-permeability beds. Aquifer C (47–81 m [155–265 ft] bgs) is

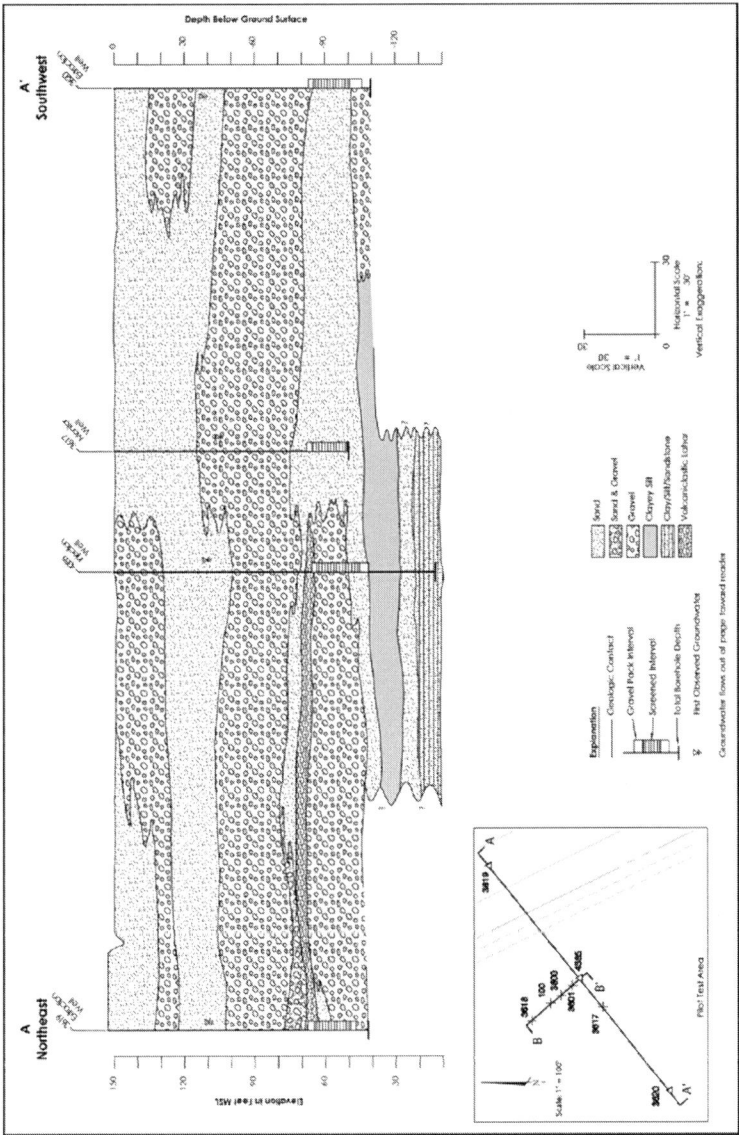

Figure 6.11(a). Geologic cross sections of the PTA oriented perpendicular to groundwater flow.

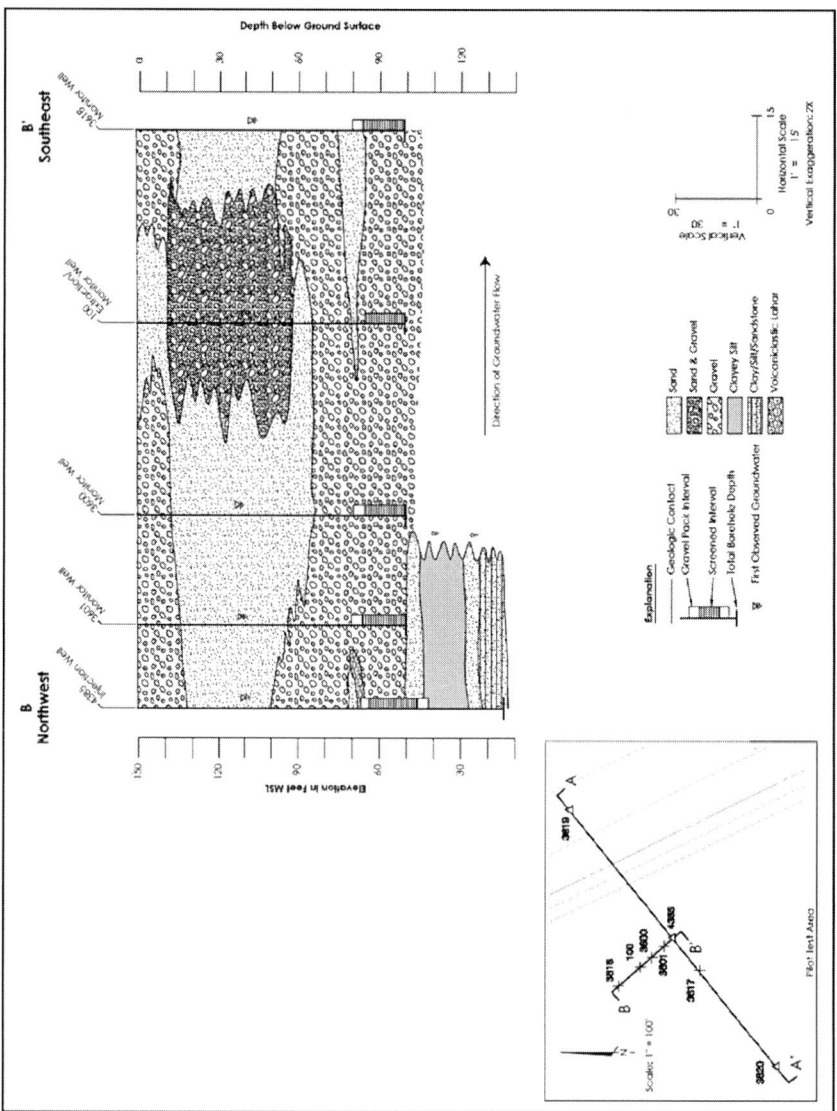

Figure 6.11(b). Geologic cross sections of the PTA oriented parallel to groundwater flow.

predominantly sand and silty sand with gravel and some sandstone. Aquifer D (>81 m [265 ft] bgs) is located well below the area of influence of the pilot test (and perchlorate impacts).

Figures 6.11a and 6.11b present geologic cross sections of the PTA, oriented perpendicular and parallel to groundwater flow, respectively. Perchlorate impacts in the PTA are primarily confined to Aquifer B. Accordingly, all PTA wells were screened within Aquifer B at a depth interval of about 26 to 30 m (85 to 100 ft) bgs. Groundwater in the PTA was determined to flow to the west-northwest with a horizontal hydraulic gradient of about 0.008 m/m (0.008 ft/ft).

6.6.3 Pilot Test Design

The pilot test involved the use of an active ER system, whereby groundwater was extracted using two pumping wells, amended with ethanol as an electron donor, and then reinjected into the aquifer via a single injection well. The system was designed to be operated as a single-pass biobarrier, rather than as a recirculation loop (see Figure 6.3c). To improve well coverage for this pilot test (within the project budget), infrastructure built for a previous bioremediation pilot test was used. This pilot test was conducted from May 2000 to April 2001. The original PTA consisted of a single electron donor injection well (designated 4385) and 3 downgradient monitoring wells (designated 3601, 3600 and 100) located along the prevailing groundwater flowpath at distances of 4.6, 10.7 and 19.8 m (15, 35 and 65 ft) downgradient from the injection well. For the previous pilot test, the PTA was bioaugmented with dehalorespiring bacteria (mixed culture KB-1 containing *Dehalococcoides* spp.) in December 2000 to evaluate the ability of bioaugmentation to improve the rate and extent of TCE dechlorination to ethene in Aerojet groundwater. Therefore, while the focus of this pilot test was to demonstrate perchlorate biodegradation, the fate of TCE in the groundwater was also assessed.

To evaluate the locations of the extraction wells and several new monitoring wells, a simplified numerical groundwater flow and transport model was developed (using MODFLOW) for the PTA. The domain was 915 m by 915 m (3,000 ft by 3,000 ft) to ensure that the model boundaries were far enough away from the PTA to have no significant effect on the simulations. The vertical dimension of the domain was from ground surface (about 46 m (150 ft) above mean sea level [amsl]) down to sea level. Groundwater flow across a hydraulic gradient of 0.008 m/m (0.008 ft/ft) was simulated with constant head boundaries at either end of the domain. The aquifer was simulated to have a horizontal hydraulic conductivity of 0.01 centimeters per second (cm/sec) (30 ft/day) and a vertical hydraulic conductivity of 0.001 cm/sec (3 ft/day). Pumping wells were simulated with 6 m (20 ft) screened intervals in the B Aquifer.

Results of the modeling indicated that two new extraction wells, each spaced at a distance of ~61 m (200 ft) from the single central injection well and pumping at 37.8 L/min (10 gpm) each, would be capable of capturing the core of the perchlorate plume in the area (Figure 6.12). The extracted groundwater would be combined and recharged (at 76 L/min [20 gpm]) via the single injection well. Figure 6.12 shows the results of the groundwater flow simulation at PTA scale. The particle tracking simulations show the area influenced by the extraction wells (about 185 m [600 ft] width). Each arrowhead represents a groundwater travel time of 2 weeks. The particle tracks were used to confirm the suitability of using existing monitoring wells for pilot test performance monitoring. According to particle tracks, groundwater recharged via well 4385 would be expected to reach monitoring wells 3601, 3600, 100 and 3618 within about 2, 6, 21 and 56 days, respectively, while particle arrival at transgradient well 3617, located 15 m (50 ft) from the injection well (4385), was estimated to be 35 days. Based on these travel estimates, the system residence time was expected to be long enough to allow complete biodegradation of perchlorate in the single-pass biobarrier system. Bromide tracer test results, as described in Section 6.6.6, were subsequently used to validate and refine the PTA model.

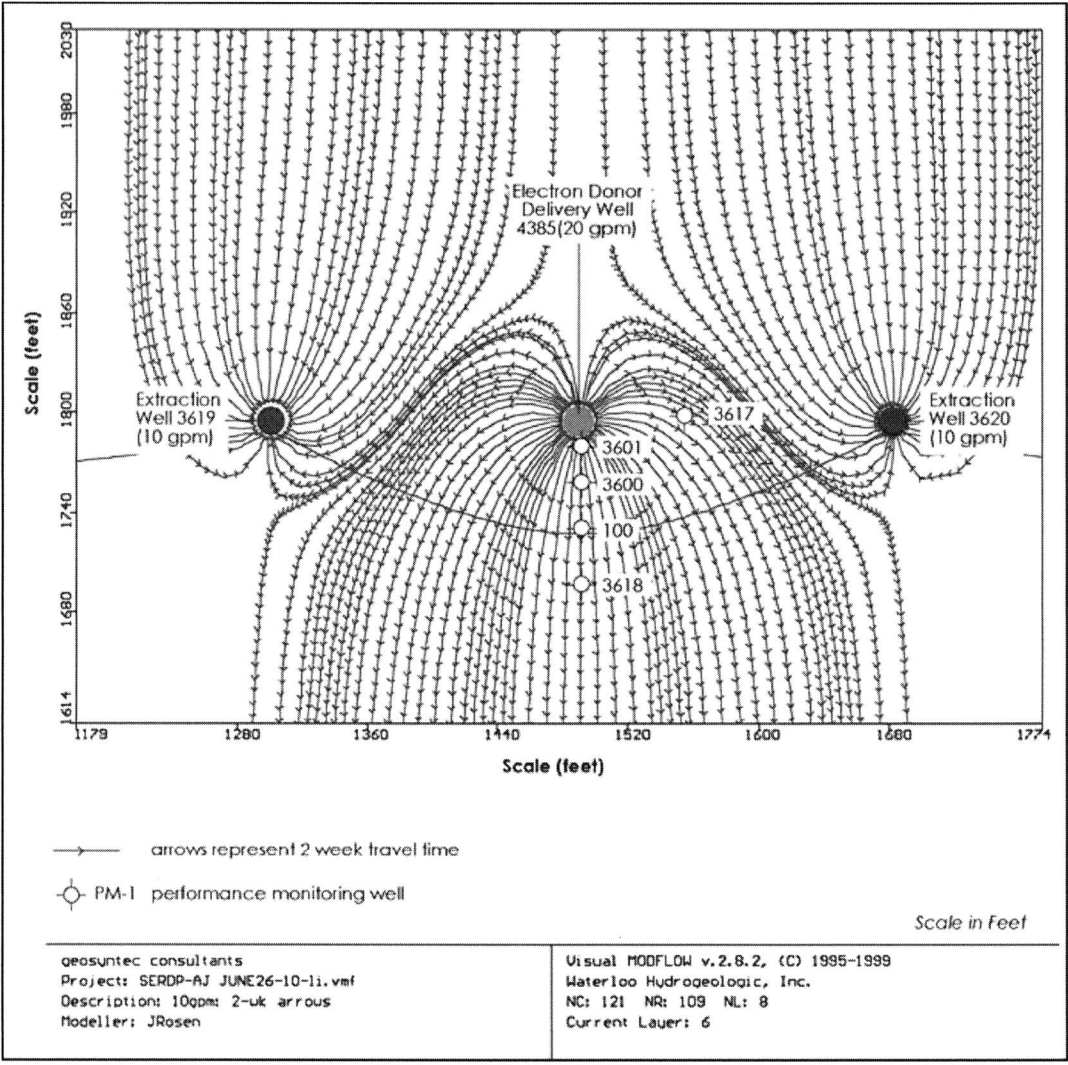

Figure 6.12. Groundwater flow simulation at PTA scale. The particle tracking simulations show the area influenced by the extraction wells. Each arrowhead represents a groundwater travel time of 2 weeks (values are in feet).

6.6.4 PTA Installation, Instrumentation and Operation

Figure 6.13 presents the layout of the groundwater extraction, injection and monitoring wells in the PTA in plan view. The two 15-cm (6-inch) diameter groundwater extraction wells were designated 3619 (east) and 3620 (west). Well 4385 (20-cm [8-inch] diameter) was used for electron donor delivery/injection, and wells 3601, 3600 and 100 were used as downgradient monitoring wells. Two additional wells (5-cm [2-inch] diameter) were installed to improve the coverage of the monitoring well network: well 3618, located 30.5 m (100 ft) downgradient from injection well 4385; and well 3617, located 15 m (50 ft) from injection well 4385.

Active Bioremediation 119

A schematic of the electron donor delivery system is given in Figure 6.14. Groundwater was extracted from wells 3619 and 3620 at a rate of approximately 38 L/min (10 gpm) per well. The extracted water was directed through a filter system to remove particulates, followed by a series of in-line meters to measure pH, ORP and perchlorate concentration in the re-circulating groundwater. An in-line flow meter was used to provide feedback control to the extraction well pumps to maintain steady extraction rates. The re-circulating groundwater was amended with electron donor (see Section 6.6.7) using a metering pump and then re-injected to Aquifer B via well 4385. System operation was controlled and monitored using a programmable logic controller and personal computer.

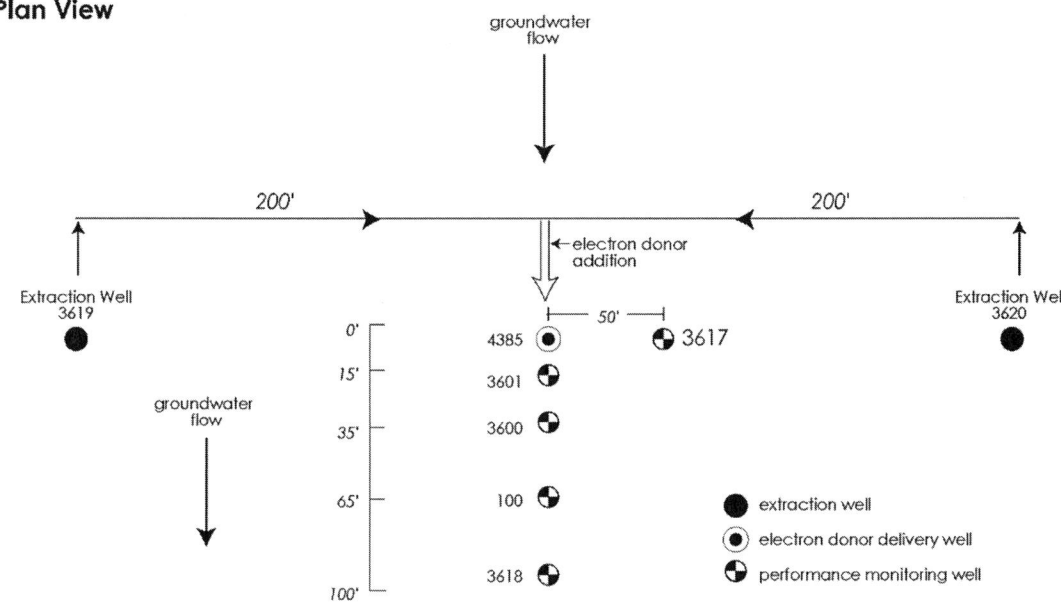

Figure 6.13. Plan view of the groundwater extraction, injection and monitoring wells in the PTA.

6.6.5 Baseline Geochemical Characterization

Several baseline groundwater sampling events were conducted in the PTA including: (1) May 2000, prior to the initial Aerojet pilot testing activities at the test site; (2) October 2001, prior to bromide tracer testing for the pilot test; and (3) November 2001, following the bromide tracer test and immediately prior to electron donor addition. At these events, samples were collected for baseline analysis of:

- Field parameters (DO, ORP, pH and temperature)
- Perchlorate and associated degradation products (e.g., chlorate, chloride)
- Volatile organic compounds (VOCs)

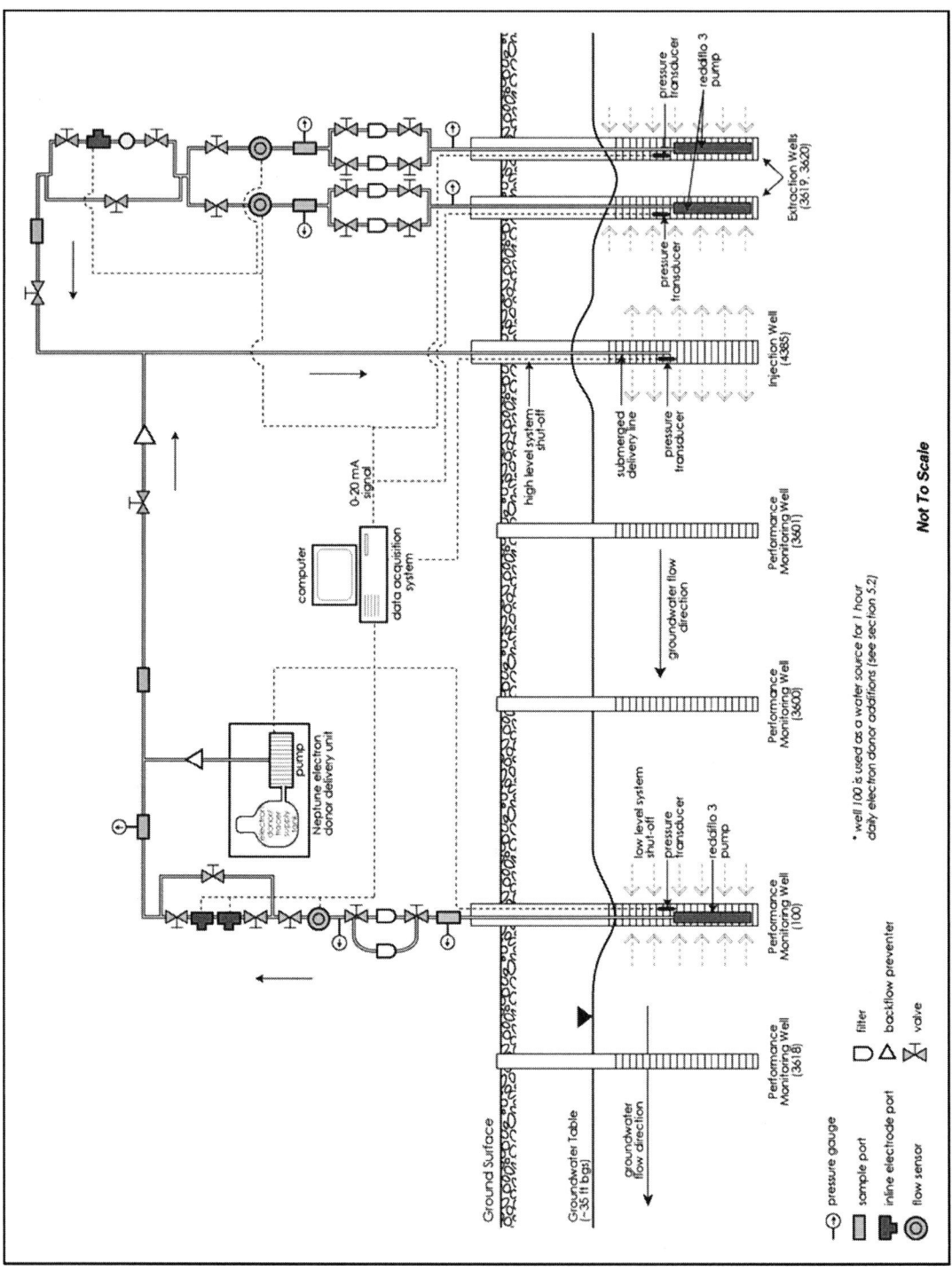

Figure 6.14. Schematic of the active treatment system. Groundwater extracted from wells 3619 and 3620 was passed through an in-line filter system, followed by a series of in-line electrodes to measure pH, ORP, and perchlorate concentration, then amended with ethanol and re-injected into well 4385.

Active Bioremediation 121

- Anions (bromide, nitrate, nitrite and sulfate)
- Dissolved hydrocarbon gases (DHGs; methane, ethane, ethene)
- Dissolved metals
- Biological oxygen demand and chemical oxygen demand
- Sulfide
- Volatile fatty acids (VFAs; acetate, propionate)

Table 6.4 summarizes the results of the baseline geochemical characterizations. For most parameters, concentrations measured in November 2001 (prior to electron donor addition) were similar to those measured in May 2000 prior to initiating bioremediation pilot testing activities at the site.

Table 6.4. Baseline Geochemical Parameters and Contaminant Concentrations at the Aerojet Pilot Test Area

Well ID:	3619	3620	4385	3601	3600	100	3618	3617
Date Sampled:	19-Nov-01	19-Nov-01	19-Nov-01	19-Nov-01	19-Nov-01	19-Nov-01	19-Nov-01	19-Nov-01
Field Parameters								
pH	7.10	6.88	6.99	7.02	7.29	7.22	7.26	7.01
Conductivity (mmhos)	n/a	n/a	n/a	n/a	n/a	n/a	n/a	n/a
Temperature (°C)	19.4	19.1	19.1	18.8	18.1	19.2	18.5	18.3
Oxidation Reduction Potential (mV)	96	125	119	140	136	-52	121	132
Dissolved Oxygen (mg/L)	5.4	4.9	4.4	3.9	5.9	3.0	1.5	2.8
Perchlorate (mg/L)	2.1	13	7.7	7.7	7.8	6.3	3.9	8.0
VOC (µg/L)								
Tetrachloroethene	n/a	n/a	35	34	24	36	28	32
Trichloroethene	n/a	n/a	1,500	1,600	1,400	1,500	1,600	1,500
1,2-Dichloroethene (total)	n/a	n/a	23	24	26	79	110	23
1,1-Dichloroethene	n/a	n/a	55	59	58	64	47	61
Vinyl Chloride	n/a	n/a	< 0.50	< 0.50	0.51	8.8	27	< 0.50
Dissolved Hydrocabon Gases (mg/L)								
Methane	n/a	n/a	< 15	< 15	< 15	15	180	< 15
Ethane	n/a	n/a	< 1	< 1	< 1	< 1	1.8	< 1
Ethene	n/a	n/a	< 1	< 1	< 1	4.5	60	< 1
Organic Carbon Indicators (mg/L)								
Biochemical Oxygen Demand (BOD)	n/a	n/a	n/a	n/a	n/a	n/a	n/a	n/a
Chemical Oxygen Demand (COD)	n/a	n/a	< 3.0	12	10	12	< 3.0	9.2
Inorganics (mg/L)								
Chlorate	n/a	n/a	< 1.0	< 1.0	< 1.0	< 1.0	< 1.0	< 1.0
Chloride	11	57	34	34	37	36	42	35
Nitrite as N	< 0.05	< 0.05	< 0.05	0.62	3.2	0.32	0.36	0.56
Nitrate as N	28	20	25	21	6.0	5.1	0.78	21
Phosphate	0.32	0.20	0.26	0.15	< 0.10	< 0.10	0.16	0.20
Sulfate	10	15	13	14	14	14	11	14

Note:
n/a = not analyzed.

6.6.6 Hydraulic Characterization (Tracer Testing)

Tracer testing was initiated in November 2001 to calibrate the PTA numerical model and refine estimates of PTA residence time and breakthrough at each monitoring well, and to estimate the perchlorate biodegradation rates. For the tracer test, sodium bromide (stock solution in deionized water) was added as a daily one-hour pulse (same method as electron donor addition) for 14 consecutive days to achieve a target time-weighted average concentration of 100 mg/L as bromide. Breakthrough of the conservative tracer at the monitoring and extraction wells was monitored by collecting samples on a daily to semi-weekly basis from each of the wells. Samples were analyzed onsite (for screening purposes) using a bromide ion-specific electrode and were submitted for laboratory confirmation analysis by ion chromatography.

Maximum breakthrough concentrations of bromide in wells 3600, 3617 and 100 (10.7, 15.2 and 19.8 m [35, 50 and 65 ft] from the injection well) represented about 100%, 76% and 72% of the injected concentrations, respectively (Figure 6.15). The maximum bromide concentration at well 3618 (30.5 m [100 ft] from the injection well) represented about 40% of the injected concentration, confirming that this well was also on the flowpath, but the breakthrough curve indicated that significant dispersion of the added bromide pulse occurred over the 30.5 m (100 ft) distance. Bromide concentrations at well 3601, located 4.6 m (15 ft) from the injection well, showed significant variability related to the pulse addition methodology. The data suggest that this well was located too close to the injection well (at the injection rate of 76 L/min [20 gpm]) to provide useful performance data, and therefore, well 3601 was not used for subsequent biodegradation performance assessment.

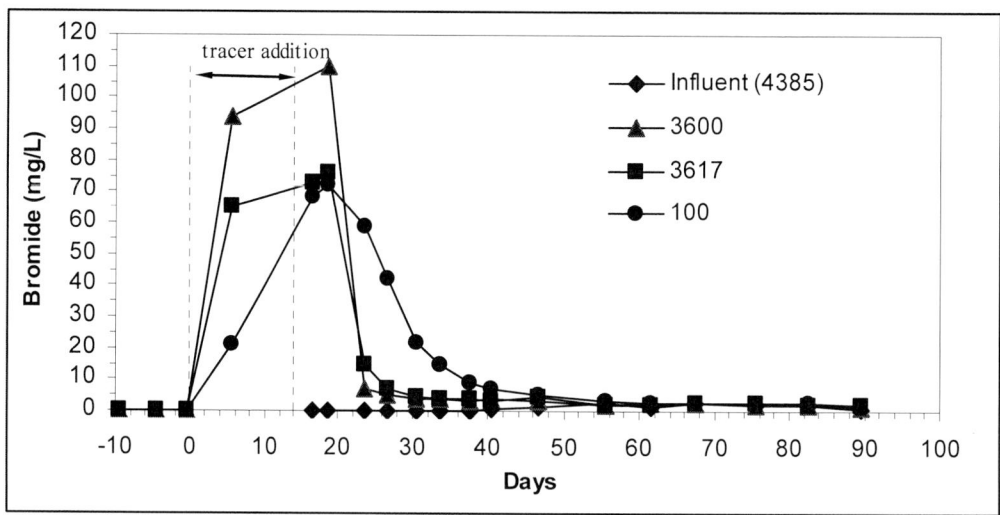

Figure 6.15. Bromide concentrations in downgradient monitoring wells 3600, 3617 and 100 (11, 15, and 20 m from the injection well, respectively) with time.

Based on the bromide tracer breakthrough curves, the average travel times for non-retarded particles to reach downgradient monitoring wells 3600, 100 and 3618 were estimated to be 5, 10 and 38 days, respectively, which is reasonably consistent with the travel times predicted by the model (5, 21 and 56 days, respectively). By comparison, the average travel time for non-retarded particles to reach transgradient monitoring well 3617 was only 5 days, as compared to 35 days predicted by the model. Therefore, the PTA model was revised to provide a "best-fit" to the observed travel times.

In addition to bromide breakthrough at the monitoring wells, bromide was detected at extraction well 3620. The maximum breakthrough concentration observed was about 5% of the injected concentration. This result confirmed that the biobarrier provided effective capture/treatment over the full 61-m (200-ft) western portion of the PTA, from well 4385 to 3620. It should be noted that significant bromide levels were not expected in extraction wells for this one-pass biobarrier design. However, detection of dilute bromide (and comparison of tracer results with those predicted by the calibrated site model) provided an indication of capture zone width effectiveness. Bromide was not detected significantly above background concentrations at extraction well 3619 in the eastern portion of the biobarrier, suggesting that the groundwater extraction rate at well 3619 was not high enough during the pilot test to provide complete 61-m (200-ft) lateral coverage on the eastern portion of the biobarrier. A review of the site geology (see Figure 6.11a; data were not available prior to pilot testing) indicated that the aquifer materials differed in the eastern and western portions of the biobarrier. Specifically, in the western portion (from well 4385 to 3620), the target aquifer consisted largely of sands, with some gravels, whereas in the eastern portion (from well 4385 to 3619), the geology consisted predominantly of gravels. The apparent lack of bromide capture by well 3619 was likely related to the presence of higher permeability materials in this eastern portion of the biobarrier and an insufficient extraction rate at this well to achieve capture.

6.6.7 System Operation

6.6.7.1 Electron Donor Addition

Ethanol was selected as the electron donor for the pilot test because: (1) it has been determined to be one of the most cost-effective electron donors for large-scale use (see Table 6.1); and (2) it does not adversely impact groundwater quality other than ORP and alkalinity. Other electron donors have been shown to have the potential to contribute metals (e.g., molasses) or cations (e.g., sodium lactate or sodium citrate) to groundwater. Since the PTA at the Aerojet site is located in proximity to a drinking water supply aquifer, the introduction of non-native constituents, such as metals or cations (e.g., sodium), was deemed unacceptable by local regulatory agencies.

Ethanol was pulsed into the groundwater one time per day (rather than added continuously) to minimize microbial fouling. Balanced oxidation-reduction reactions were used to determine the amount of ethanol required to promote complete reduction of perchlorate and TCE in the extracted groundwater. Specifically, sufficient ethanol was added to reduce oxygen (5 mg/L), nitrate (23 mg/L), perchlorate (8 mg/L), sulfate (13 mg/L) and TCE (1.7 mg/L) in the groundwater. Based on these average influent concentrations, the ethanol demand was estimated to be 17 mg/L. To account for uncertainty and biomass production, a 3-fold safety factor was applied to this concentration, and therefore the time-weighted average ethanol addition concentration was ~50 mg/L.

6.6.7.2 Biofouling Control

The injection well was periodically treated with chlorine dioxide (ClO_2) to prevent microbial fouling. This chemical biocide is commonly used to disinfect drinking water and to prevent biofilm formation in *ex situ* treatment systems, cooling towers and industrial applications. However, installation of the pilot-scale chlorine dioxide generator was not completed in time to meet the schedule for electron donor delivery, and therefore, electron donor addition was conducted for approximately one month without biofouling control. In the absence of chlorine dioxide addition, biofouling was quickly evident, resulting in rising water-levels in the injection well. The rate of rise was subsequently reduced by chlorine dioxide addition, but groundwater extraction and recharge rates were reduced on several occasions to maintain system operation during the course of the test.

6.6.8 Demonstration Results

Monitoring of groundwater chemistry consisted of weekly measurements of field parameters, and weekly to bi-weekly collection of groundwater samples from the influent (4385) and PTA wells 3600, 3617, 100 and 3618 for laboratory analyses of perchlorate, anions, VFAs, VOCs and DHGs. Field parameter measurements were conducted onsite.

6.6.8.1 Oxidation-Reduction Potential (ORP) and Dissolved Oxygen (DO)

Figures 6.16a and 6.16b present ORP and DO data, respectively, from the PTA groundwater over the duration of the pilot test. Initial redox conditions in the groundwater were generally aerobic and oxidizing, with DO concentrations exceeding 1.5 mg/L and ORP values exceeding 100 millivolts (mV). Over the duration of the pilot test, the influent DO concentrations (well 4385) were consistently aerobic, ranging from 3.4 to 7.4 mg/L, while the ORP was consistently oxidizing, ranging from 56 to 258 mV.

Following addition of electron donor, ORP values in monitoring wells 3601, 3600 and 100 quickly became reducing (within 6 to 9 days), with ORP values declining to and stabilizing between −50 and −100 mV through the remainder of the pilot test. ORP values also declined at transgradient well 3617, and became reducing by Day 16. While ORP values at downgradient well 3618 declined, they generally remained oxidizing through the end of the pilot test. DO concentrations in wells 3601 and 3600 generally declined below 1 mg/L, suggesting the development of anoxic conditions. While declines in DO concentrations were observed in groundwater at downgradient well 3618, concentrations generally remained above 1 mg/L, reflecting the return to background redox conditions downgradient from the PTA.

6.6.8.2 Perchlorate

Perchlorate concentrations in the extraction wells differed in magnitude, with concentrations at well 3619 ranging from 2.1 to 3.0 mg/L during the test, while concentrations at well 3620 ranged from 12 to 14 mg/L. As a result, the influent perchlorate concentration (a 50:50 blend from wells 3619 and 3620) ranged from 7.1 to 8.3 mg/L over the duration of the study.

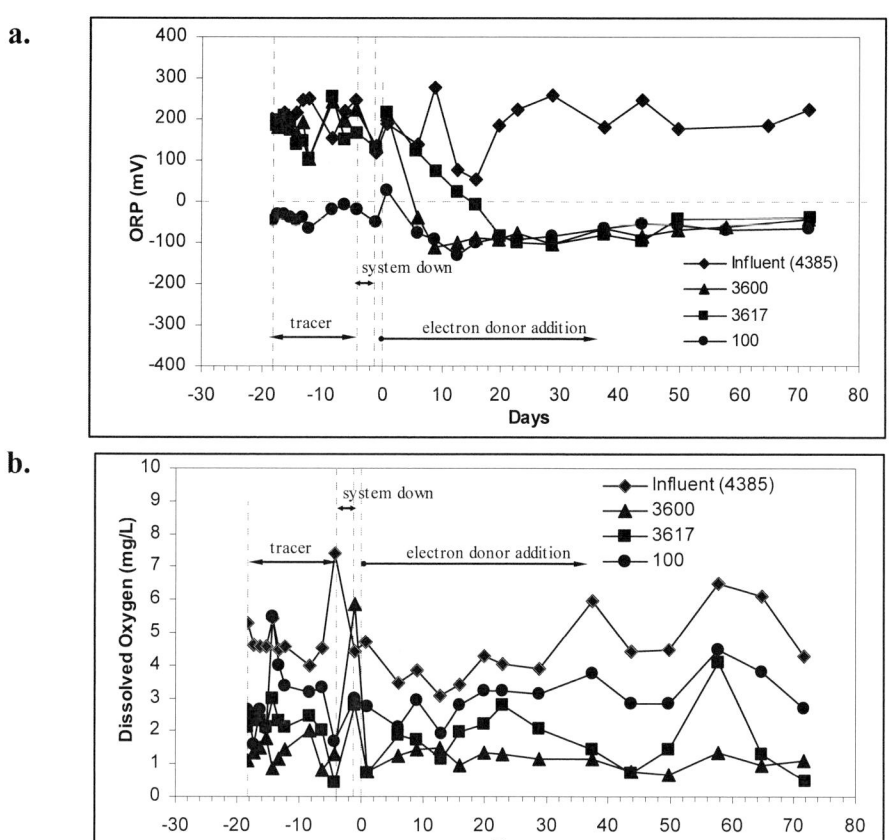

Figure 6.16. (a) Oxidation-reduction potential (ORP) and (b) dissolved oxygen (DO) concentrations in the influent water and in monitoring wells 3600, 3617, and 100 over the duration of the pilot test. ORP values are reported in mV.

Following electron donor addition, perchlorate concentrations declined rapidly, with little to no acclimation period (Figure 6.17). For example, perchlorate concentrations at well 3600 declined from 7.8 mg/L to <4 µg/L (the PQL for this study) within 9 days from the start of electron donor addition, and remained below 4 µg/L (with only one exception) for the remainder of the study. Similarly, at wells 3617 and 100, perchlorate concentrations declined to <4 µg/L within 20 and 29 days, respectively, and remained below detection. Perchlorate concentrations at downgradient well 3618 declined from 3.9 mg/L to 0.15 mg/L by the end of the pilot test (72 days), and concentrations continued to decline (data not shown). Given the degree of dispersion observed for bromide at this well, it was expected that perchlorate concentrations would have eventually declined to below the PQL, but that this would have required several additional months. Perchlorate concentrations showed an overall declining trend at well 3601, but as discussed in Section 6.6.6, the data from this well showed significant variability related to the short travel time and the electron donor pulse addition methodology.

To estimate perchlorate biodegradation rates, first-order degradation half-lives were approximated using the influent data (well 4385) as the initial concentration and data from monitoring wells 3600, 3617 and 100 as the final concentrations. Travel times of 5, 5 and 10 days, respectively, were used as the elapsed time for the respective wells, based on the results from the bromide tracer test. Using these data, the perchlorate biodegradation half-lives during the early portion of system operation at 76 L/min (20 gpm) were estimated to be in the range of 0.5 to 1.2 days. These rates are consistent with half-lives calculated from previous pilot tests at the Aerojet facility (0.2 to 1.8 days). It should be noted that these values are a rough approximation, assuming first-order degradation kinetics. A coupled site model can be used to provide a more detailed analysis (see Section 6.5.2). As a final note related to perchlorate reduction, chlorate (a potential perchlorate degradation intermediate) was not detected above its PQL (1 mg/L) in any of the monitoring wells during perchlorate reduction.

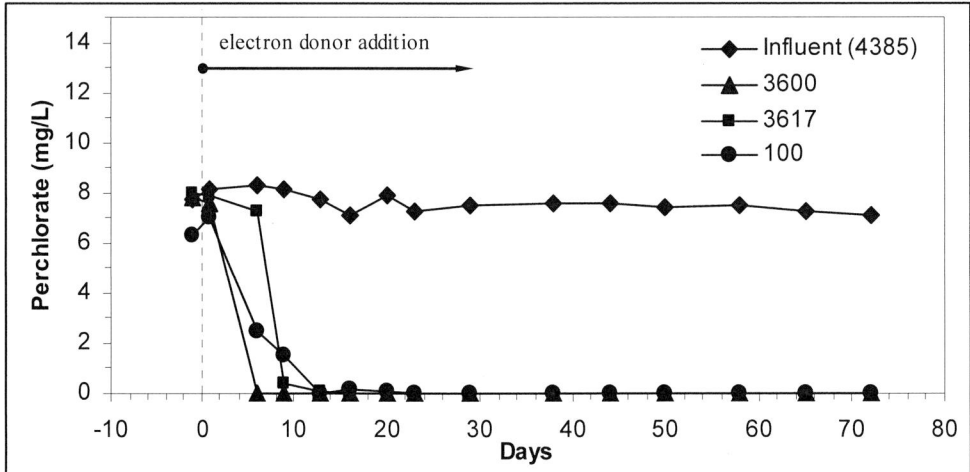

Figure 6.17. Perchlorate concentrations in the influent water and in monitoring wells 3600, 3617, and 100 during the pilot test. Values after Day 29 were below the PQL of 4 µg/L for the three monitoring wells.

6.6.8.3 VOCs

Coincident with perchlorate reduction, the addition of ethanol to the groundwater promoted rapid and complete dechlorination of TCE (1.7 mg/L) to ethene within 10.7 to 19.8 m (35 to 65 ft) from the injection well. Figure 6.18 provides a comparison of the relative proportions (in micromoles per liter [µmoles/L]) of TCE, *cis*-1,2-dichloroethene (1,2-DCE), vinyl chloride (VC), and ethene at the start of the pilot test and at Days 44, 58 and 72 following initiation of ethanol addition. At the start of the demonstration (Day -1), TCE was the dominant VOC in the biobarrier influent and at all downgradient and transgradient monitoring wells. Dechlorination products present in wells 100 and 3618 were present from a previous pilot test in this area. By Day 58, ethene was the dominant product at wells located 10.7 to 19.8 m (35 to 65 ft) downgradient, within the portion of the PTA that was previously bioaugmented with strain KB-1. By Day 72, TCE and 1,2-DCE concentrations were below their respective maximum contaminant levels at wells 3600 and 100, while VC concentrations had declined to 12 µg/L at well 100, and were continuing to fall. VOC concentrations were also declining at downgradient well 3618, the furthest downgradient well in the PTA. Based on the data summarized above, the calculated half-life for TCE dechlorination to cis-1,2-DCE under steady state conditions ranged between 1.3 to 3.7 days, while the half-life for complete TCE dechlorination to ethene ranged between 4.1 to 11 days.

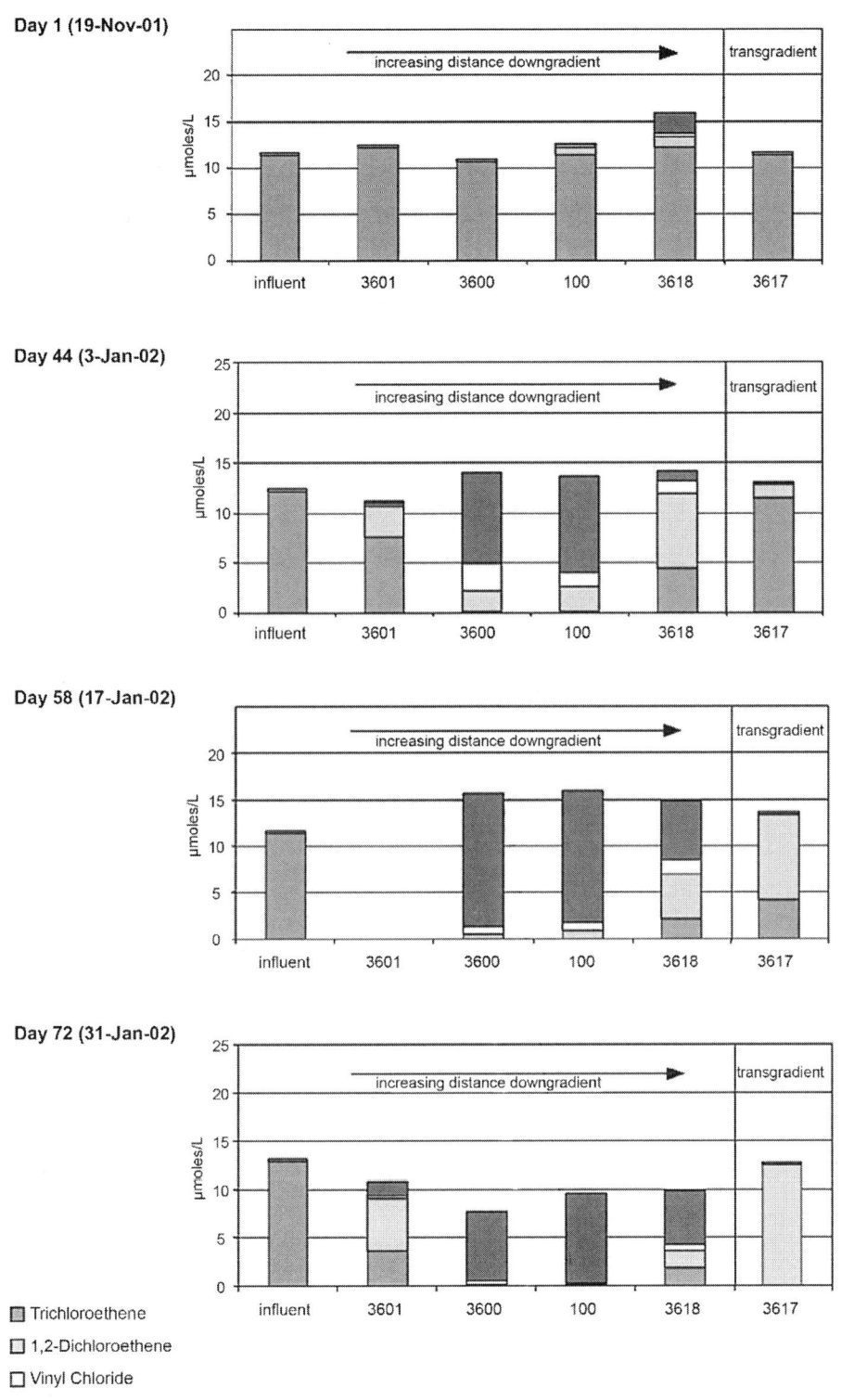

Figure 6.18. Concentrations of trichloroethene and daughter products cis-1,2-dichloroethene, vinyl chloride and ethene as a function of time in the influent water and in monitoring wells 3601, 3600, 100, 3618, and transgradient well 3617.

6.6.8.4 Nitrate

Influent nitrate concentrations averaged approximately 23 mg/L during the pilot test. Following electron donor addition, nitrate concentrations in wells 3600, 100 and 3617 declined to less than the PQL (0.05 mg/L) within 6 to 9 days (data not shown). Nitrate concentrations also declined at well 3618, reaching non-detect (<0.05 mg/L) by Day 44. Based on these data, the calculated biodegradation half-life for nitrate in the PTA was 0.6 to 0.7 days.

6.6.8.5 Sulfate and Sulfide

Because of the intent to treat VOCs as well as perchlorate during the pilot test, enough ethanol was added to the aquifer to promote significant sulfate reduction. In the absence of VOCs (and of the requirement for a low ORP to promote dechlorination of these compounds), less ethanol would have been required and sulfate reduction minimized.

Influent sulfate concentrations averaged approximately 14 mg/L during the pilot test. Following electron donor addition, sulfate concentrations in wells 3600, 100 and 3617 declined to less than the PQL (0.05 mg/L) within 60 days (data not shown). Sulfate concentrations also declined at well 3618, reaching 2.1 mg/L by Day 72. Coincident with sulfate reduction, sulfide concentrations increased in wells 3600, 100 and 3617 to maximum values of 1.1, 1.2 and 2.5 mg/L, respectively, at Day 30, but then declined to below method detection limits by Day 72 at each well. It is unclear whether sulfide was produced and precipitated with dissolved metals or whether there was analytical variability between sampling events.

6.6.8.6 Ethanol and Degradation Intermediates

Ethanol was rarely detected in groundwater samples within the PTA (PQL of 0.5 mg/L). However, the metabolism of ethanol by acetogenic bacteria in aquifers often results in the production of acetate and propionate, which can subsequently be used by a variety of bacteria as electron donors for the reduction of nitrate, perchlorate and/or chlorinated solvents. Acetate was observed within the initial 10.7 to 19.8 m (35 to 65 ft) of the injection well, as reflected by the increasing concentrations over time at wells 3600, 100 and 3617 (Figure 6.19). Acetate concentrations were consistently below detection (3 mg/L) at downgradient well 3618. Similarly, propionate concentrations showed increasing trends at wells 3600 and 3617 but tended to decrease thereafter (data not shown). As with acetate, propionate concentrations were consistently below detection (3 mg/L) at downgradient well 3618. These data suggest that ethanol is rapidly metabolized to acetate and propionate, which subsequently serve as secondary electron donors along the groundwater flowpath.

6.6.8.7 Methane

Methane was detected in the PTA, but concentrations generally remained below 100 µg/L (data not shown). The low levels of methanogenesis reflect the addition of stoichiometric concentrations of electron donor, with consumption being primarily directed toward the desired degradation reactions for perchlorate, nitrate and TCE, as well as significant sulfate reduction.

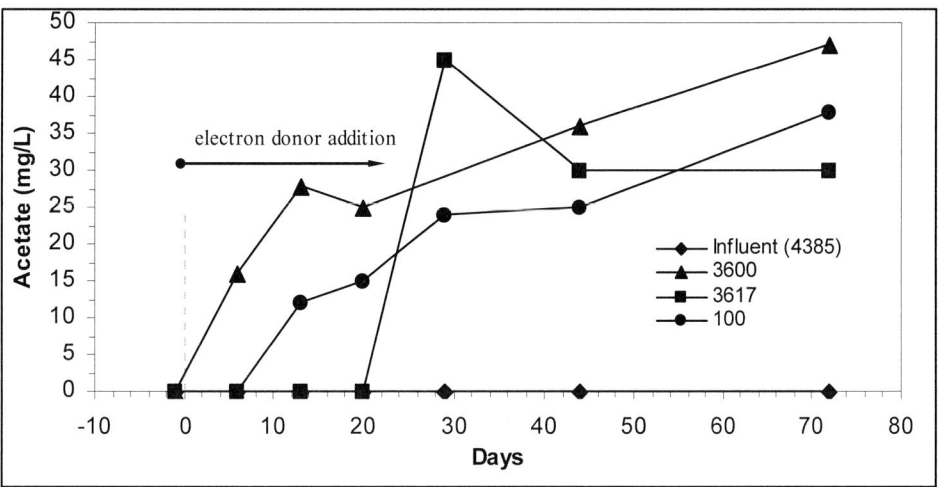

Figure 6.19. Acetate concentrations in the influent water and in monitoring wells 3600, 3617, and 100 during the pilot test. Acetate is derived from ethanol oxidation.

6.6.8.8 Dissolved Metals

The addition of electron donor to the PTA groundwater promoted bacterial reduction and mobilization of iron and manganese but not any other metals. Dissolved iron concentrations increased at wells 3600 and 100 to maximums of 0.8 and 4.6 mg/L, respectively (Figure 6.20). However, dissolved iron was not detected during the pilot test at downgradient well 3618, suggesting that the dissolved iron precipitated within 19.8 to 30.5 m (65 to 100 ft) from the injection well. Interestingly, dissolved iron was not detected at transgradient well 3617 during the pilot test. By comparison, dissolved manganese was produced in the PTA and detected at wells 3600, 100 and 3617 (Figure 6.21). Once produced, concentrations did not decline before reaching downgradient well 3618. These data are consistent with previous pilot tests at the Aerojet site which show that manganese tends to persist in groundwater once formed. The best solution for manganese control is to optimize electron donor addition further to prevent or limit its formation.

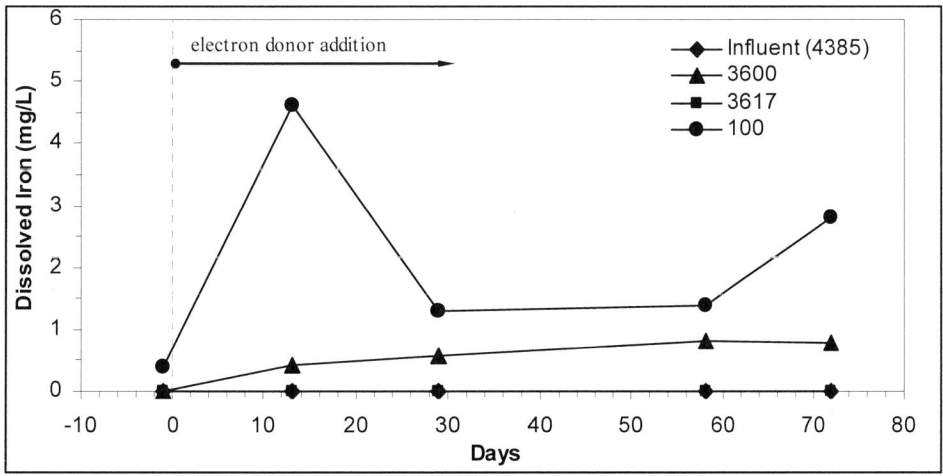

Figure 6.20. Dissolved iron (Fe) concentrations in the influent water and in monitoring wells 3600, 3617, and 100 during the pilot test.

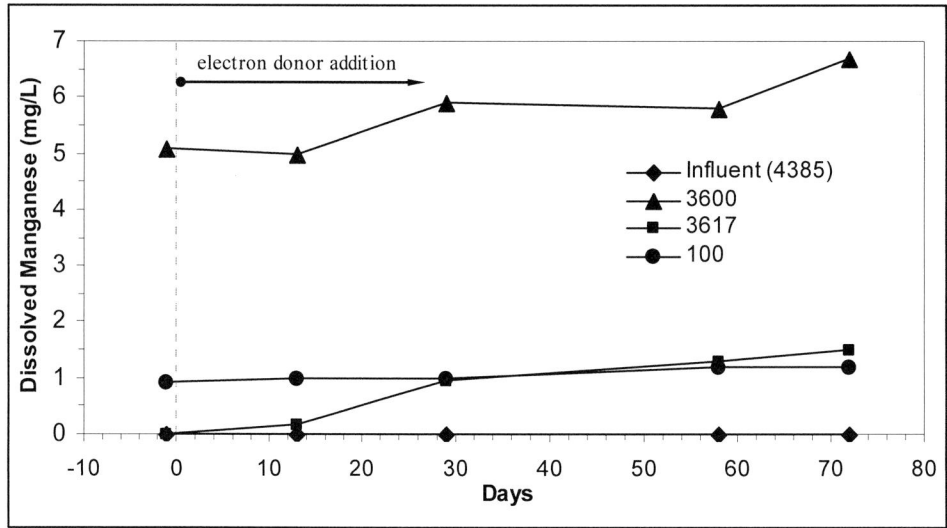

Figure 6.21. Dissolved manganese (Mn) concentrations in the influent water and in monitoring wells 3600, 3617, and 100 during the pilot test.

6.6.9 Pilot Test Conclusions

Summary conclusions from the pilot test include:

1. Perchlorate biodegradation was readily initiated, without acclimation period, through the addition of ethanol as electron donor. Perchlorate concentrations in excess of 8 mg/L were consistently reduced to less than the PQL of 4 µg/L within 10.7 m (35 ft) of the injection well. Perchlorate biodegradation half-lives were rapid, ranging from 0.5 to 1.2 days.

2. Coincident with perchlorate reduction, TCE (~2 mg/L) was also dechlorinated to ethene in a portion of the aquifer that was previously bioaugmented with a dehalorespiring microbial consortium. The calculated half-life for TCE dechlorination to ethene ranged from 4.1 to 11 days.

3. Nitrate concentrations in the influent (~23 mg/L) were routinely reduced to <0.05 mg/L within 10.7 m (35 ft) of the injection well. The half-lives for nitrate were in the range of 0.6 to 0.7 days.

4. Ethanol was a highly effective and efficient electron donor for perchlorate, nitrate and TCE reduction when added at 3x stoichiometric requirements. The data suggest that ethanol is rapidly metabolized to acetate and propionate *in situ*, which were subsequently used as electron donors and depleted within 30.5 m (100 ft) of the injection well.

5. Pulsed addition of chlorine dioxide in the injection well as a biocide was effective for slowing the progress of well biofouling. Biofouling control measures are necessary to maintain system operation.

6. The sole groundwater impact created by *in situ* bioremediation appeared to be mobilization and persistence of low levels (about 1 mg/L) of dissolved manganese.

Overall, the results of the field pilot test show that an active treatment approach can be used to effectively treat perchlorate, nitrate and VOCs in groundwater without causing appreciable secondary impacts to aquifer geochemistry. The most significant impact was the mobilization and transport of manganese downgradient. Well fouling appears to be the most important O&M issue. The data suggest that this approach should be considered as a potential groundwater remediation technology for both source remediation and control of perchlorate plumes.

6.7 SUMMARY

This chapter describes active *in situ* treatment systems for perchlorate and common co-contaminants in groundwater, and summarizes previous field trials of this approach. Active treatment systems, which include ER and HFTW approaches, utilize pumping wells to capture perchlorate-contaminated groundwater and meter and mix soluble electron donor with the captured water.

The key advantages of these systems include (1) applicability over a broad range of hydrogeologic, geochemical and contaminant conditions; and (2) minimization of secondary groundwater impacts that are typical of passive approaches. System disadvantages include (1) the necessity for increased infrastructure compared to passive approaches; and (2) O&M costs and issues associated with biofouling.

Several field demonstrations of active treatment approaches have been conducted, and one full-scale ER system is currently operating at the former PEPCON site in Henderson, Nevada. A plant located at this site manufactured perchlorate salts from the mid 1950s until 1988, when it was destroyed by an explosion. Significant contamination of groundwater extends from the region of the former plant downgradient nearly 4.8 km (3 mi) toward the Las Vegas Wash. The full-scale biobarrier consists of 9 groundwater extraction wells in two separate well fields (~950 L/min [250 gpm] total flow rate), 6 electron donor delivery/aquifer recharge wells, a permanent electron donor dosing facility, and nearly 8,230 m (27,000 linear ft) of conveyance piping to connect the extraction wells via the electron donor dosing station to the barrier recharge area. Sodium benzoate is used as the electron donor, added continuously at a concentration equal to 1.25 times the stoichiometric amount needed to reduce the concentration of oxygen, nitrate, chlorate and perchlorate in the influent groundwater. Short duration applications of chlorine dioxide are also applied to the operating recharge wells to control biofouling and maintain recharge capacity. Additional details concerning the installation and operation of this system are provided in ITRC, 2008.

Based on several years of laboratory and field pilot testing, improved techniques to mitigate biofouling and successful field application, it is anticipated that active systems will play an important role in perchlorate clean-up in the years to come.

REFERENCES

Chopra G, Dutta L, Nuttall E, Anderson W, Hatzinger P, Goltz MN. 2005. Investigation of methods to control biofouling during in situ bioremediation. In Alleman BC, Kelley ME, eds, In Situ and On-Site Bioremediation—2005, Proceedings of the Eighth International In Situ and On-Site Bioremediation Symposium (on CD), Battelle Press, Columbus, OH, USA, Paper # A-22.

Chopra G, Dutta L, Nuttall E, Hatzinger P, Goltz MN. 2004. Investigation of biomass and their mitigation for in-situ bioremediation. In Gavaskar AR, Chen ASC, eds, Remediation of Chlorinated and Recalcitrant Compounds—2004, Proceedings of the Fourth International

Conference on Remediation of Chlorinated and Recalcitrant Compounds (on CD), Battelle Press, Columbus, OH, USA, Paper # 4F-02.

Clement TP. 1997. RT3D—A Modular Computer Code for Simulating Reactive Multi-Species Transport in 3-Dimensional Groundwater Aquifers. PNNL 11720. Pacific Northwest National Laboratory, Richland, WA, USA.

Coates JD, Achenbach LA. 2004. Microbial perchlorate reduction: rocket-fuelled metabolism. Nature Rev Microbiol 2:569–580.

Coates JD, Michaelidou U, Bruce, RA, O'Conner SM, Crespi JN, Achenbach LA. 1999. Ubiquity and diversity of dissimilatory (per)chlorate-reducing bacteria. Appl Environ Microbiol 65:5234–5241.

Costerton JW, Lewandowski Z, Caldwell DE, Korber DR, Lappin-Scott HM. 1995. Microbial biofilms. Annu Rev Microbiol 49:711–745.

Cox EE, McMaster M, Neville SL. 2001. Perchlorate in groundwater: scope of the problem and emerging remedial solutions. In Proceedings 36th Annual Engineering Geology and Geotechnical Engineering Symposium, Las Vegas, NV, USA, pp 27–32.

Cullimore R. 2000. Microbiology of Well Fouling. Lewis Publishers, NY, USA. 435 pp.

Donlan RM. 2002. Biofilms: Microbial life on surfaces. Emerg Infect Dis 8:881–890.

ESTCP (Environmental Security Technology Certification Program). 2005. A Review of Biofouling Controls for Enhanced *In Situ* Bioremediation of Groundwater. ESTCP Technical Report. ESTCP, Arlington, VA, USA. http://www.estcp.org/viewfile.cfm?Doc=ER%2D0429%2DWhtPaper%2Epdf. Accessed April 15, 2008.

Fetter CW. 1999. Contaminant Hydrogeology, 2nd Ed. Prentice Hall, Upper Saddle River, NJ, USA. 500 pp.

Gandhi RK, Hopkins GD, Goltz MN, Gorelick SM, McCarty PL. 2002a. Full-scale demonstration of *in situ* cometabolic biodegradation of trichloroethylene in groundwater, 1: Dynamics of a recirculating well system, Water Resour Res, 38:10.1–16.

Gandhi RK, Hopkins GD, Goltz MN, Gorelick SM, McCarty PL. 2002b. Full-scale demonstration of *in situ* cometabolic biodegradation of trichloroethylene in groundwater, 2: Comprehensive analysis of field data using reactive transport modeling, Water Resour Res 38:11.1–19.

GeoSyntec Consultants. 2002. In Situ Bioremediation of Perchlorate Impacted Groundwater. Project ER-1164 Final Technical Report. DoD Strategic Environmental Research & Development Program (SERDP), Arlington, VA, USA. http://www.serdp.org/Research/upload/CU-1164-FR-01.pdf. Accessed April 15, 2008.

Hatzinger PB. 2005. Perchlorate biodegradation for water treatment. Environ Sci Technol 39:239A–247A.

Hatzinger PB, Diebold J, Yates CA, and Cramer RJ. 2006. Field demonstration of *in situ* perchlorate bioremediation in groundwater. In Gu B, Coates JC, eds, Perchlorate: Environmental Occurrence, Interactions, and Treatment, Springer, New York, USA, pp 311–341.

Hatzinger PB, Whittier MC, Arkins MD, Bryan CW, Guarini WJ. 2002. In-situ and ex-situ bioremediation options for treating perchlorate in groundwater. Remediation 12:69–85.

ITRC (Interstate Technology & Regulatory Council). 2008. Remediation technologies for perchlorate contamination in water and soil. PERC-2. ITRC Perchlorate Team, Washington DC, USA. http://www.itrcweb.org/Documents/PERC-2.pdf. Accessed April 20, 2008.

Knarr MR. 2003. Optimizing an In situ Bioremediation Technology to Manage Perchlorate-Contaminated Groundwater, MS Thesis, AFIT/GEE/ENV/03-14, Graduate School of Engineering and Management, Air Force Institute of Technology, Wright-Patterson AFB, OH, USA.

McCarty P, Goltz MN, Hopkins GD, Dolan ME, Allan JP, Kawakami BT, Carrothers TJ. 1998. Full-scale evaluation of *in situ* cometabolic degradation of trichloroethylene in groundwater through toluene injection. Environ Sci Technol 32:88–100.

McDonald MG, Harbaugh AW. 1988. A modular three-dimensional finite-difference groundwater flow model. In U.S. Geological Survey Techniques of Water-Resources Investigations, Book 6, Chapter A1, 586 pp.

Parr JC. 2002. Application of Horizontal Flow Treatment Wells for In Situ Treatment of Perchlorate Contaminated Groundwater. MS Thesis, AFIT/GEE/ENV/02-M08, Graduate School of Engineering and Management, Air Force Institute of Technology, Wright-Patterson AFB, OH, USA.

Parr JC, Goltz MN, Huang J, Hatzinger PB, Farhan YH. 2003. Modeling *in situ* biodegradation of perchlorate-contaminated groundwater. In Magar VS, Kelley ME, eds, In Situ and On-Site Bioremediation—2003, Proceedings of the Seventh International In Situ and On-Site Bioremediation Symposium, Orlando, FL (on CD), Battelle Press, Columbus, OH, USA.

Secody RE. 2007. Modeling In Situ Bioremediation of Perchlorate-Contaminated Groundwater. MS Thesis, AFIT/GEM/ENV/07-M13, Graduate School of Engineering and Management, Air Force Institute of Technology, Wright-Patterson AFB, OH, USA.

Waddill DW, Widdowson MD. 1998. Three-dimensional model for subsurface transport and biodegradation. J Environ Eng 124:336–344.

Wallace W, Beshear S, Williams D, Hospadar S, Owens M. 1998. Perchlorate reduction by a mixed culture in an up-flow anaerobic fixed bed reactor. J Ind Microbiol Biotechnol 20:126–131.

Waller AS, Cox EE, Edwards EA. 2004. Perchlorate-reducing microorganisms isolated from contaminated sites. Environ Microbiol 6:517–527.

Weight WD, Sondregger JL. 2001. Manual of Applied Field Hydrogeology. McGraw Hill, New York, USA, 608 pp.

Woods K. 2006. Perchlorate: the new scourge. The Weekend Pinnacle. http://www.pinnaclenews.com/news/contentview.asp?c=180410. Accessed April 15, 2008.

Wu J, Unz RF, Zhang HS, Logan BE. 2001. Persistence of perchlorate and the relative numbers of perchlorate- and chlorate-respiring microorganisms in natural waters, soils, and wastewater. Bioremediation J 5:119–130.

Xu J, Song Y, Min B, Steinberg L, Logan BE. 2003. Microbial degradation of perchlorate: Principles and applications. Environ Eng Sci 20:405–422.

Zheng C, Wang PP. 1999. MT3DMS—A modular three-dimensional multispecies transport model for simulation of advection, dispersion and chemical reactions of contaminants in groundwater systems, University of Alabama, AL, USA. http://hydro.geo.ua.edu/mt3d/. Accessed April 15, 2008.

CHAPTER 7

SEMI-PASSIVE *IN SITU* BIOREMEDIATION

Thomas A. Krug[1] and Evan E. Cox[1]

[1]Geosyntec Consultants, Inc., Guelph, ON, Canada

7.1 BACKGROUND

7.1.1 What is a Semi-Passive Approach

Semi-passive enhanced *in situ* bioremediation (EISB) of perchlorate involves the addition of electron donor on a periodic basis to stimulate natural microbiological populations. Semi-passive EISB approaches are similar to active approaches in that groundwater is recirculated between injection and extraction wells; however, with the semi-passive approach, groundwater is recirculated for an "active phase" of a limited duration (e.g., several days to several weeks) to distribute the electron donor, and then the recirculation system is shut off for a "passive phase" of longer duration (e.g., several weeks to several months).

Figure 7.1 shows the induced and natural groundwater flow patterns during the active and passive phases of a semi-passive system. In this case, the injection and extraction wells are configured to create a biobarrier perpendicular to groundwater flow. Groundwater extracted from the central well is amended with electron donor and injected into the wells on either side of the extraction well during the active phase. Some of the injected water flows back to the central extraction well and some water moves out in other directions from the injection wells. Some of the ambient flow of groundwater from upgradient of the biobarrier is collected in the central extraction well and some of the flow is diverted around the ends of the biobarrier. During the passive phase, ambient groundwater flow patterns are reestablished, and the natural groundwater gradient directs groundwater through the area where the electron donor has been added to the subsurface.

The semi-passive approach also can be used to distribute electron donor in source areas or throughout other target treatment zones. The semi-passive approach differs from the passive approach in that it relies on some recirculation of groundwater to distribute electron donor, and it differs from the active approach in that the recirculation of groundwater is conducted on a periodic and not a continuous basis. The equipment used to implement the semi-passive approach may be mobile and moved from one area to another as required or may be permanent installations operated on an intermittent basis.

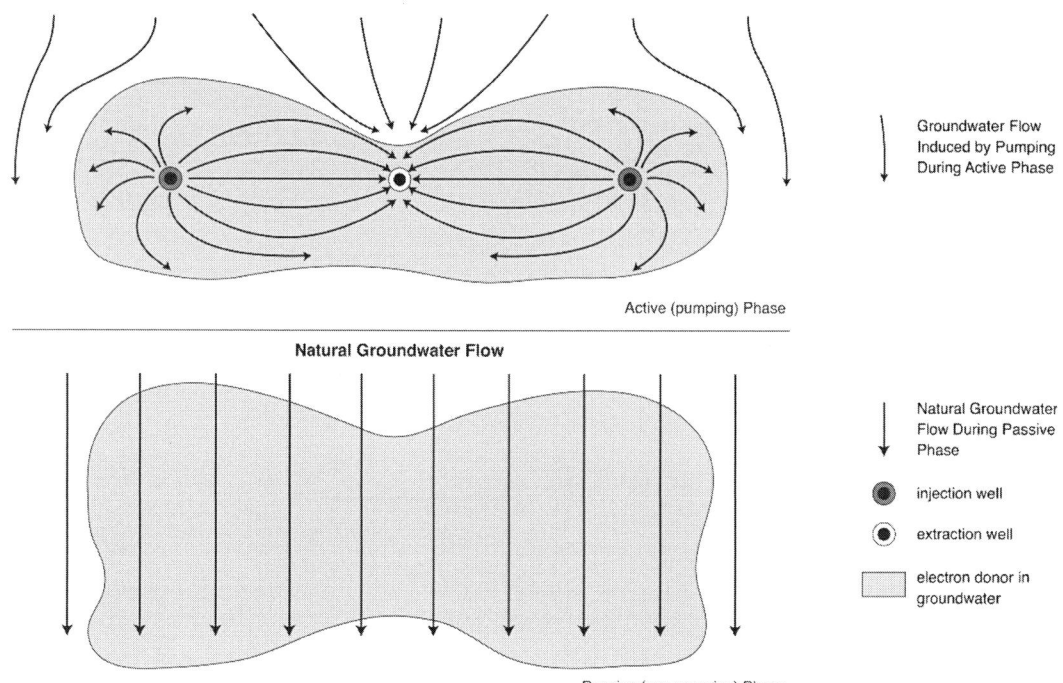

Figure 7.1. Plan view of groundwater flow for semi-passive biobarrier system

As with the active remediation approaches, the electron donor used for the semi-passive approach must be sufficiently mobile to travel some distance between the injection and extraction wells in order to achieve the desired electron donor coverage. Soluble electron donors, such as sodium lactate, have been used in field applications, and it may be possible to use mobile forms of emulsified vegetable oil, methyl esters and other slower release forms of electron donor as well. Biomass grows rapidly during the active phase when high concentrations of electron donor are present. During the passive phase, some of this biomass dies off, providing a source of electron donor to promote microbial degradative activity until the next addition of electron donor. The high level of biological activity also reduces natural minerals in the subsurface, leaving behind a more reduced geochemistry.

Semi-passive approaches are similar to "passive" bioremediation approaches in that electron donor is added to the subsurface and the system is allowed to operate predominantly under natural groundwater flow conditions. The "active phase" of the semi-passive approach can allow for a better distribution of electron donor than with the "passive" approach because electron donor is pushed from the injection wells and pulled towards the extraction wells of the groundwater recirculation system. In addition, because the amount of electron donor injected at any one time using the semi-passive approach is typically less than is used in passive systems, there are generally less impacts to secondary water quality. As with any bioremediation approach, groundwater quality may be adversely impacted by trace constituents present in the electron donors injected. Care must be taken in the selection of electron donors to avoid those that could cause increases in concentrations of dissolved metals or other undesirable constituents.

The semi-passive approach, with periodic operation of a groundwater recirculation system, is less expensive to operate than the active approach because the recirculation system is not operated on a continuous basis. Periodic operation of the recirculation system will also result in less biofouling of the injection wells than with continuous recirculation. The semi-passive approach also allows for the use of simple equipment such as a trailer-mounted recirculation system that can be moved from one area to another in sequence.

7.1.2 When to Consider a Semi-Passive Approach

The characteristics of sites which tend to favor a semi-passive approach over either the passive or active approaches include:

- Deep sites, >12 or 15 meters (m) (40 or 50 feet [ft]) below ground surface (bgs), where injection of electron donor to create a passive biobarrier or installation of a mulch biobarrier system would be difficult and expensive;
- Wide plumes, where the number and cost of electron donor injection points would be prohibitive;
- Sites where impacts to secondary water quality characteristics (such as increasing the concentrations of iron and manganese, sulfide or methane) due to the injection of a large quantity of electron donor at one time, as in a passive approach, would create problems; and
- Sites where the high capital and the on-going operating costs of an active recirculation system would make an active bioremediation system uneconomical.

7.1.3 Advantages and Limitations Relative to Other Approaches

The semi-passive approach, with periodic operation of a groundwater recirculation system, has the following advantages over passive approaches:

- Semi-passive systems require fewer wells or injection points because the groundwater recirculation during injection provides an induced flow of groundwater to distribute electron donor across the natural flow of groundwater in the subsurface or across greater distances in an area to be treated. This factor is particularly relevant when the target treatment zone is deep and the costs to install wells or injection points are high.
- Semi-passive systems do not inject unduly high concentrations of electron donor at one time as is typical with passive systems. The more moderate concentration of electron donor added to semi-passive systems reduces the impacts to secondary water quality characteristics (such as increasing the concentrations of iron and manganese, sulfide and methane) and reduces the tendency for electron donor to be consumed in biological pathways that will not contribute to perchlorate reduction (i.e., methane generation).
- Semi-passive systems do not inject large volumes of oil emulsion that can reduce the hydraulic conductivity of the treatment zone and cause diversion of groundwater around the treatment zone.

The semi-passive approach has the following limitations relative to passive approaches:

- Semi-passive systems normally require the installation of permanent injection wells to allow for periodic amendment of electron donor. Passive systems can use direct push injection points rather than permanent wells.
- Semi-passive systems require periodic re-amendment of the subsurface with electron donor on a more frequent basis than most passive approaches.

The semi-passive approach, with periodic operation of a groundwater recirculation system rather than continuous operation, has the following advantages over active approaches:

- The groundwater recirculation equipment of a semi-passive system does not need to be dedicated to a specific set of injection and extraction wells. The equipment may operate for a few weeks, then be shut off for several months at any specific set of wells. The semi-passive approach can also allow for the use of simple equipment, such as a trailer-mounted recirculation system that is moved from one area to another in sequence, thus avoiding significant capital costs.

- The operating costs for a semi-passive system are significantly less than for an active system because: (1) the system is not operated on a continuous basis and therefore does not incur costs for labor and power during the long "passive phase" of operation; and (2) the injection wells are less susceptible to biofouling because the injection of electron donor is not conducted on a continuous basis.

- The equipment required for semi-passive operation can be significantly less complex and is less likely to require complex controls and permitting because of the relatively short duration of the operation of the equipment.

The semi-passive approach has the following limitations relative to active approaches:

- The semi-passive approach results in greater variations in the concentration of electron donor in the subsurface than active systems. The variations in concentration with the semi-passive approach are not as great as with the passive approach but are greater than with the active approach. As discussed earlier, variations in the concentration of electron donor can impact secondary water quality characteristics.

In theory, the semi-passive bioremediation approach can integrate the best aspects of both the active approach (wider well spacings and less impact on secondary water quality characteristics) and the passive approach (minimal permanent *ex situ* infrastructure, lower operations and maintenance [O&M]), in order to optimize the balance of capital and O&M costs.

7.1.4 Technology Maturity

The semi-passive approach is a variation of *in situ* bioremediation systems which have been used for a number of years. The semi-passive approach was discussed as early as 1994 (Devlin and Barker, 1994), and a field demonstration was conducted in the late 1990s (Devlin and Barker, 1996; Devlin and Barker, 1999). Researchers from the University of Waterloo conducted a field demonstration of the approach for the biological treatment of nitrate (Gierczak et al., 2006; Gierczak et al., 2007). Other researchers used a semi-passive approach for *in situ* bioremediation of carbon tetrachloride and nitrate in groundwater at the Schoolcraft site in Michigan (Hyndman et al., 2000; Dybas et al., 2002; Phanikumar et al., 2005). Design issues related to the semi-passive approach were evaluated by the researchers working on the Schoolcraft site (Phanikumar et al., 2002a; Phanikumar et al., 2002b; and Phanikumar and Hyndman, 2003).

The semi-passive approach was used for a demonstration of *in situ* bioremediation for perchlorate-impacted groundwater funded by the Department of Defense (DoD) Environmental Security Technology Certification Program (ESTCP) at the Longhorn Army Ammunitions Plant (LHAAP) in Karnack, Texas. This demonstration is discussed further in the Case Study section of this chapter (Section 7.3).

The components of the semi-passive approach (injection wells, groundwater recirculation pumps, electron donor amendment systems) have all been used extensively for the EISB of perchlorate and other compounds.

7.2 SYSTEM DESIGN, OPERATION, AND MONITORING

7.2.1 Typical System Design

Semi-passive bioremediation systems typically include the following components:

- Groundwater recirculation wells (including injection and extraction wells that may be used for either injection or extraction of groundwater)
- Groundwater recirculation system, including pumps, to extract groundwater and transfer it to the injection wells, interconnecting piping, and valves
- Electron donor amendment system including the tanks, pumps and piping needed to add electron donor to the groundwater as it is being injected into the injection well
- System instrumentation and controls

7.2.1.1 Recirculation Wells

A series of groundwater recirculation wells are required to deliver electron donor and to distribute the electron donor in the subsurface. The pattern of injection and extraction wells could consist of a line of wells perpendicular to the natural groundwater flow direction or could consist of other patterns of wells to distribute electron donor across a source area or other target treatment zone. A line of wells may include a single line or several lines of wells, each perpendicular to the direction of groundwater flow. A good understanding of the geology and hydrogeology at the site and a clear understanding of the objectives of the remedial program are important in developing an appropriate pattern of recirculation wells. The screened intervals of the wells must coincide with the target treatment interval, and the system may include separate wells with screens in distinct aquifer layers in the subsurface, if present.

In certain situations where the hydraulic conductivity of the subsurface is low or the geology is heterogeneous, it may be appropriate to consider the use of a groundwater recirculation trench, similar to a French Drain. Recirculation trenches can foster more rapid and uniform distribution of electron donors in the subsurface (Devlin and Barker, 1999). Such trenches typically contain inert media (i.e., coarse sand or fine gravel) with a hydraulic conductivity greater and more uniform than the native geological material, and are installed perpendicular to the direction of groundwater flow. Recirculation wells in such trenches can be used to circulate groundwater at a much faster rate than would be possible in low hydraulic conductivity native material, and electron donors can be distributed quickly across the target treatment area. Such an approach avoids potential difficulties in distributing electron donor in low hydraulic conductivity geological units.

7.2.1.2 Groundwater Recirculation System

The groundwater recirculation system serves to extract groundwater and direct it to the groundwater injection wells. The system must include pumps to extract groundwater and pump it to injection wells, interconnecting piping and valves. Groundwater extraction pumps are typically submersible, multi-staged centrifugal pumps, sized based on the flowrate and head (discharge) pressure required to lift the groundwater from the extraction well and direct the flow into the injection well(s). Semi-passive systems normally operate for a limited period of time, and the pumps and interconnecting piping need not be a permanent installation. In many situations it may be most cost effective to construct a simple portable, trailer-mounted system that can be used at one set of wells and subsequently transported and set up at another set of wells for electron donor injection.

7.2.1.3 Electron Donor Amendment System

An electron donor amendment system is used to add electron donor to the groundwater as it is being pumped into the injection well or to add electron donor directly into the injection well or other wells within the influence of the groundwater recirculation system. The dosage of electron donor should be adjusted to provide sufficient donor to support biodegradation of perchlorate but low enough to minimize the mobilization of certain metals and creation of secondary water quality issues. The dosage of electron donor must be sufficient to reduce any dissolved oxygen and nitrate as well as perchlorate and its intermediate byproducts (chlorate and chlorite).

In situ anaerobic biological treatment systems also can be used to degrade co-contaminants such as nitrate and chlorinated solvents. Nitrate will be reduced under conditions very similar to those that are necessary for perchlorate biodegradation, and no special measures are necessary to treat nitrate. Chlorinated solvents, such as chlorinated ethenes, may also be treated using anaerobic biological treatment, but the reduction of chlorinated solvents will require a lower oxidation reduction (redox) environment than is required for perchlorate biodegradation, and may require bioaugmentation to provide appropriate organisms required to degrade chlorinated solvents effectively. The dosage rates for electron donors will need to be higher if chlorinated solvents are present and are targeted for treatment.

The system for adding electron donors can be as simple as manual addition of electron donor to the injection wells or other wells within the influence of the groundwater recirculation system. However, it may be more complicated, or may include an electron donor storage tank with an automated electron donor dosing pump to automatically provide the target dose of electron donor based on the actual flow of groundwater being recirculated during the active phase of operation. Complex systems may require less operating labor but are more costly to construct and may require significant amounts of labor to maintain. Simple systems may require more labor and have less equipment to maintain, but they can also provide maximum flexibility for desired changes in operation and are significantly less expensive.

The case study presented later in this chapter describes work done at the Longhorn Army Ammunitions Plant where groundwater was recirculated on a continuous basis and electron donor was added to the injection wells and wells between the injection and extraction wells on a manual basis three times a week for three weeks. This simple system required several hours of labor to add the electron donor but avoided the costs of storage tanks and dosing pumps as well as the labor costs for maintaining such equipment.

7.2.1.4 Instrumentation and Controls

Instrumentation and controls are required to control the groundwater recirculation system and may be required to control the electron donor addition system. Basic instrumentation and controls for the groundwater recirculation system typically include: (1) flowmeters to monitor the rates of groundwater extraction and injection; (2) flow control valve or pump controller to adjust the rate of groundwater extraction and injection; (3) water level sensor in the extraction well to shut off the groundwater extraction pump if the water level in the extraction well drops too low; and (4) water level sensor in the injection well to shut off the groundwater extraction pump if the water level in the injection well rises too high. More complex instrumentation and controls may be used to increase the level of automation, or allow for remote data acquisition or remote control.

Instrumentation and controls for electron donor addition systems may be very simple for systems where manual addition of electron donor is used, but will be more complex for systems where automated control is required. More complex systems may include instrumentation and controls to: (1) monitor the supply of electron donor in the storage tanks; (2) meter the appropriate dose of electron donor to the recirculating groundwater; (3) adjust the dose of electron donor based on real time measurements of the groundwater recirculation rate; and (4) shut down the system in response to varying upset conditions.

7.2.2 Site Assessment Needs

As with any *in situ* groundwater remediation technology, it is critical to have a good understanding of the contaminant distribution, hydrogeology and groundwater geochemistry in the target treatment area. Pre-design investigations are likely to include: (1) a site assessment to define the extent (both horizontal and vertical) of contamination; (2) multiple sets of water level measurements to identify groundwater flow directions and magnitude and possible seasonable variations; (3) aquifer testing to assess hydrogeological characteristics of the subsurface geology; and (4) an assessment of the geochemistry of the groundwater.

A good understanding of the horizontal and vertical distribution of target contaminants is obviously important in the design of any remediation system as it will impact the extent and location of the treatment area and the appropriate configuration. It is also important to understand the location and nature of the source of perchlorate contributing chemicals to groundwater as it will impact the configuration, the objectives and the required duration of operation of the groundwater remediation system. If perchlorate is leaching from soil above the water table, the concentrations of perchlorate in groundwater may be sustained for a considerable period of time. Treatment of perchlorate in groundwater often will be integrated with treatment of chemicals present in source areas above the water table. Perchlorate concentrations may be sustained for a considerable period of time if perchlorate has diffused into low permeability geological media in the subsurface due to perchlorate back diffusion.

Design of the groundwater recirculation systems (wells and pumps) should be supported with a site-specific hydrogeological model using pump test data. Long-term pump tests may add significantly to pre-design costs but will decrease uncertainty in system design. Data obtained during the early stages of operation will help guide modifications in operating parameters such as recirculation rates and groundwater recirculation patterns.

The geochemistry of groundwater at a site will determine the appropriate dosage of electron donor and other amendments. Electron donor delivery must meet electron donor demand from all sources including reduction of perchlorate. Electron demand can be exerted by dissolved oxygen (DO), nitrate, sulfate and subsurface minerals, especially iron. Sites with high DO and nitrate concentrations require additional electron donor beyond that needed for reduction of perchlorate. Sulfate reducing bacteria that reduce sulfate to sulfide consumes electron donor; however, the oxidation-reduction potential (ORP) for this process is significantly lower than for perchlorate reduction. Hence, sulfate reduction is usually discounted in determining electron donor dosage except in the vicinity of electron donor injection where electron donor concentrations in excess of that needed for perchlorate reduction are unavoidable. In addition, in the early phase of operation of a semi-passive bioremediation system, some electron donor will be consumed by iron reducing bacteria that reduce iron minerals in geological media common in aquifers.

As a result of the difficulty in accounting for all possible electron donor demands, in practice it is common to estimate electron donor demand based on the concentrations of DO, nitrate and perchlorate and then apply a safety factor of at least ten to twenty. Monitoring data obtained during the initial stages of operation should be used to optimize successive additions of electron donor to achieve reduction of perchlorate without overdosing and creating secondary water quality issues.

Some electron donors (e.g., sodium lactate) can cause an increase in inorganic constituents such as sodium. Potential impacts can be evaluated by modeling during the design phase. If predicted impacts on secondary groundwater quality are significant, alternate electron donors, such as citric acid or ethanol, should be considered.

7.2.3 Groundwater Modeling

Groundwater modeling is an important step in the design of any *in situ* groundwater remediation system. Groundwater models also can be useful in evaluating and interpreting operating data to determine if modifications should be made to any of the operating parameters. Information on site geology and hydrogeology is required to develop a groundwater model. The accuracy and usefulness of the groundwater model will be only as good as the input data available.

Once a basic groundwater model has been developed for the target treatment area, various groundwater recirculation scenarios can be evaluated. Model simulations can be used to evaluate different well configurations, well spacings, and groundwater recirculation rates to determine a range of travel times between injection and extraction wells. The results of the groundwater modeling, along with knowledge of the costs for various components of the system, will support development of an optimal remediation system design.

The groundwater model also can be very useful in evaluating the results of system operation. Data from an initial tracer test and from the initial stages of operation can be used to calibrate or otherwise modify the groundwater model to provide more accurate representation of the characteristics of the site. This model, as it is improved with site-specific data, can be used to evaluate potential modifications in operating parameters such as the groundwater recirculation rates.

7.2.4 Tracer Testing

A groundwater tracer test can be an important step in confirming the results of groundwater modeling used to design the groundwater recirculation system. The tracer test can be conducted using conservative tracers, such as bromide or iodine, to evaluate groundwater flow directions and velocity under active groundwater recirculation conditions and under passive natural gradient conditions. Operating parameters for the system can be modified, if necessary, based on analysis of the results of the tracer testing.

7.2.5 Operation and Maintenance

Active operation of the semi-passive *in situ* bioremediation system is required only during the groundwater recirculation and electron donor amendment phase. The system must be operated to recirculate groundwater and to add electron donor. The amount of operator attention required will depend upon the degree of automation built into the system and the degree to which unexpected operating issues arise with the system. The groundwater recirculation component of the system is fundamentally very simple and often requires very little operator attention given the short duration of the operation, typically several days to several weeks at a time.

If a simple manual electron donor system is being used, operation may simply involve addition of electron donor to injection wells on a predetermined cycle such as once per day or once every several days. Addition of electron donor may involve manually pouring electron donor into injection wells or operating electron donor dosing pumps for a short period of time on a predetermined cycle. If equipment for electron donor addition is automated, then operation may simply involve checking to make sure that the supply of electron donor is sufficient and that the system is operating properly. Given the short duration of the active phase of operation relative to the passive phase, the overall operating requirements of a semi-passive system can be small relative to systems with active full-time groundwater recirculation.

7.2.6 Monitoring

Monitoring of semi-passive systems involves: (1) monitoring the aboveground components of the groundwater recirculation and electron donor amendment systems to assure they are operating as intended; and (2) monitoring the groundwater to determine if the electron donor is being distributed as intended and that the electron donor is creating appropriate conditions for biodegradation of the target compounds. Monitoring of the aboveground components of the systems will include activities such as ensuring that: (1) the groundwater recirculation flow rates are within the target range; (2) the water levels in the extraction wells are not dropping close to the extraction well pump intakes; (3) the water levels in the injection wells are not increasing (indicating potential fouling of the injection wells); (4) the pressure drop across any filters is not increasing, indicating the need to replace or backwash the filters; (5) the supply of electron donor is sufficient for the automatic dosing systems; and (6) the electron donor is being added at the target dosage rate.

Monitoring of the groundwater will include sampling from extraction and monitoring wells for: (1) field parameters including ORP, pH, DO, and specific conductance; (2) perchlorate; (3) electron donor such as volatile fatty acids (VFAs); (4) other anions including nitrate, nitrite, sulfate, sulfide and chloride; and (5) dissolved metals including iron and manganese.

Measurement of ORP in groundwater can provide a simple method of monitoring the impact of the addition of electron donor. The ORP generally will drop quickly in response to the addition of adequate quantities of electron donor, and a negative ORP is required to promote degradation of perchlorate. ORP is a simple measurement, and frequent measurements can provide valuable data on impacts within the subsurface. If the ORP is not sufficiently reduced in the target injection areas, then injection of higher concentrations of electron donor or modifications to the groundwater recirculation pattern or system may be required to improve distribution of the electron donor. The ORP is typically low in proximity to areas where electron donor has been added to the subsurface and where perchlorate degradation may be occurring. The ORP may increase back to ambient levels within a short distance downgradient of a biobarrier.

Measurements of perchlorate in groundwater provide the actual demonstration that the *in situ* bioremediation system is working as designed. Perchlorate degradation should occur in the areas with reduced ORP within the biobarrier or treatment zone. As treated water flows downgradient of a biobarrier system, concentrations of perchlorate will also decline. The rate of decline in the concentrations of perchlorate downgradient of a biobarrier may be slowed by the release of perchlorate from the geological media downgradient, but concentrations should decline as treated groundwater flows from the biobarrier and perchlorate is flushed from the areas downgradient of the biobarrier.

Measurements of VFAs may provide another indication of the distribution of electron donor in the subsurface. It is necessary to analyze for a range of VFAs, such as acetate and propionate, as natural biological activity in the subsurface can convert added electron donor (e.g., lactate or ethanol) to metabolic intermediates (e.g., acetate) that can also serve as effective electron donors.

Measurements of anions and metals will provide an indication of the impacts of the addition of electron donor on groundwater quality. The addition of electron donor will promote the reduction of nitrate in addition to the reduction of perchlorate. The addition of significant quantities of electron donor may also promote the reduction of sulfate to sulfide. Elevated concentrations of electron donor may also create conditions that mobilize certain metals such as iron and manganese. Monitoring of iron and manganese concentrations in groundwater within the biobarrier or treatment zone and downgradient of the biobarrier or treatment zone will provide an indication of the degree to which metals are mobilized and the degree to which they are subsequently removed from the groundwater as the geochemical conditions, such as ORP, are re-established downgradient of the biobarrier or treatment zone.

7.2.7 Health and Safety

Health and safety issues with semi-passive EISB systems are not significantly different than for passive or active EISB systems. Standard precautions are required for installation of extraction and injection wells and construction and operation of groundwater extraction pumps at sites with contaminated groundwater. Special safety precautions will need to be taken with handling and storage if flammable electron donors, such as ethanol or methanol, are used. The semi-passive EISB systems are unlikely to require aggressive and potentially harmful chemicals that may be needed for biofouling control or rehabilitation of injection wells such as might be required for active EISB systems.

7.3 CASE STUDY: SEMI-PASSIVE BIOREMEDIATION OF PERCHLORATE AT THE LONGHORN ARMY AMMUNITIONS PLANT

7.3.1 Demonstration Test Procedures

A demonstration of the semi-passive bioremediation approach was conducted at the LHAAP (the "Site") in north-eastern Texas starting in 2004. A semi-passive biobarrier was constructed downgradient of a former landfill, where earlier investigations identified a 76-m (250-ft) wide perchlorate plume with concentrations exceeding 2,000 micrograms per liter (μg/L). A shallow aquifer extends to a depth of approximately 10 m (~35 ft) bgs within interbedded sand, silt and clay, with a groundwater velocity in the treatment zone of approximately 11 m/yr (~37 feet per year [ft/yr]). Nitrate concentrations varied across the treatment area with some concentrations as high as 8 to 15 milligrams per liter (mg/L) to the north and generally less than 0.4 mg/L to the south. Sulfate concentrations were generally very high and varied significantly across the treatment area. Some areas had sulfate concentrations as high as 3,000 to 4,000 mg/L and others as low as 200 mg/L. Background DO concentrations were typically between 1 and 4 mg/L, and the background ORP was typically between +50 and +150 millivolts (mV).

Five 10-centimeter (cm) (4-inch [in]) diameter groundwater recirculation wells, each with 3 m (10 ft) long screens, were installed on 10 m (35 ft) centers along a line perpendicular to the direction of groundwater flow. The wells were designed to be used as either injection or extraction wells, depending upon the selected groundwater recirculation pattern. The top of each well screen coincides with the area's water table, at about 4.6 m (15 ft) bgs. Two 5-cm (2-in) diameter intermediate injection wells with 4.6 m (15 ft) screens were installed between each of the adjacent recirculation wells. The intermediate wells were installed to reduce the time needed to operate the groundwater recirculation system by allowing electron donor to be added at more locations along the biobarrier. Figure 7.2 shows the locations of the recirculation wells, intermediate injection wells and performance monitoring wells in the vicinity of the demonstration area at the Site.

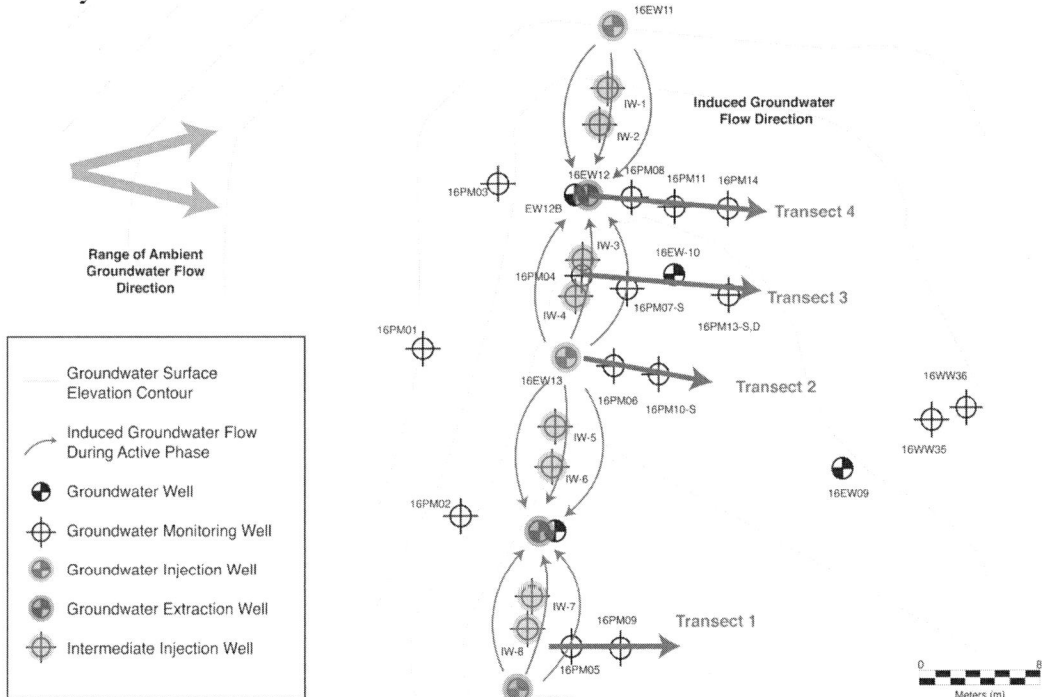

Figure 7.2. Layout of demonstration test area at LHAAP

A groundwater recirculation system, including two extraction pumps, flowmeters and piping, to split the flow from the extraction wells into the injection well, was constructed at the site. The groundwater extraction pumps were set to extract groundwater at the maximum sustainable yield (about 1 to 2 gallons per minute [gpm] or 3.8 to 7.6 liters per minute [L/min]) of the two extraction wells initially used for extraction of groundwater. These extraction well pumping rates were lower than initially predicted based on available hydraulic data for the Site but were sufficient to provide recirculation of groundwater for the distribution of electron donor in the biobarrier.

Hydraulic data from the Site were used to develop a simplified numerical groundwater flow and transport model (using VisualMODFLOW) for the demonstration area. The model allowed for simulation of a variety of operating scenarios (groundwater flow patterns which could be induced by extraction and injection of groundwater from different recirculation wells) to confirm that groundwater could be recirculated in the subsurface between injection and extraction wells and to provide an estimate of the travel time between adjacent wells.

The groundwater recirculation system was started in March 2004 by extracting groundwater from EW-12B and EW-14B at rates of 1.0 gpm (3.8 L/min) and 1.7 gpm (6.4 L/min), respectively. Groundwater was injected into EW-11, EW-13 and EW-15 at rates of 1.0 gpm (3.8 L/min), 0.85 gpm (3.2 L/min) and 0.85 gpm (3.2 L/min), respectively.

A tracer test was conducted using both bromide and iodine for two reasons: (1) to confirm the transport velocities of amended water from the injection wells along the biobarrier during the active recirculation phase; and (2) to determine the time needed to achieve lateral coverage across the entire electron donor delivery zone. The tracer test was conducted using bromide in the two injection wells furthest from the center of the biobarrier and iodide in the injection well in the center of the biobarrier. The use of different tracers made it possible to determine the travel times from different injection wells to the intermediate and extraction wells.

The first cycle of electron donor addition was initiated on March 25, 2004, with the groundwater recirculation system in operation. The dosage of electron donor used was calculated based on the geochemistry of the groundwater including the DO, nitrate, and perchlorate concentrations. The amount of electron donor added in the first cycle was calculated based on the amount of electron donor required to reduce DO, nitrate and perchlorate in the groundwater moving into the biobarrier for a period of eight months with a safety factor of 28. A higher than normal safety factor was used to account for electron donor consumed by: (1) the demand of non-target compounds including the very high concentrations of sulfate; (2) the demand of minerals present in the native geological material; and (3) normal microbiological metabolic processes.

A solution containing 60% by weight of sodium lactate was added to the injection wells and each of the eight intermediate injection wells by pouring the sodium lactate solution directly into the wells three times per week for a period of three weeks, from March 25, 2004 until April 14, 2004. A total of 273 gallons (gal) (1,033 L) (2,980 pounds [lb]) (1,352 kilograms [kg]) of 60% sodium lactate was added to the wells during the first injection cycle. Following the addition of sodium lactate, the recirculation system was turned off and the system was allowed to operate in a passive mode for 7.5 months.

The recirculation system was activated again in early December 2004, using the same extraction and injection configuration used during the first cycle of electron donor amendment. During the second cycle of amendment, a total of 443 gal (1,677 L) (4,830 lb, or 2,190 kg) of 60% sodium lactate was added to the two injection wells and eight intermediate wells three times per week over a three-week period. A larger dose of electron donor was used in the second cycle of addition in order to achieve a greater and more sustained reduction in the ORP in the second cycle. The recirculation system was shut off after the addition of electron donor and the system was again allowed to operate in a passive mode.

The third amendment cycle was conducted in November and December 2005. The groundwater recirculation pattern was modified during the third amendment cycle to provide higher quantities of electron donor to one segment of the biobarrier which appeared to have received less than the target dosage of electron donor during the first and second amendment cycles. Groundwater was extracted from EW-14B at a rate of 1.7 gpm (6.4 L/min), and the entire flow was injected into EW-12B. A second tracer test was conducted to evaluate the forced gradient groundwater flow conditions of the modified recirculation pattern. During the third cycle of amendment, a total of 1,110 gal (4,202 L) (12,000 lb, or 5,443 kg) of 60% sodium lactate was added to the one injection well, eight intermediate wells and three recirculation wells, which were not being used for extraction or injection of groundwater, three times per week over a three-week period. The recirculation system was shut off after the addition of electron donor and the system was again allowed to operate in a passive mode.

7.3.2 Demonstration Test Results

The results of groundwater flow modeling based on the recirculation scenario used for the first and second injection cycles showed that groundwater could be recirculated in the subsurface from the injection wells to the extraction wells perpendicular to the natural flow of groundwater. The model outputs for the first and second injection cycles show a high density of flow lines between extraction and injection wells in most areas of the biobarrier, but a lower density of flow lines was seen in one area of the biobarrier as a result of the diversion of some portion of the groundwater injected into EW13 to the south, towards EW-14B which was able to operate at a higher extraction rate (~1.7 gpm or 6.4 L/min) than EW-12B (~1.0 gpm or 3.7 L/min). The travel time for groundwater between adjacent electron donor amendment wells, including intermediate injection wells in most areas of the biobarrier, was approximately one to three weeks. The travel time for groundwater between injection wells in the area where the model showed a lower density of flow lines was approximately two to four weeks, suggesting that the distribution of electron donor in this segment would not be as effective as in other areas of the biobarrier.

The results of groundwater flow modeling, based on the recirculation scenario used for the third injection cycle, showed that groundwater could be recirculated in the subsurface from a single injection location to the two extraction wells oriented perpendicular to the natural flow of groundwater. The model output shows a high density of flow lines between the extraction and injection wells. The travel times for groundwater between the adjacent electron donor amendment wells (intermediate injection wells and recirculation wells not used for groundwater recirculation) during the third cycle of electron donor amendment ranged between from less than one week to about three weeks.

The results of the first tracer test showed travel times of approximately one to two weeks in most areas of the biobarrier between the injection well and the first intermediate well (located 4.6 m [15 ft] from the injection well). These results were consistent with the groundwater modeling results. The tracer results from the area of the biobarrier, which modeling suggested would have slower groundwater flow, showed slower movement of the tracer. The tracer test and groundwater modeling both indicated slower movement of groundwater in this area as a result of some of the water injected into EW-13 being pulled back towards the south into the higher pumping EW-14B because EW-12B could not sustain as high a yield. The results of the second tracer test, conducted prior to the third cycle of electron donor amendment, showed travel times between the electron donor addition points across the biobarrier to be approximately one to two weeks, consistent with the results of groundwater modeling of this recirculation scenario.

Perchlorate concentrations in groundwater samples collected prior to addition of electron donor ranged from non-detect up to 2,000 µg/L in the upgradient monitoring well PM-03. The ORP of groundwater samples collected prior to addition of electron donor were generally high (greater than +150 mV).

Significant reductions in perchlorate concentrations were achieved across the line of recirculation wells in the semi-passive biobarrier (Table 7.1). The concentrations of perchlorate in some of the monitoring wells further downgradient of the biobarrier were not reduced to the same extent as in monitoring wells closer to the biobarrier in the first and second injection cycle. This may be a result of perchlorate diffusion out of low hydraulic conductivity units downgradient of the biobarrier, insufficient delivery of electron donor to this area or mixing with untreated water containing perchlorate. Perchlorate concentrations were reduced more significantly following the third cycle of electron donor addition as a result of the increased dose and better distribution of electron donor.

Table 7.1. Summary of Groundwater Monitoring Results at Site 16 Landfill, LHAAP, Karnack, Texas

Well ID	Date	ORP (mV)	Perchlorate (µg/L)	Well ID	Date	ORP (mV)	Perchlorate (µg/L)
Upgradient				Downgradient			
16PM03	23-Mar-04	643	1,690	16PM13-S	23-Mar-04	-	< 4.0
16PM03	10-Mar-05	66	1,180	16PM13-S	14-Mar-06	47	< 4.0
Downgradient				16PM13-D	23-Mar-04	-	220
16EW12B	24-Mar-04	223	1,040	16PM13-D	14-Mar-06	16	< 4.0
16EW12B	14-Mar-06	-32	< 4.0	16PM06	23-Mar-04	-	968
16PM08	23-Mar-04	132	129	16PM06	14-Mar-06	-7	7
16PM08	14-Mar-06	-	< 4.0	16PM10-S	24-Mar-04	227	669
16PM11	23-Mar-04	216	161	16PM10-S	14-Mar-06	-62	8
16PM11	14-Mar-06	-	< 4.0	16EW14B	24-Mar-04	206	1,000
16PM14	23-Mar-04	250	428	16EW14B	09-Mar-05	-178	< 4.0
16PM14	14-Mar-06	-	4	16PM05	24-Mar-04	216	883
16PM04	23-Mar-04	417	286	16PM05	14-Mar-06	-37	< 4.0
16PM04	14-Mar-06	-111	< 4.0	16PM09	24-Mar-04	206	918
16PM07-S	23-Mar-04	-	39	16PM09	14-Mar-06	75	< 4.0
16PM07-S	14-Mar-06	-9	10	16EW09	24-Mar-04	108	749
16EW10	23-Mar-04	-	111	16EW09	14-Mar-06	60	< 4.0
16EW10	14-Mar-06	20	< 4.0				

Notes: ORP—Oxidation-Reduction Potential; mV—millivolt; µg/L—micrograms per liter
    "-" indicates that no measurement was taken

Concentrations of perchlorate in Transect 1 monitoring wells 16PM05 and 16PM09 (Figure 7.3) were in the range of 900 to 1,100 µg/L before the addition of electron donor. Following the addition of electron donor, the concentrations of perchlorate decreased rapidly (over about 1 month) to less that 200 µg/L and continued to decline over the following two months. Low concentrations of perchlorate were maintained through the beginning of December 2004 when the second amendment of electron donor was conducted. Concentrations of perchlorate remained generally low following the second addition of electron donor, then increased significantly immediately following the start up of the groundwater recirculation system for the third addition of electron donor. It is believed that the operation of the groundwater recirculation system began to draw groundwater containing very high concentrations of perchlorate into the vicinity of Transect 1. The concentrations of perchlorate dropped to non-detect in the sampling event following the third addition of electron donor and operation under passive conditions.

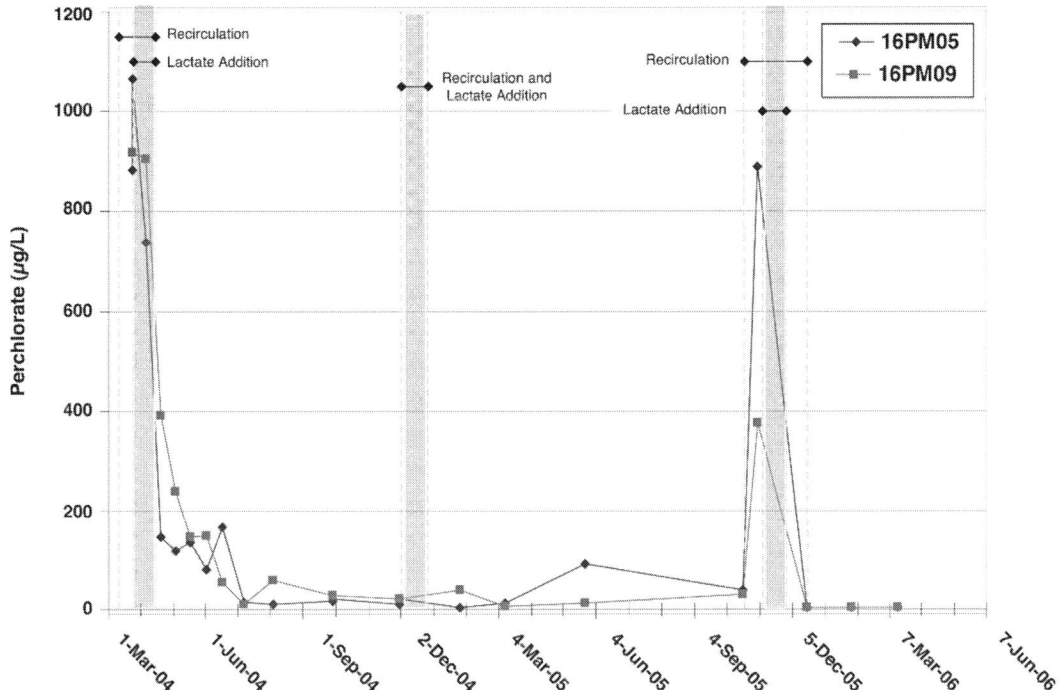

Figure 7.3. Perchlorate concentrations over time in monitoring transect one (modified from Krug et al., 2007)

Concentrations of perchlorate in Transect 4 monitoring wells EW-12B, PM-08 and PM-11 are shown in Figure 7.4. The perchlorate concentration in the extraction well (EW-12B) was in the range of 1,000 to 1,100 µg/L before and during the initial addition of electron donor in April 2004. The perchlorate concentrations in the monitoring wells (PM-08 and PM-11) were in the range of 100 to 200 µg/L before and during the initial addition of electron donor in April 2004. Following the first addition of electron donor, the concentration of perchlorate in monitoring wells in this transect decreased significantly, and the concentrations were reduced even further following the second addition of electron donor, but the concentrations achieved were not as low or as consistent as was seen in the other transects.

Transect 4 is located directly downgradient of extraction well EW-12B and at the greatest distance from an electron donor injection well compared to the other monitoring well transects. It is believed that the amount of electron donor added to the biobarrier in the vicinity of this transect was insufficient to obtain the target treatment concentration of perchlorate. The injection pattern was therefore modified during the third addition of electron donor to provide better distribution of electron donor in this area. Following the third addition of electron donor, the concentrations of perchlorate in monitoring wells were reduced more significantly and consistently than had been observed previously. The perchlorate data support other data in demonstrating that the modified injection pattern was effective in achieving distribution of electron donor throughout the biobarrier.

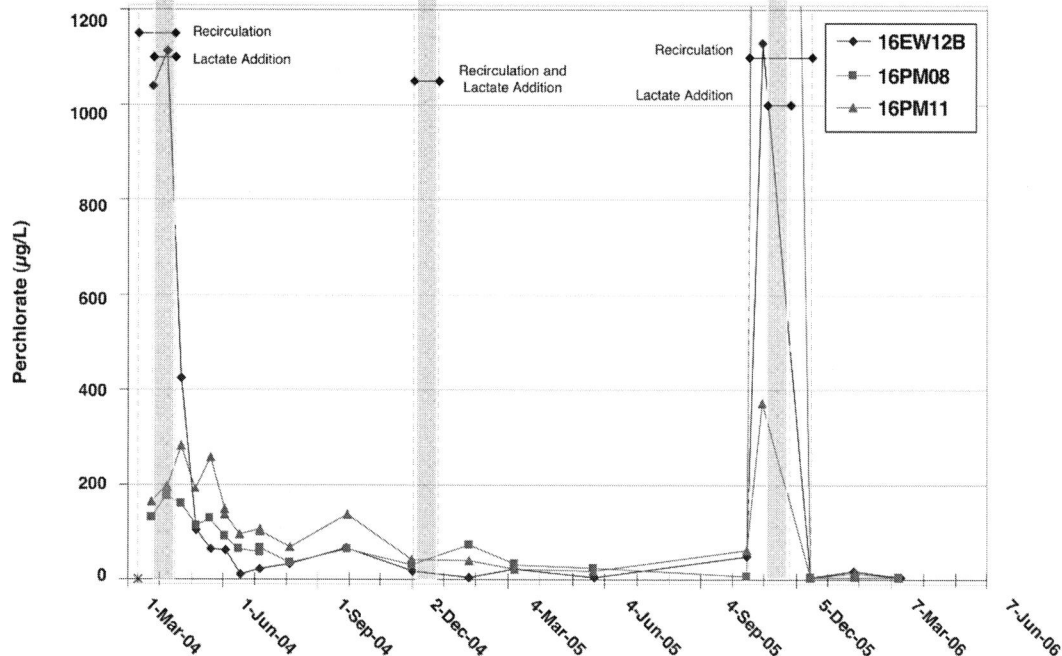

Figure 7.4. Perchlorate concentrations over time in monitoring transect four (modified from Krug et al., 2007)

The ORP was reduced significantly following addition of electron donor. Following the first and second additions of electron donor, significant reductions were observed in Transects 1 and 2, whereas less significant reductions were seen in Transects 3 and 4. Following the third addition of electron donor, significant reductions in ORP were sustained or achieved in all monitoring transects.

The ORP in Transect 1 monitoring wells PM-05 and PM-09 (Figure 7.5) were in the range of +200 mV before the addition of electron donor. Following the addition of electron donor, the ORP decreased rapidly (over about 1 month) to about 0 mV then rose slowly over the next few months to a level of about +100 mV. The ORP remained at about 100 mV until after additional electron donor was added to the wells in December 2004. Following the second addition of electron donor in December 2004, the ORP declined sharply to about -20 mV in PM-05 and +20 mV in PM-09. As can be seen in Figure 7.5, the drop in ORP was greater and was sustained longer following the second amendment with electron donor. It is believed that electron donor from the first amendment cycle reduced minerals present in the geological media and produced additional biomass that allowed for the second addition of electron donor to produce a greater and more sustained reduction in ORP. The concentration of electron donor added during the second amendment cycle was also 60% greater than during the first amendment cycle. The reduction in ORP was sustained following the third addition of electron donor.

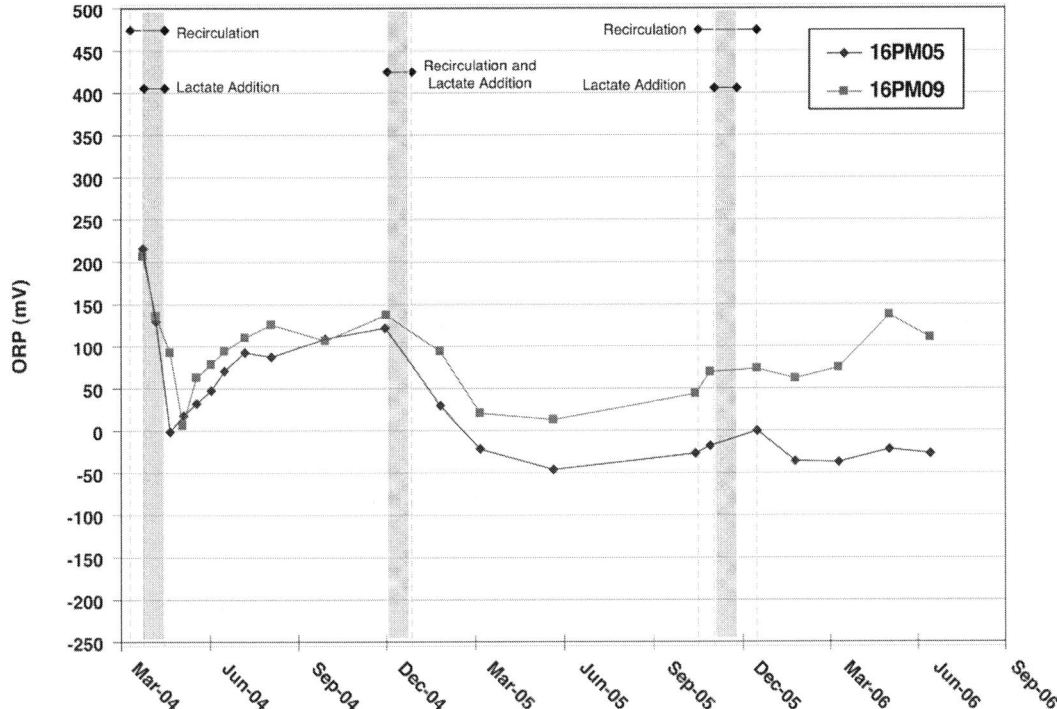

Figure 7.5. ORP over time in monitoring transect one (modified from Krug et al., 2007)

The ORP in Transect 4 extraction well EW-12B (Figure 7.6) appeared to rise to +450 mV immediately following the addition of electron donor although this high a value is not consistent with baseline ORP measurements in the vicinity of this well which were generally in the range of +200 mV. Shortly thereafter, the ORP dropped into the negative range and remained negative until December 2004 when it increased slightly into the positive range (10 mV). Following the second addition of electron donor in December 2004, the ORP declined to -200 mV. The ORP in monitoring wells downgradient of the centerline of the biobarrier in Transect 4 had an ORP generally in the range of 150 mV to 250 mV before addition of electron donor, and little change was observed following the first addition of electron donor. Following the second addition of electron donor in December 2004, the ORP in PM-11 declined slightly to about 55 mV. Following the third addition of electron donor, there was a significant and sustained drop in the ORP in injection well EW-12B and significant declines in the ORP in the two downgradient monitoring wells PM-11 and PM-08. The results demonstrate that the injection pattern used during the third addition of electron donor was much more effective in distributing electron donor in the vicinity of Transect 4 than the injection pattern used during the first and second amendment cycle.

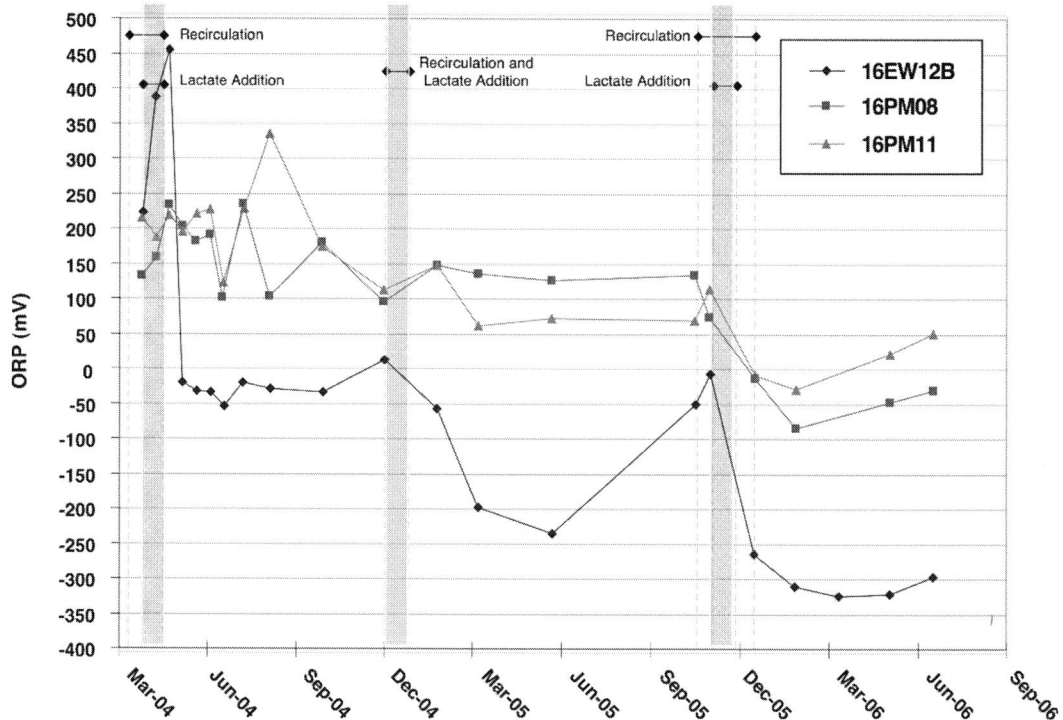

Figure 7.6. ORP over time in monitoring transect four (modified from Krug et al., 2007)

7.3.3 Conclusions of Case Study

The following conclusions can be drawn based on the results of the demonstration to date:

- Microorganisms capable of perchlorate degradation responded quickly following addition of sodium lactate as electron donor.

- Significant reductions in perchlorate concentrations were measured within and downgradient of the biobarrier. The concentrations of perchlorate were reduced following each of the three cycles of addition of electron donor. Following the third cycle of electron donor addition, perchlorate concentrations were reduced from levels in excess of 800 μg/L to below detection levels of <4.0 μg/L in 10 of 14 monitoring wells, with a maximum concentration in the other monitoring wells of only 10 μg/L.

- The ORP in the groundwater in the treatment zone decreased from oxidizing conditions (values were > +150 mV) to reduced conditions (<0 mV) following addition of the electron donor. ORP can provide a simple real-time field measurement of the extent and distribution of electron donor influence. The reduction in ORP increased in magnitude following each successive addition of electron donor, probably because of the reduction of geological materials, the growth of biomass in response to the electron donor additions and the higher concentrations of electron donor used.

- Dissolved iron, manganese and arsenic concentrations increased in the biobarrier but decreased to ambient levels generally within 9 m (30 ft) downgradient of the center line of the injection and extraction wells.

- No indications of biofouling were observed during groundwater recirculation and addition of electron donor, and no measures were necessary to maintain the ability of the injection wells to accept groundwater.

7.4 SUMMARY

Semi-passive enhanced *in situ* bioremediation of perchlorate involves groundwater recirculation and the addition of electron donor on a periodic basis to stimulate natural microbiological populations. The approach is similar to active EISB in that groundwater is recirculated between injection and extraction wells. However, with the semi-passive approach, groundwater is recirculated for an "active phase" of a limited duration to distribute the electron donor and then the recirculation system is shut off for a "passive phase" of longer duration. The extraction and injection wells may be designed to create a biobarrier perpendicular to the direction of groundwater flow or to distribute electron donor throughout source areas or other target treatment zones.

The semi-passive strategy is also similar to "passive" bioremediation in that the system is allowed to operate under natural groundwater flow conditions for most of the time. The "active phase" of the semi-passive approach can, however, allow for a better distribution of electron donor than with the "passive" approach, and semi-passive approaches are less likely to cause undesirable impacts on secondary water quality because a smaller amount of electron donor is added during each electron donor amendment event than is typically used with the passive approach. Semi-passive systems can be significantly less expensive than continuous active recirculation because the pumping system is not operated on a continuous basis and biofouling of the injection wells is not generally a problem.

The semi-passive bioremediation approach can integrate the best aspects of both the active approach (wider well spacings and less impact on secondary water quality characteristics) and the passive approach (minimal permanent *ex situ* infrastructure, lower O&M) in order to optimize the balance of capital and O&M costs.

The semi-passive approach was successfully demonstrated at the LHAAP site. At this site, intermittent injections of sodium lactate were used to create a biobarrier that reduced perchlorate within and downgradient of the barrier to below regulatory criteria. The flexibility of the semi-passive approach proved valuable at the LHAAP site, as modifications to the injection system over time improved the distribution of the donor injections.

REFERENCES

Devlin JF, Barker JF. 1994. A semipassive nutrient injection scheme for enhanced in situ bioremediation. Ground Water 32:374–380.

Devlin JF, Barker JF. 1996. Field investigation of nutrient pulse mixing in an in situ biostimulation experiment. Water Resour Res 32:2869–2877.

Devlin JF, Barker JF. 1999. Field demonstration of permeable wall flushing for biostimulation of a shallow sandy aquifer. Ground Water Monitor Remediat 19:75–83.

Dybas MJ, Hyndman DW, Heine R, Tiedje J, Linning K, Wiggert D, Voice T, Zhao X, Dybas L, Criddle CS. 2002. Development, operation, and long-term performance of a full-scale biocurtain utilizing bioaugmentation. Environ Sci Tech 36:3635–3644.

Gierczak R, Devlin JF, Rudolph D. 2006. Combined use of laboratory and in situ hydraulic testing to predict preferred flow paths of solutions injected into an aquifer. J Contam Hydrol 82:75–98.

Gierczak R, Devlin JF, Rudolph D. 2007. Field test of a nutrient injection wall for stimulating *in situ* denitrification near a municipal water supply well. J Contam Hydrol 89:48–70.

Hyndman DW, Dybas L, Forney L, Heine R, Mayotte T, Phanikumar MS, Tatara G, Tiedje J, Voice T, Wallace R, Zhao X, Criddle CS. 2000. Hydraulic characterization and design of a full-scale biocurtain. Ground Water 38:462–474.

Krug T, Cox EE, Bertrand DM, Harre B. 2007. Demonstration of Active and Semi-Passive Electron Donor Addition for In Situ Bioremediation of Perchlorate. Presented at the Partners in Environmental Technology Technical Symposium & Workshop, Washington, DC, USA, December 4-6, 2007, poster #119. Abstract available at http://www.serdp.org/Symposium/upload/F_posters_2007_Wednesday_ALL_NEW.pdf. Accessed May 5, 2008.

Phanikumar MS, Hyndman DW, Criddle CS. 2002a. Biocurtain design using reactive transport models. Ground Water Monit Remediat 22:113–123.

Phanikumar MS, Hyndman DW, Wiggert D, Dybas L, Witt ME, Criddle CS. 2002b. Simulation of microbial transport and carbon tetrachloride biodegradation in intermittently-fed aquifer columns. Water Resour Res 38:1033–1046.

Phanikumar MS, Hyndman DW. 2003. Interactions between sorption and biodegradation: Exploring bioavailability and pulsed nutrient injection efficiency. Water Resour Res 39:1122–1135.

Phanikumar MS, Hyndman DW, Zhao X, Dybas L. 2005. A three-dimensional model of microbial transport and biodegradation at the Schoolcraft, Michigan, site. Water Resour Res 41:WO5011.

CHAPTER 8

PASSIVE BIOREMEDIATION OF PERCHLORATE USING EMULSIFIED EDIBLE OILS

Robert C. Borden[1] and M. Tony Lieberman[2]

[1]North Carolina State University, Raleigh, NC 27695; [2]Solutions-IES, Inc., Raleigh, NC 27607

8.1 INTRODUCTION

A variety of passive treatment approaches for managing perchlorate-contaminated sites have been developed in recent years (Lieberman et al., 2004; Knox et al., 2005). These approaches range from simply monitoring and documenting natural physical, chemical and biological processes as they occur (monitored natural attenuation [MNA]) to creating conditions in the contamination zone that will enhance biodegradation rates. Both MNA and passive bioremediation rely on natural biological activity in an aquifer to reduce perchlorate concentrations. However, unlike MNA, passive bioremediation involves the addition of a slow release organic substrate to stimulate the indigenous microbial population.

Biologically active treatment zones can be formed in an aquifer to treat source areas or control downgradient migration of dissolved perchlorate by injecting organic substrates through injection wells in a grid formation or through lines of temporary or permanent injection wells to form permeable reactive barriers (PRBs). This approach can be implemented with a variety of different hydrogen releasing substrates, including polymerized lactate products, neat edible oil, and emulsified edible oil. Solid substrates (e.g., chitin, cellulose, mulch) also can be emplaced in the subsurface to create a PRB or biowall. As groundwater is transported through the PRB or biowall by the natural hydraulic gradient, perchlorate is degraded by naturally occurring bacteria using the organic substrate as an electron donor. Although a variety of substrates are available for conducting passive bioremediation of perchlorate, this chapter focuses on the use of emulsified edible oils as an example of one strategy currently being applied.

After emplacement of substrate into the subsurface, the primary advantages of passive bioremediation systems are:

- Rapid establishment of reducing conditions *in situ*
- Long-lasting *in situ* treatment (depending on substrate used)
- No permanent aboveground remediation equipment
- Low operation and maintenance costs

The effectiveness of passive bioremediation systems is potentially limited by the absence of appropriate microorganisms, impacts on groundwater geochemistry, depth to groundwater and subsurface heterogeneity. Fortunately, in the case of perchlorate, many microbial genera with diverse metabolic capabilities are present in the environment. Perchlorate-reducing microorganisms are widespread in pristine and hydrocarbon-contaminated soils, aquatic sediments, and industrial and agricultural waste sludges (Gingras and Batista, 2002), and no site has been reported where the presence of perchlorate-reducing microorganisms could not be shown (Coates et al., 1999; Logan, 2001). Thus, at perchlorate-contaminated sites, microorganisms typically are not an important design consideration, and the need for bioaugmentation, often considered for other contaminants of concern, is not anticipated. However, proper design can account for and address the other issues of paramount importance for implementing passive bioremediation technology.

In order for site conditions to be suitable for passive bioremediation, it must be possible to effectively distribute the organic substrate in the subsurface and generate reducing conditions. Substrate distribution depends primarily on the hydraulic conductivity of the aquifer, the depth to groundwater and the groundwater flow direction and velocity. In many cases, systems can be designed to overcome difficult hydrogeologic conditions; however, doing so may increase the implementation costs. The ability to generate reducing conditions conducive to perchlorate biodegradation is dependent on the aquifer geochemistry. Specific factors to consider include dissolved oxygen (DO), oxidation reduction potential (redox), iron, nitrate, sulfates/sulfides, pH, and alkalinity. Sections 8.2 and 8.3 provide additional information on how these hydrogeologic and geochemical factors impact the use of emulsified oils.

8.2 DESIGN OF PASSIVE BIOREMEDIATION SYSTEMS

8.2.1 Treatment System Configurations

Passive bioremediation systems can be designed in different configurations to treat perchlorate-contaminated aquifers. The most common approaches are for source area treatment and plume control, using injections in a grid formation, a single PRB, or multiple PRBs (Figure 8.1). In choosing a treatment approach for a given site, it is important to understand the overall objectives of the project. The objectives may be to reduce contaminant concentrations below the maximum contaminant levels (MCLs), to reduce mass flux as part of an overall risk reduction approach or to limit plume migration.

8.2.1.1 Source Area Treatment

Emulsified oils can be distributed throughout a source area to reduce contaminant mass flux from the source area and to eventually treat the source zone. Oil injection will stimulate microbial activity, generating strongly reducing conditions and promoting anaerobic biodegradation of the target contaminants. Biodegradation of the aqueous phase contaminants is enhanced by dissolved organic carbon (DOC) released during the biodegradation of the edible oils. DOC is released when soybean oil (a triglyceride) hydrolyzes, releasing glycerol (an alcohol) and long-chain fatty acids (LCFAs). Glycerol is very soluble and relatively easy to biodegrade, so this material will be quickly consumed or will migrate downgradient with the ambient groundwater flow. The LCFAs are much less soluble in water and are hypothesized to initially sorb to sediment surfaces. Over time, the LCFAs are slowly fermented via beta oxidation, releasing acetate and H_2. The acetate and H_2 are then used as electron donors for biodegradation of perchlorate and other anaerobically biodegradable contaminants. The time required for complete contaminant biodegradation should be considered in the design of the emulsified oil application. Methods for estimating the volume of injected oil required are included in Section 8.2.2.

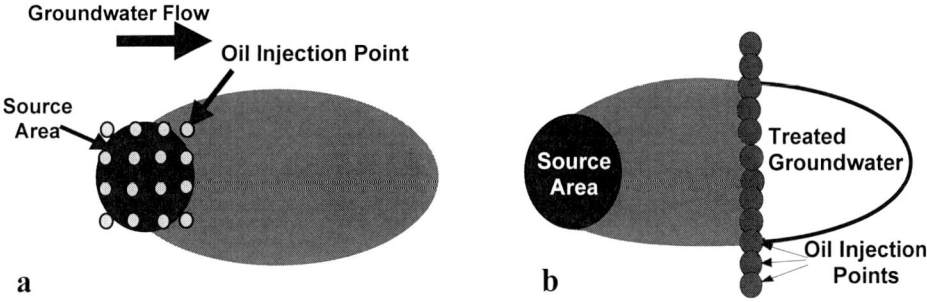

Figure 8.1 Use of emulsified oils to treat contaminated groundwater in (a) source areas using grid injection and (b) plumes using permeable reactive barriers.

A variety of different injection patterns can be used to treat source areas including uniform grids of injection wells, grids of injection and temporary extraction wells, or a series of PRBs spaced to repeatedly treat contaminated water as it flows through the area. Where the thickness of the contaminated zone is substantially greater than a typical well screen length (e.g., 1.5 to 6 meters [m] or 5 to 20 feet [ft]), injection wells can be constructed with screened intervals staggered at variable depths. For example, injection into a contaminated aquifer approximately 18 m (~60 ft) in thickness can be accomplished using three 6-m (20-ft) injection screens at staggered depths. All subsurface injection strategies should consider the distribution of contaminants in relation to subsurface heterogeneity. The presence of more permeable aquifer sediment layers or zones next to less permeable layers will lead to preferential flow into more permeable strata and could result in less than optimal substrate-to-contaminant contact (Figure 8.2).

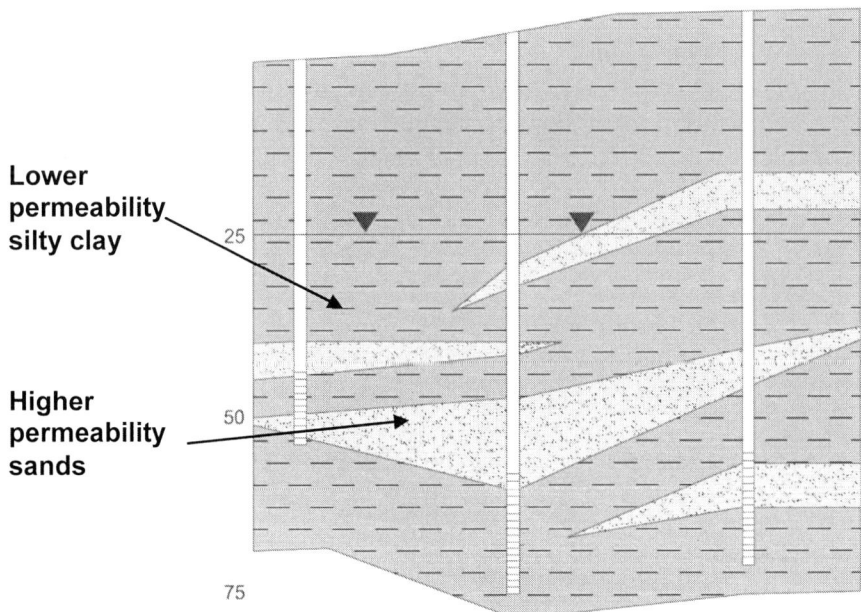

Figure 8.2 Cross-section showing zones of higher permeability. Emulsion injected into this aquifer will preferentially flow through the higher permeability zones leaving a portion of the lower permeability zones untreated.

8.2.1.2 Permeable Reactive Barriers

In cases where the contaminant source is poorly defined, source treatment may not be feasible. In other cases, it may be desirable to intercept a contaminant plume upgradient of a boundary or potential receptor. Under these conditions, emulsified oils can be injected in a reactive barrier configuration for plume containment through enhanced biodegradation. As with any permeable barrier configuration, the reaction zone must be uniformly distributed and an effort made to maintain the permeability of the reactive barrier. Edible oil emulsions can be effectively used to create permeable reactive barriers.

Residence time within the reactive zone of the barrier is controlled by the groundwater flow velocity and length of the oil treated zone along the direction of groundwater flow. At present, there is no reliable, all-inclusive method for determining the required contact time for effective treatment. Laboratory column and microcosm experiments and limited field studies suggest that perchlorate can degrade within 1 to 2 weeks in the presence of donor (ESTCP, 2006b). However, when emulsified oil barriers are used to treat mixtures of chlorinated solvents and perchlorate, longer contact times are desirable (Borden, 2007b).

Barriers are typically installed across the plume, perpendicular to groundwater flow. The barrier width (perpendicular to groundwater flow) should be wider than the contaminant plume to allow for uncertainties in the actual plume dimensions, variations in groundwater flow direction and some permeability loss. When using edible oil emulsions, the permeability loss associated with the actual emulsion injection is expected to be minor. However, biomass growth and gas production may result in up to an order-of-magnitude reduction in permeability (Long and Borden, 2006). Common groundwater flow and transport models (MODFLOW and MT3D) can be used to assess the impacts of permeability loss on barrier performance and determine the required barrier width to prevent contamination from bypassing the barrier. In most cases, up to a factor of ten reduction in permeability in a 6- to 12-m (20- to 40-ft) thick barrier is not a significant issue. However, the barrier width must be increased somewhat (typically 10 to 30%) to prevent a portion of the flow from bypassing around the edges of the barrier. Both the Interstate Technology & Regulatory Council (ITRC) and the Air Force Center for Environmental Excellence (AFCEE) have published documents providing in-depth discussions of permeable reactive barrier applications and design considerations (ITRC, 2005; AFCEE, 2007).

8.2.2 Planning and Design of Passive Bioremediation Systems

8.2.2.1 Amount of Substrate Required

Two main issues to consider in determining how much oil emulsion to inject into the subsurface are:
- Consumption of oil during biodegradation of the contaminants, reduction of competing electron acceptors (e.g., oxygen, nitrate, sulfate), and the downgradient release of dissolved organic carbon and methane; and
- Retention of emulsified oil by aquifer materials.

To determine the amount of oil required, the oil requirement for both biodegradation and retention by the aquifer materials should be calculated. To ensure sufficient substrate is applied to meet the desired design life under the site-specific biogeochemistry of the aquifer, the oil required will be the larger of the two amounts. When designing barriers using emulsions, oil retention by the subsurface matrix controls in lower velocity environments, while substrate demand for biodegradation commonly controls in very high flow rate systems with large amounts of competing electron acceptors.

Oil Consumption during Contaminant Biodegradation

The amount of oil required to support contaminant biodegradation will be a function of: (1) treatment zone dimensions, (2) site hydrogeology, (3) system design life, (4) amount of electron acceptors entering the treatment zone (both contaminants and naturally occurring electron acceptors), and (5) additional hydrogen demands and release of dissolved organic carbon to the downgradient aquifer. The "Protocol for Enhanced *In Situ* Bioremediation Using Emulsified Edible Oil" (ESTCP, 2006a) provides detailed information on the various calculations and potential safety factors that should be considered to estimate the amount of substrate required using site-specific data and design criteria. Based on these calculations the practitioner can determine the amount of substrate needed for a given site. Other approaches may be available, and as the science and engineering behind the edible oils technology evolves, new and improved tools will likely become available.

Oil Retention by Aquifer Materials

For effective treatment, edible oil emulsions must be distributed throughout the targeted treatment zone. There are a number of commercially available emulsions. For best performance, the emulsion should be composed of small oil droplets (~1 micrometer [μm] diameter) with a negative surface charge. The small oil droplets will attach to positively charged sites on surfaces of the aquifer solids. Oil retention for one commercially available product (EOS®) is between 1.6 to 0.16 kilograms per cubic meter (kg/m^3) (0.1 to 0.01 pound oil per cubic foot [lb/ft^3]) of treated material. Unfortunately, measured oil retention values are not available for other commercial products. Table 8.1 illustrates the range of emulsion retained in a variety of aquifer solids.

Table 8.1 Observed Oil Droplet Retention by Aquifer Solids

Site-Specific Aquifer Material	Maximum Retention*	Average Retention*
Blended sand (7% silt + clay) (Coulibaly and Borden, 2004)	0.0054 g/g or 9.2 kg/m^3 (Lab Column)	0.0066 g/g or 11.2 kg/m^3 (Sandbox)
Blended sand (9% silt + clay) (Coulibaly and Borden, 2004)	0.0061 g/g or 10.4 kg/m^3 (Lab Column)	0.0035 g/g or 6.0 kg/m^3 (Sandbox)
Blended sand (12% silt + clay) (Coulibaly and Borden, 2004)	0.0095 g/g or 16.2 kg/m^3 (Lab Column)	0.0037 g/g or 6.3 kg/m^3 (Sandbox)
Aluvium (clayey sand) treated with EOS® (ESTCP, 2006b) **	0.0037 g/g or 6.3 kg/m^3 (Lab Column)	0.0013 g/g or 2.2 kg/m^3 (Field)
Low K, weathered rock treated (sandy clay with remnant fractures) with EOS® (Borden et al., 2007) ***	—	0.0028 g/g or 4.8 kg/m^3
High K, gravelly sand with EOS® (Kovacich et al., 2007) ****	—	0.0003 g/g or 0.5 kg/m^3

* Assumed soil density of 1,700 kg/m^3
** Treatability Study, manufacturing site in Elkton, MD (ESTCP, 2006b)
*** EOS® remediation at former Army site in Burlington, NC (Borden et al., 2007)
**** EOS® remediation at active commercial site in Connersville, IN (Kovacich et al., 2007)

The minimum amount of oil required to treat a certain volume of aquifer is calculated as

$$\text{Oil Requirement} = V_t OR \qquad \text{Eq. 8.1}$$

where V_t is the volume to be treated (m³ or ft³) and OR is the average oil retention (kg/m³ or lb/ft³). In many aquifers, OR (the amount of oil required to adequately coat the aquifer material) will be much greater than the amount of oil required for biodegradation, and will therefore determine the total amount of oil that must be injected.

8.2.2.2 Amount of Water Required

The distribution of edible oil emulsions during the injection process is controlled by several factors: the subsurface permeability, the natural groundwater flow within the immediate injection zone and the injection process. The injected emulsified oil will migrate with groundwater until all droplets have adhered to aquifer surfaces. Consequently, to achieve a greater radius of influence around an injection point, water must be injected to transport the oil droplets throughout the target treatment zone. Common procedures used include: (1) injecting a concentrated emulsion followed by chase water to distribute the oil, (2) continuous injection of a more dilute emulsion, and (3) recirculation of emulsion through the treatment zone.

Modeling studies conducted by Borden (2007a) for homogeneous aquifers indicate that injection flow rate and emulsified oil concentration have essentially no effect on the final oil distribution in the sediment. The only factors that significantly influence the final oil distribution are: (1) the total amount of oil injected, and (2) the total amount of water injected. Procedures for estimating the amount of emulsified oil to inject are described in Section 8.2.2.1, above. The current best recommendation is for the total fluid injection volume (including emulsion and chase water) to equal the effective pore volume of the target treatment zone. When installing an edible oil barrier using injection wells, the water volume injected per well can be calculated as (ESTCP, 2006a):

$$V_F = (\pi D^2/4)(Z) n_e \qquad \text{Eq. 8.2}$$

where: V_F is the volume of injection fluid per well
D is the injection well spacing
Z is the effective vertical height of the treatment zone
n_e is the effective porosity

Research is underway to develop more effective methods for distributing emulsions in heterogeneous aquifers (Borden, 2007c). Preliminary results suggest that contact efficiency can be increased by injecting excess emulsion and/or excess water. However, the benefits of the "over treatment" are still not well understood.

Example calculations illustrating the amount of emulsified oil and water required to treat a 30-m x 30-m (100-ft x 100-ft) area that is 3 m (10 ft) thick are provided in Table 8.2. Edible oils, when purchased in bulk, are relatively inexpensive (see Chapter 10). However, potential treatment volumes required at some contaminated sites are large. Thus, process designers should carefully consider the size of the source treatment zone and/or barrier location required to meet remediation objectives.

Table 8.2 Estimated Volumes of Oil and Water Required for Treatment of 30-m x 30-m (100-ft x 100-ft) Area

Injection Point Spacing	1.5 m	3.0 m	5.0 m	7.5 m	10.0 m
Vertical Injection Interval	3.0 m	3.0 m	3.0 m	3.0 m	3.0 m
Porosity (n)	0.25	0.25	0.25	0.25	0.25
Total Volume of 30-m x 30-m Treatment Zone	2,700 m^3	2,700 m^3	2,700 m^3	2,700 m^3	2,700 m^3
Pore Volume (PV) of 30-m x 30-m Treatment Zone	675 m^3	675 m^3	675 m^3	675 m^3	675 m^3
Number of Injection Wells to Treat 30-m x 30-m Area	400	100	49	25	16
Injection Volume per Well	1.7 m^3	6.8 m^3	18.8 m^3	42.2 m^3	75.0 m^3
Time to Inject One PV at 4 L/min/well (~1 gpm per well)*	7 hr	28 hr	78 hr	176 hr	313 hr
Emulsified Oil required for average retention of 0.002 g/g	9,200 kg (20200 lb)	9,200 kg (20200 lb)	9,200 kg (20200 lb)	9,200 kg (20200 lb)	9,200 kg (20200 lb)

* See Section 8.2.2.3 for a discussion on constraints on the rate of injection.

8.2.2.3 Injection Point Spacing

The injection point spacing is primarily a trade-off between the costs for installing injection points/wells and labor costs. Wider spacing of the injection points or wells can reduce installation costs, but will increase the time and labor required for injection. The injection point installation costs are affected by the geology and depth to groundwater, while the labor costs are determined by the time required to inject the oil, which is largely a function of the aquifer permeability and the available water supply. If the aquifer has a high permeability, the oil will be easier to inject and the injections will take less time. Often, multiple wells can be injected simultaneously to reduce the amount of time required to complete the injections. Injection tests are often done to help determine the anticipated injection flow rates and pressures and the approximate time it will take to complete the injections. Installation and labor costs associated with injection of oil should be evaluated on a site-specific basis to determine the appropriate injection point spacing. Figure 8.3 illustrates the competing effects of well spacing on costs for well installation, injection labor and substrate. Design tools for planning aqueous amendment injection systems currently in the development phase will aid users in conducting more cost-effective and successful *in situ* application of emulsified oil and other substrates (Borden, 2006).

Injections are typically designed to provide 100% coverage of a targeted treatment zone. However, subsurface heterogeneities will affect the distribution of the oil in the subsurface. Permeability differences will cause some zones to be over-treated and some zones to be insufficiently treated. Groundwater flow and dispersion will provide some spreading of aqueous organic carbon to increase the reactive zone. However, design factors are often used to provide overlap between the injections and minimize the potential for untreated or insufficiently treated zones. The need for a design factor will depend on hydrogeologic complexities, the amount of available site characterization data, and site-specific concerns such as sensitive downgradient receptors.

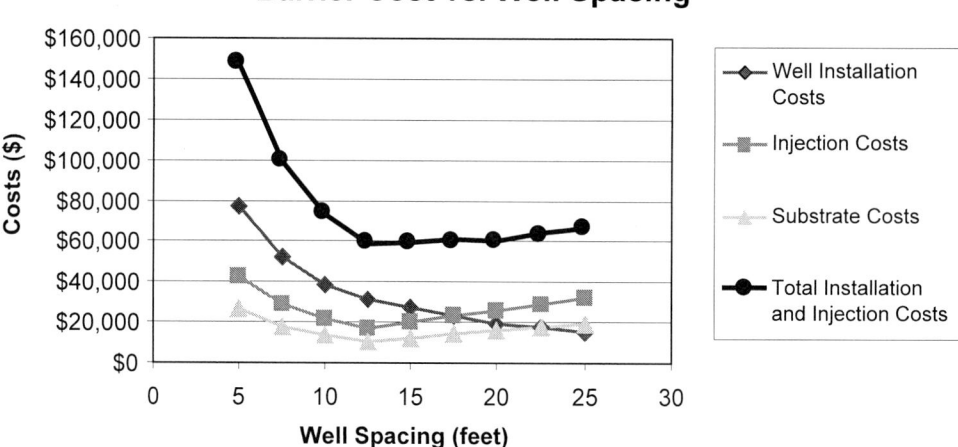

Figure 8.3 Example cost analysis for a PRB with various injection well spacings (1 m = 3.28 ft) (ESTCP, 2006a).

8.2.2.4 Additional Planning Considerations

While emulsified oil injection can enhance biodegradation of perchlorate, there are some secondary effects of oil injection that need to be considered, including secondary water quality issues and soil gas emissions.

Secondary Water Quality Issues

The term "secondary water quality" is used in this chapter to refer to water quality issues or concerns, apart from the primary contaminants being treated, that result from substrate addition. Degradation of secondary water quality can occur as a result of mobilization of formerly insoluble forms of metals that occur naturally in the aquifer matrix. Other secondary water quality parameters that may be affected include chemical oxygen demand (COD), biochemical oxygen demand (BOD), total dissolved solids (TDS) and sulfides that affect taste and odor. These parameters should be monitored if of concern and/or regulated at the site.

When planning an edible oils project, an *in situ* "bioreactor" approach should be adopted. In this approach, organic substrate is added to a specific reactive zone, generating strongly reducing conditions and stimulating biodegradation of the target contaminants. Within the anaerobic reactive zone, intermediate degradation products may temporarily accumulate before subsequent treatment in the downgradient aquifer. Within the reactive zone, the taste and odor of groundwater may be impacted due to elevated levels of COD, BOD, TDS, sulfides, and/or fatty acids. In addition, the reduced groundwater environment in the reactive zone may increase the mobility of some naturally occurring, but regulated, metals (e.g., iron, manganese, and arsenic). While these metals are more soluble under reducing conditions, migration of metals out of the reactive zone is often substantially retarded by adsorption to the aquifer matrix and/or precipitation as insoluble metal sulfides (Butler and Hayes, 1999). Mobilized metals can also precipitate once again when redox conditions increase. Careful calculation of the amount of substrate employed (Section 8.2.2.1) may help minimize the development of some undesirable secondary water quality characteristics.

In naturally aerobic aquifers, groundwater typically returns to near background conditions within a relatively short distance downgradient of the induced anaerobic reactive zone (ESTCP,

2006a). As groundwater migrates downgradient, any excess substrate will be consumed by biological processes and the anaerobic groundwater will mix with background aerobic groundwater, resulting in precipitation/immobilization of dissolved metals. In naturally anaerobic aquifers, secondary water quality impacts may extend farther downgradient. However, the groundwater in naturally anaerobic aquifers is generally not of drinking water quality.

Based on petroleum hydrocarbon plume studies, dissolved organic reactive zones are not likely to extend more than 60 to 100 m (200 to 330 ft) downgradient of the oil injection zone (Rice et al., 1995; Mace et al., 1997; Newell and Connor, 1997). Impacted zones downgradient of emulsified oil injections are expected to be even more limited because emulsified oils are much more biodegradable than petroleum hydrocarbons. Monitoring data from existing emulsified oil sites indicate that DOC and secondary groundwater quality parameters are not affected more than 15 m (49 ft) downgradient of emulsified oil injection zones. To provide a substantial factor of safety, a typical recommendation is to maintain a minimum distance of 75 m (246 ft) between injection locations and critical downgradient receptors.

In summary, the potential for degradation of secondary water quality should be considered when working in close proximity to drinking water supplies. It also should be noted that these changes in water quality, and those discussed under generation of noxious gases (see Section 8.2.2.4.2), are not easily reversed and, in the case of a slow release carbon source, it may take many years for the effects of the substrate addition to diminish. These secondary water quality issues should be carefully considered before proceeding with an enhanced anaerobic bioremediation project. Specific groundwater quality goals should be established for wells upgradient of sensitive areas that will allow for temporal increases in breakdown products or byproducts within the reactive zones.

Soil Gas Generation

There is a potential for methane production as a result of emulsified oil injection. Highly elevated methane concentrations could potentially pose a problem when found near buildings. Therefore, soil gas monitoring should be conducted when emulsified oils are applied near the water table surface and in close proximity to buildings. Methane will rapidly biodegrade in the presence of oxygen, and soil gas oxygen concentrations should be measured to determine if methane is likely to be biodegraded *in situ*. Soil gas carbon dioxide concentrations should also be measured, because elevated carbon dioxide levels often correlate with methane generation.

8.2.3 Site Characterization Requirements

A conceptual site model (CSM) should be developed to determine if passive bioremediation is suitable for a site. Most sites being evaluated for enhanced anaerobic bioremediation generally have been investigated and characterized to some extent, and a limited assessment of remedial alternatives will have been conducted. An assessment of the potential to stimulate anaerobic biodegradation is based upon a review of site-specific data including hydrogeology, contaminant distribution and trends and biogeochemical conditions (electron donors, electron acceptors, metabolic byproducts, and general geochemical indicators). A CSM summarizes the fate and transport of contaminants, migration pathways, exposure mechanisms and potential receptors. Site characterization considerations for selection, development and evaluation of a passive bioremediation system are described in the following subsections and discussed in detail in available guidance documents (ESTCP, 2006a; AFCEE, 2007).

8.2.3.1 Hydrogeology

Subsurface hydrogeology must be considered in site selection and process design, as inadequate characterization of the site hydrogeology can lead to system failure. In many cases, the remediation system can be designed to mitigate difficult hydrogeologic conditions. Depth to water and the depth of the contaminant plume primarily impact the capital cost of drilling and delivering the substrate to the intended treatment zone. Where possible, installation of injection wells using direct push equipment will result in a less costly installation. Direct push equipment also may be used to inject the emulsion directly, which may further reduce cost for materials. However, drilling costs for direct push injection can be higher since only one or two points can be injected at a time. In addition, it may be difficult to uniformly distribute emulsion (or any reagent) using long well screens, and practitioners should consider using multiple vertical injection points when treating a large saturated thickness.

Hydraulic conductivity is a primary factor in effective distribution of substrate in the subsurface. In general, hydraulic conductivities greater than 4×10^{-3} centimeters per second (cm/s) (~10 ft/day), are best for effective distribution of emulsified oils away from the injection points (AFCEE, 2007). It is generally not cost effective to distribute substrates in zones having a hydraulic conductivity less than 10^{-4} cm/s. Although alternate injection techniques, such as pneumatic fracturing, have been used to inject emulsified oil, these techniques often result in a much less uniform oil distribution and may not bring the oil into direct contact with the contaminant, reducing treatment effectiveness. Strongly heterogeneous sites present special challenges for achieving uniform substrate distribution. Any injected fluid will preferentially flow into more permeable materials. Distribution of emulsified oil in more permeable materials may be very effective in reducing the mass flux of contaminants out of a source area since the contaminants will be treated as they pass through these higher permeability, emulsion treated zones. However, if the majority of the contaminant mass has partitioned into less permeable clays, silts or bedrock, then overall biodegradation rates will be slow and will be controlled by slow diffusion of the contaminants out of these lower permeability layers.

Groundwater velocity, flow direction, and horizontal and vertical gradients will impact the effectiveness of emulsified oil addition. Excessively high groundwater flow rates (greater than 1.5 meters per day (m/d) [4.9 ft/d]) may require large amounts of substrate to overcome a large influx of competing electron acceptors migrating into the reactive zone. A substantially larger treatment zone also may be required to maintain sufficiently reducing conditions in high-flow aquifers. Where groundwater flow rates are very low (less than 1 m/yr [3.3 ft/yr]), the timeframe for remediation may be extended due to reduced mixing of substrate and contaminants.

8.2.3.2 Contaminant Distribution

Emulsified oils can provide a long-lasting substrate to support anaerobic biotransformation processes. Emulsified oils will be most cost-effective for treatment of small to midsize source areas. For very large sources, it may be more cost-effective to contain the source using either an impermeable barrier or possibly a biologically active barrier created by injecting emulsified oil into points surrounding the source.

For large plumes, it may not be economically feasible to remediate the entire plume at one time due to the relatively high cost of installing injection wells. As in treating the source area, oil emulsions can be used to generate a larger radius of influence around each injection point. However, a much more cost-effective approach is to install barriers at several different points along the plume. For example, if the barriers are spaced 1 to 2 years travel time apart, the entire plume should be treated by passage of contaminated groundwater through one or more barriers within 5 years.

8.2.3.3 Geochemistry

Geochemical evaluations are focused on determining the prevailing redox conditions and demonstrating that the "footprints" of the expected degradation processes are present. Characterizing the initial geochemical and redox conditions is useful to determine the prevailing terminal electron acceptor processes. This also can be used to evaluate the changes in redox conditions required for optimal contaminant biodegradation. High levels of alternate electron acceptors (e.g., DO, nitrate, or sulfate), and possibly concentrations of co-contaminants, should be taken into account when determining substrate demand. Electron donor supply is often measured and tracked by measuring parameters such as total organic carbon (TOC) or metabolic volatile fatty acids (VFAs).

8.2.4 Monitoring

Biodegradation of perchlorate stimulated by substrate addition brings about measurable changes in the chemistry of groundwater in the treated area. By measuring these changes, it is possible to document and quantitatively evaluate the effect of adding substrate to the subsurface to enhance anaerobic biodegradation at a site. Ongoing process monitoring of key contaminant and biogeochemical characteristics of the site is critical to evaluating the effectiveness of the system to meet remedial objectives. Primary groundwater parameters that should be sampled regularly for process monitoring include contaminants and daughter products, biogeochemical indicators of redox conditions, and the strength and distribution of organic substrate. These parameters provide basic information on the efficacy of substrate delivery to the treatment zone and the prevailing redox conditions.

8.2.4.1 Contaminants and Biodegradation Products

The effectiveness of the bioremediation of contaminants of concern can be monitored by simply assessing changes in concentration of the parent compounds. For many compounds, the formation and then subsequent degradation of intermediate metabolic products is corroborating evidence that complete degradation is occurring. The target analytes and metabolic daughter products are used to determine the type, concentration, and distribution of contaminants and degradation products in the aquifer. In addition, the ratio of the parent and daughter compounds should change as biodegradation is stimulated.

The microbial degradation of perchlorate proceeds enzymatically through a pathway leading from perchlorate (ClO_4^-), to chlorate (ClO_3^-), to chlorite (ClO_2^-), and finally to chloride (Cl^-) and oxygen (O_2). Both chlorate and chlorite are rapidly biodegraded under anaerobic conditions. As a consequence, these byproducts are rarely detected. In most cases, demonstrations of perchlorate remediation rely on the observed disappearance of the parent compound. In some cases, increases in chloride concentrations can be detected. However, this is only possible when high levels of perchlorate are degraded in aquifers with a low background chloride concentration. New advances in measurements of stable isotopes of perchlorate ($\delta^{18}O/\delta^{17}O$ and $\delta^{35}Cl/\delta^{37}Cl$) may also prove valuable in demonstrating the biodegradation process (Sturchio et al., 2003; Bao and Gu, 2004; Böhlke et al., 2005; Lieberman et al., 2006).

8.2.4.2 Biogeochemistry

Biogeochemical parameters are measured to determine whether conditions are suitable for enhanced anaerobic biodegradation to occur. Profound changes in redox processes may occur as a result of substrate addition, and the predominant electron acceptor being utilized by microbial activity often varies in zones across the site. Addition of emulsified oil is intended to deplete competing electron acceptors and to maintain anaerobic conditions that are optimal for high rates of anaerobic biodegradation to occur. Excessive levels of competing electron acceptors (e.g., DO, nitrate and sulfate) may limit the effectiveness of substrate addition. Nitrate, in particular, has been shown to compete with or inhibit perchlorate biodegradation. Perchlorate reduction and denitrification occur under similar redox conditions (Coates et al., 1999; Logan et al., 2001). Laboratory evidence has shown that low nitrate concentrations or processes that effectively remove nitrate (such as presence of electron donors) allow perchlorate biodegradation to proceed efficiently (Tan et al., 2004). Therefore, groundwater geochemical conditions across the site should be measured in order to identify any undesirable geochemical conditions. At a minimum, parameters that should be measured include DO, redox, nitrate, Fe(II), sulfate, methane, alkalinity, and pH.

8.2.4.3 Indicators of Organic Carbon

Indicators of organic substrate available for biodegradation processes include TOC (for unfiltered samples), DOC (for filtered samples), VFAs, and edible oil fatty acids (EOFA). Elevated levels of TOC in groundwater can be used as an indicator of emulsion distribution. Elevated DOC levels are an indicator of soluble organic products produced from oil fermentation. Elevated levels of total inorganic carbon (TIC) are an indicator of organic carbon that has been degraded to inorganic byproducts. However, this analysis may not be useful in carbonate aquifers where background TIC levels are high.

TOC (or DOC) and VFAs should be monitored over time to evaluate longevity of the edible oil. VFAs are produced by bacterial fermentation of edible oils and serve as biomarkers of anaerobic metabolism. Levels of TOC and VFAs should be expected to decline over time as the substrate is consumed.

Emulsified oil distribution in an aquifer can be evaluated by analyzing aquifer solid samples for TOC or EOFAs. In general, TOC is a lower cost and more readily available analysis. However, many aquifers naturally contain significant amounts of TOC, and it can be difficult to distinguish injected oil from background TOC. In contrast, EOFA analysis targets the long-chain fatty acids present in vegetable oil (e.g., oleic and linoleic acids) and can be used to conclusively demonstrate emulsion distribution in an aquifer.

8.3 CASE STUDY

The cost and performance of edible oil emulsions for passive bioremediation of perchlorate plumes has been evaluated at the laboratory and field-scale level (ESTCP, 2006a, 2006b; Borden, 2001a; Borden, 2007a; Borden, 2007b). A demonstration was conducted at a site in Maryland with a mixed perchlorate and 1,1,1-trichloroethane (1,1,1-TCA) groundwater plume. The performance of the field pilot test was evaluated by monitoring the distribution of the oil emulsion in the aquifer, the impact of the oil injection on the aquifer permeability and groundwater flow paths, and the changes in contaminant concentrations and biodegradation indicator parameters both upgradient and downgradient of the PRB. Data obtained during the pilot test were used to demonstrate the cost-effectiveness of emulsified edible oils for remediation of perchlorate and chlorinated ethanes in groundwater through enhanced biodegradation. This section provides a brief overview of the demonstration activities and results. Additional details can be found in the final technical report (ESTCP, 2006b).

8.3.1 Demonstration Design

Groundwater beneath the manufacturing portion of the Maryland project site was contaminated with concentrations up to 72,000 micrograms per liter (µg/L) perchlorate and 19,000 µg/L 1,1,1-TCA. The shallow aquifer at the site consists of silty sand and gravel to a depth of approximately 5 m (16 ft) below ground surface (bgs). The aquifer was impacted by a former lagoon that received ammonium perchlorate and waste solvent. The water table is approximately 1.5 m (4.9 ft) bgs. The demonstration activities included both laboratory studies using site aquifer solids and a field pilot test involving injection of emulsified oil substrate (EOS®) to form a PRB. The objective of the pilot test was to evaluate the performance of an emulsified oil PRB for treatment of groundwater contaminated with high levels of perchlorate and 1,1,1-TCA, and to determine if the PRB could effectively reduce perchlorate to a concentration less than the laboratory detection limit of 4 µg/L.

The laboratory microcosm study was conducted first to evaluate the effectiveness of emulsified oil for remediating perchlorate and 1,1,1-TCA, and a column study was performed concurrently to assess emulsified oil distribution in site sediments. Figure 8.4 shows measured perchlorate concentrations in microcosms at 2 and 14 days after construction. Perchlorate was reduced to below the analytical detection limit of 8 µg/L in triplicate microcosms amended with emulsified liquid soybean oil and emulsified solid hydrogenated soybean oil, but was not significantly depleted in killed controls and microcosms without added carbon. In microcosms constructed with aquifer material, groundwater and emulsified soybean oil, 1,1,1-TCA was degraded to 1,1-dichloroethane (1,1-DCA) (Figure 8.5). 1,1-DCA subsequently degraded with only minor production of chloroethane (CA). These results indicated that bioaugmentation would not be required to achieve complete dechlorination of 1,1,1-TCA and other chlorinated compounds (data not shown) to non-toxic end products. The column study results indicated that emulsified oil could be effectively distributed in aquifer material from the Maryland site (ESTCP, 2006b).

The subsequent field demonstration consisted of a one-time injection of emulsified oil and chase water to create a 15-m (50-ft) long PRB. The PRB was located approximately 15 m (50 ft) upgradient of an existing interceptor trench that is part of an on-site groundwater treatment system. Historically, groundwater from the interceptor trench has been treated via an air stripper and re-injected via an upgradient infiltration gallery. The air stripper system effectively treated the 1,1,1-TCA and other volatiles, but did not impact perchlorate in the waste stream.

Ten 2.5-cm (1-inch) diameter direct-push injection wells were installed 1.5 m (4.9 ft) on center perpendicular to groundwater flow, and monitoring wells were installed upgradient and downgradient of the injection wells to evaluate changes in groundwater concentrations over time. In October 2003, approximately 416 L (110 gallons [gal]) of EOS® and 7,835 L (2070 gal) of water were injected into the subsurface. Five wells were injected simultaneously using a manifold injection system to decrease the time required to complete the injection activities. The injection was completed in 2 days by a 2-person field team. Additional details on the layout of monitor and injection wells is provided in the final technical report for the site (ESTCP, 2006b).

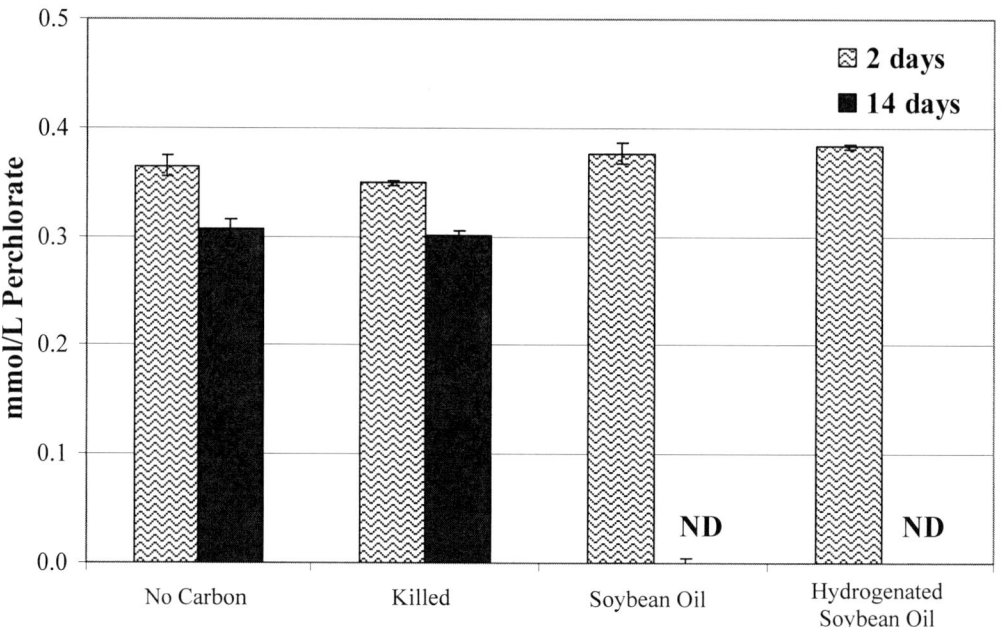

Figure 8.4 Comparison of perchlorate concentrations in microcosms after 2 and 14 days illustrating the effect of different treatments on perchlorate biodegradation (ESTCP, 2006b). ND indicates non-detect in triplicate incubations at 14 days.

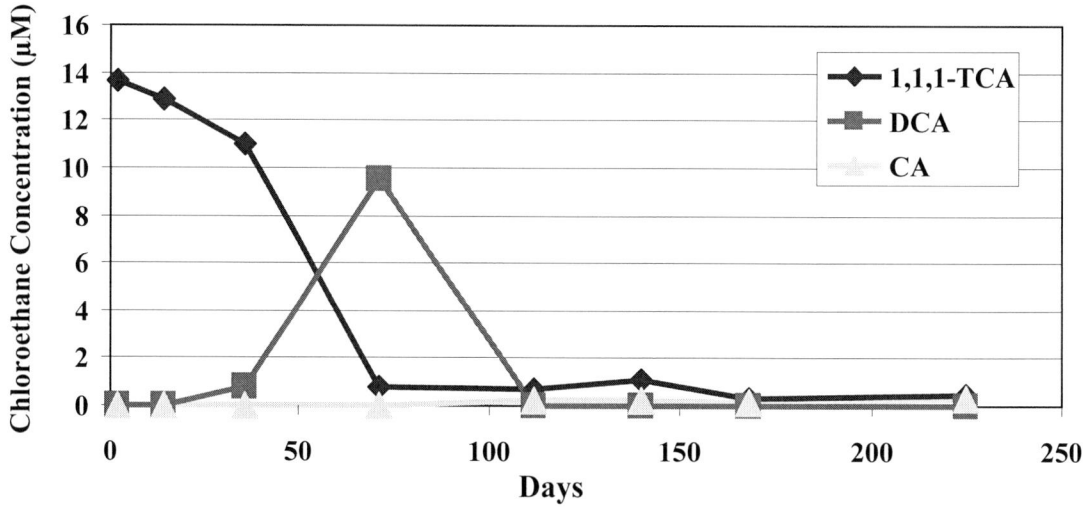

Figure 8.5 Reductive dechlorination of 1,1,1-TCA to 1,1-Dichloroethane (1,1-DCA) and Chloroethane (CA) in microcosms constructed with aquifer material, groundwater and emulsified soybean oil (ESTCP, 2006b).

8.3.2 Monitoring

Monitoring activities were conducted over an 18-month period to evaluate performance of the PRB. Groundwater samples were analyzed for perchlorate, volatile organic compounds (VOCs) (including chlorinated aliphatic hydrocarbons and trihalomethanes), electron acceptors (oxygen, nitrate, sulfate, phosphate), electron donors (TOC and VFAs), indicator parameters (pH, redox potential), metals (Fe^{+2} and Mn^{+2}), light hydrocarbon gases (ethene, ethane, methane), and chloride. Additional monitoring activities included hydraulic conductivity testing, soil sampling, and pre- and post-injection tracer tests.

8.3.3 Results

Emulsified oil injection resulted in substantial reductions in perchlorate and 1,1,1-TCA concentrations within and downgradient of the PRB. Perchlorate concentrations in all of the injection wells were non-detect (<4 µg/L) within 5 days of injection. Eighteen months post-injection perchlorate removal rates remained greater than 99% in the downgradient monitor wells compared to pre-injection levels. The perchlorate results are presented in Figure 8.6.

In general, 1,1,1-TCA concentrations decreased in the downgradient monitoring wells during the pilot test with subsequent increases in 1,1-DCA and chloroethane. Eighteen months post-injection, 1,1,1-TCA was reduced by 94 to 98% 6 m downgradient of the barrier. Figure 8.7 shows the changes in 1,1,1-TCA and its daughter products for SMW-6, located approximately 6 m downgradient of the PRB.

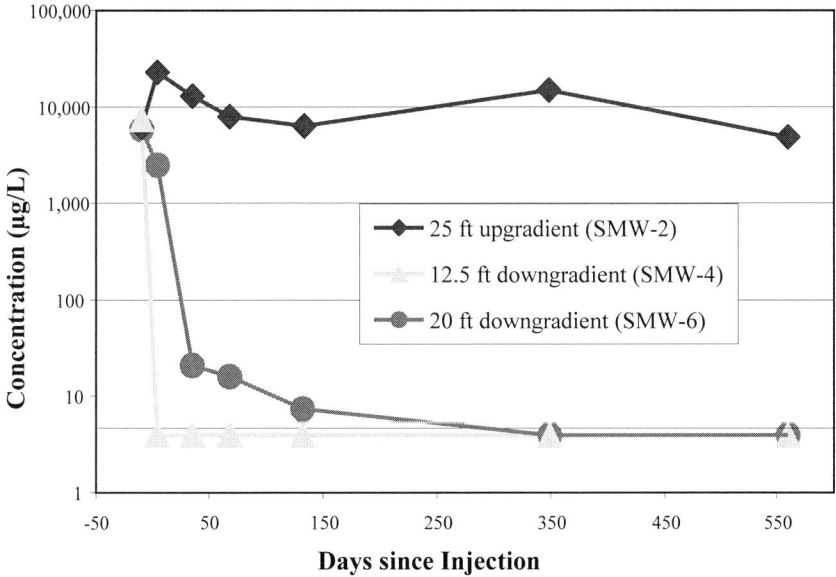

Figure 8.6 Change in perchlorate concentration versus time in monitoring wells upgradient and downgradient of the emulsified oil barrier (ESTCP, 2006b).

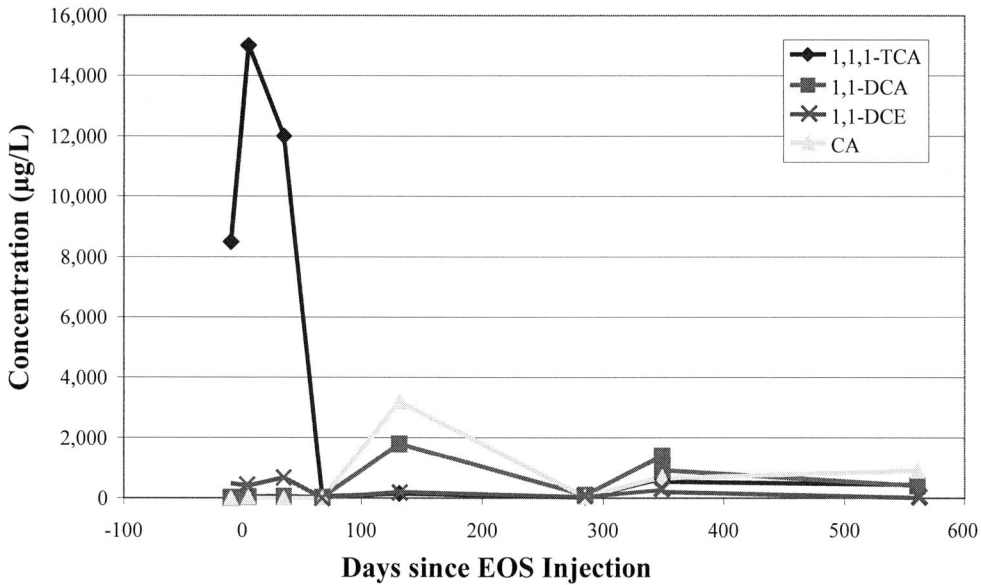

Figure 8.7 Change in 1,1,1-TCA and degradation products versus time in Monitoring Well SMW-6 showing effect of emulsified oil addition in stimulating reductive dechlorination (ESTCP, 2006b).

Geochemical data collected at the site confirmed that anaerobic conditions favorable for biodegradation of perchlorate and 1,1,1-TCA were established in the treatment area. In general, nitrate and sulfate concentrations decreased with time in the injection and downgradient wells, indicating nitrate and sulfate reduction, while dissolved iron and manganese concentrations increased with time, indicating iron and manganese reducing conditions. Methane concentrations increased in the injection wells, suggesting methanogenic conditions within the PRB. No significant changes were observed in the upgradient monitor wells. After about one year of activity, floccular material appeared in the groundwater interceptor trench downgradient of the barrier. This material increased fouling of the air stripper, possibly a result of the increased levels of dissolved iron and manganese, or due to increased biofouling as a result of enhanced microbial activity in the groundwater and/or elevated BOD in the air stripper influent.

Elevated concentrations of TOC within and immediately downgradient of the injection wells indicated good distribution of emulsified oil throughout the target zone forming a PRB. Although permeability reductions were observed in the injection wells, tracer test data indicated that groundwater flow through the barrier did not appear to have been affected by the measured changes in hydraulic conductivity. Monitoring data collected during the pilot test indicate the emulsified oil barrier performed very well for over 2.5 years. This was beyond the original design life of 1.5 years planned for the pilot demonstration. However, based on a mass balance analysis of organic and inorganic carbon released from the barrier, performance began to decline when approximately half of the injected substrate had been depleted.

8.4 TOOLS AND RESOURCES

Several documents are available to assist with design, implementation, and monitoring of passive bioremediation systems. The Department of Defense (DoD) Environmental Security Technology Certification Program (ESTCP) funded development of "A Treatability Test for Evaluating the Potential Applicability of the Reductive Anaerobic Biological *In Situ* Treatment Technology to Remediate Chloroethenes" (i.e., the RABITT Protocol), which aids users in determining the site applicability of enhanced anaerobic bioremediation for chloroethene contamination in groundwater (Morse et al., 1998). The "Principles and Practices of Enhanced Anaerobic Bioremediation of Chlorinated Solvents," published cooperatively by AFCEE, ESTCP, and the Naval Facilities Engineering Service Center (NFESC), describes the scientific basis of enhanced anaerobic bioremediation and summarizes relevant site selection, design and performance criteria for various engineered approaches to stimulate and enhance *in situ* biodegradation of chlorinated solvents (AFCEE, 2004).

The DoD Strategic Environmental Research and Development Program (SERDP) and ESTCP have funded two projects focusing on the use of emulsified edible oil to enhance *in situ* anaerobic bioremediation of groundwater contaminants: SERDP Project CU-1205, "Development of Permeable Reactive Barriers Using Edible Oils" (Borden, 2001a), and ESTCP Project ER-0221, "Edible Oil Barriers for Treatment of Chlorinated Solvent- and Perchlorate-Contaminated Groundwater" (Borden, 2001b). The information gained from these projects was used to develop the "Protocol for Enhanced *In Situ* Bioremediation Using Emulsified Edible Oil" (ESTCP, 2006a) which provides detailed information to assist base managers and project engineers in: (1) determining if the emulsified oil process is appropriate for their site, and (2) designing and implementing this technology. AFCEE has also published a "Protocol for *In situ* Bioremediation of Chlorinated Solvents using Edible Oil" (i.e., the Edible Oils Protocol, AFCEE, 2007), which focuses on the application of pure liquid edible oils and edible oil emulsions to provide a long-lived carbon source to enhance anaerobic bioremediation of groundwater contaminants.

In addition, ESTCP is funding two new projects that will provide improved information and tools for designing passive bioremediation systems. Project ER-0626, "Development of a Design Tool for Planning Aqueous Amendment Injection Systems" (Borden, 2006), will provide a set of tools to assist design engineers in developing effective, reasonably efficient systems for distributing aqueous amendments for *in situ* treatment of groundwater contaminants. Project ER-0627, "Loading Rates and Impacts of Substrate Delivery for Enhanced Anaerobic Bioremediation" (Henry, 2006), will develop practical guidelines for designing and optimizing substrate loading rates and injection scenarios for differing substrate types and for differing geochemical and hydrogeologic conditions.

8.5 FACTORS CONTROLLING COST AND PERFORMANCE

The primary costs associated with installation of passive bioremediation systems include injection point installation, substrate, and substrate injection/emplacement. These costs are affected by the mass of contaminants in the aquifer, the subsurface lithology, the depth to groundwater and the vertical extent of contamination. The amount of substrate required at a specific site depends on the amount of substrate needed for biodegradation (e.g., contaminant concentrations, competing electron acceptors). For emulsified oil, the oil retention by aquifer materials also needs to be considered. The substrate costs will depend on the material costs of the specific substrate selected (e.g., emulsified oil, polymerized lactate products, chitin, etc.). The injection costs are influenced by the number of injection points, injection point spacing, the time needed to complete the injections and how the injections are completed (i.e., direct-push points or wells). All of these factors are primarily related to the subsurface lithology and the depth to groundwater. Substrates can typically be injected through direct-push points, temporary injection wells, or conventional monitoring wells. The subsurface lithology (i.e., heterogeneity) also influences the ability to distribute substrate throughout the aquifer, which affects the number and spacing of the injection points.

The performance of a passive bioremediation system for perchlorate treatment is primarily related to the ability to distribute the substrate throughout the treatment zone, the biodegradation of the substrate after it is injected, the presence of microorganisms capable of complete biodegradation and the rate of biodegradation of the target contaminants that can be achieved *in situ*. In order to be effective as a PRB, substrate must be distributed vertically and horizontally throughout the treatment zone and must not result in an excessive decrease in the permeability of the aquifer. If the substrate is not thoroughly distributed, contaminated groundwater could short-circuit the barrier and remain untreated. If substrate injection substantially decreases the permeability of the aquifer, the contaminated groundwater may flow around the barrier instead of through the barrier. If the substrate is biodegraded too rapidly, then the barrier does not last as long as designed and re-injection could be necessary to reduce contaminant concentrations to the desired levels. This should be considered in the design when selecting an appropriate substrate. Available information suggests that perchlorate-degrading microorganisms are fairly common, if not ubiquitous. However, there is a possibility that these organisms may not be present at all sites. The presence of perchlorate-reducing microorganisms should be evaluated before implementing a passive bioremediation system.

8.6 SUMMARY

Emulsified oils can be used to enhance *in situ* anaerobic biodegradation of perchlorate. The edible oils are initially fermented to hydrogen and acetate which can then be used to support perchlorate biodegradation.

In this chapter, we focus on the use of emulsified edible oils as a low-cost, long-lasting and easy-to-distribute organic substrate. Commercially available products are typically supplied as concentrates containing 45 to 65% soybean oil, with various emulsifying agents and nutrients added. These products are typically diluted with water onsite and injected into the aquifer through permanent or temporary wells. By controlling the oil droplet size and surface charge, it is possible to distribute the oil droplets 2 to 10 m (6 to 33 ft) from the injection point. Design tools are available and under development to aid in planning an injection project.

Passive bioremediation systems can be designed in different configurations to treat perchlorate-contaminated aquifers. The most common approaches are source area treatment and plume control using injections in a grid formation, a PRB, or multiple PRBs. In choosing a treatment approach for a given site, it is important to understand the overall objectives of the project.

A case study is presented documenting the use of an emulsified oil barrier for treatment of groundwater containing perchlorate and chlorinated solvents. The 15-m (50-ft) wide barrier was formed by injecting 416 liters (110 gal) of emulsified oil concentrate (EOS®) over a two-day period through 10 direct push wells. Emulsified oil injection resulted in substantial reductions in perchlorate and 1,1,1-TCA concentrations within and downgradient of the PRB. Emulsion injection resulted in a rapid conversion to anaerobic conditions. Perchlorate concentrations were reduced from roughly 10,000 µg/L to below detection (<4 µg/L) in all injection wells within 5 days of injection. Perchlorate removal rates remained high for over

18 months, with perchlorate reduced by over 99% in the downgradient monitor wells. Chlorinated solvents were also biodegraded during migration through the emulsified oil barrier, resulting in a 94 to 98% reduction in 1,1,1-TCA, 18 months post-injection. However, reductive dechlorination degradation products (1,1-DCA and CA) were not completely degraded, presumably due to the short hydraulic residence time in the oil treated zone. Elevated concentrations of TOC within and immediately downgradient of the injection wells indicated good distribution of emulsified oil throughout the target zone forming a PRB.

REFERENCES

AFCEE (Air Force Center of Environmental Excellence). 2004. Principles and Practices of Enhanced Anaerobic Bioremediation of Chlorinated Solvents. AFCEE, Brooks City-Base, TX, USA. http://www.afcee.af.mil/shared/media/document/AFD-071130-020.pdf. Accessed June 11, 2008.

AFCEE. 2007. Protocol for *In situ* Bioremediation of Chlorinated Solvents using Edible Oil. AFCEE, Brooks City-Base, TX, USA. http://www.afcee.af.mil/shared/media/document/AFD-071203-094.pdf. Accessed June 11, 2008.

Bao H, Gu B. 2004. Natural perchlorate has a unique oxygen isotope signature. Environ Sci Technol 38:5073–5077.

Böhlke JK, Sturchio NC, Gu B, Horita J, Brown GM, Jackson WA, Batista J, Hatzinger P. 2005. Perchlorate isotope forensics. Anal Chem 77:7838–7842.

Borden RC. 2001a. Development of Permeable Reactive Barriers Using Edible Oils. Project CU-1205 Fact Sheet. Strategic Environmental Research and Development Program (SERDP), Arlington, VA, USA. http://www.serdp.org/Research/upload/CU-1205.pdf. Accessed June 11, 2008.

Borden RC. 2001b. Edible Oil Barriers for Treatment of Chlorinated Solvent- and Perchlorate-Contaminated Groundwater. Project ER-0221 Fact Sheet. Environmental Security Technology Certification Program (ESTCP), Arlington, VA, USA. http://www.estcp.org/Technology/ER-0221-FS.cfm. Accessed June 11, 2008.

Borden RC. 2006. Development of a Design Tool for Planning Aqueous Amendment Injection Systems. Project ER-0626 Fact Sheet. ESTCP, Arlington, VA, USA. http://www.estcp.org/Technology/ER-0626-FS.cfm. Accessed June 11, 2008.

Borden RC. 2007a. Effective distribution of emulsified edible oil for enhanced anaerobic bioremediation. J Contam Hydrol 94:1–12.

Borden RC. 2007b. Concurrent bioremediation of perchlorate and 1,1,1-trichloroethane in an emulsified oil barrier. J Contam Hydrol 94:13–33.

Borden RC. 2007c. Engineering delivery of insoluble amendments. Partners in Environmental Technology Technical Symposium & Workshop, Washington, DC, USA. December 4–6, 2007, Technical Session No. 1A. Abstract available at http://www.serdp-estcp.org/Symposium/upload/C_Technical_Session_All.pdf. Accessed June 11, 2008.

Borden RC, Beckwith WJ, Lieberman MT, Akladiss N, Hill SR. 2007. Enhanced anaerobic bioremediation of a TCE source area at the Tarheel Army Missile Plant using EOS. Remediat (Summer 2007):5–19.

Butler EC, Hayes KF. 1999. Kinetics of the transformation of trichloroethylene and tetrachloroethylene by iron sulfide. Environ Sci Technol 33:2021–2027.

Coates JD, Michaelidou U, Bruce RA, O'Connor SM, Crespi JN, Achenbach LA. 1999. Ubiquity and diversity of dissimilatory (per)chlorate-reducing bacteria. Appl Environ Microbiol 65:5234–5241.

Coulibaly KM, Borden RC. 2004. Impact of edible oil injection on the permeability of aquifer sands. J Contam Hydrol 71:219–237.

ESTCP (Environmental Security Technology Certification Program). 2006a. Protocol for Enhanced *In Situ* Bioremediation using Emulsified Edible Oil. ESTCP, Arlington, VA, USA. http://www.estcp.org/viewfile.cfm?Doc=ER%2D0221%20Final%20Protocol%20V2%2Epdf. Accessed June 11, 2008.

ESTCP. 2006b. Edible Oil Barriers for Treatment of Perchlorate Contaminated Groundwater. Project ER-0221 Final Report. ESTCP, Arlington, VA, USA. http://www.estcp.org/viewfile.cfm?Doc=ER%2D0221%2DFR%2D01%2Epdf. Accessed June 11, 2008.

Gingras TM, Batista JR. 2002. Biological reduction of perchlorate in ion exchange regenerant solutions containing high salinity and ammonium levels. J Environ Monit 4:96–101.

Henry B. 2006. Loading Rates and Impacts of Substrate Delivery for Enhanced Anaerobic Bioremediation. Project ER-0627 Fact Sheet. ESTCP, Arlington, VA, USA. http://www.estcp.org/Technology/ER-0627-FS.cfm. Accessed June 11, 2008.

ITRC (Interstate Technology & Regulatory Council). 2005. Permeable Reactive Barriers: Lessons Learned/New Directions. ITRC, Washington, DC, USA. http://www.itrcweb.org/Documents/PRB-4.pdf. Accessed June 11, 2008.

Knox S, Lieberman MT, Borden RC. 2005. Monitored Natural Attenuation of Perchlorate in Groundwater. Proceedings, Eighth International *In Situ* and On-Site Bioremediation Symposium, Baltimore, MD, USA, June 6–9, 2005, Paper A-20.

Kovacich MS, Beck D, Rabideau T, Pettypiece E, Smith K, Noel M, Zack MJ, Cannaert MT. 2007. Full-Scale Bioaugmentation to Create a Passive Biobarrier to Remediate a TCE Groundwater Plume. Proceedings, Ninth International *In Situ* and On-Site Bioremediation Symposium, Baltimore, MD, USA, May 7–10, 2007, Paper I-18.

Lieberman MT, Zawtocki C, Borden RC, Birk GM. 2004. Remediation of Perchlorate and Trichloroethane in Groundwater Using Edible Oil Substrate (EOS®). Proceedings, 2004 National Ground Water Association Conference on MTBE and Perchlorate: Assessment, Remediation and Public Policy, Costa Mesa, CA, USA, June 3–4, 2004, ISBN#1-56034-111-4.

Lieberman MT, Borden RC, Hatzinger PB, Sturchio NC, Böhlke JK, Gu B. 2006. Isotopic fractionation of perchlorate and nitrate during biodegradation in an EOS® biobarrier. Poster presented at Partners in Environmental Technology Technical Symposium & Workshop, Washington, DC, USA. November 28–30, 2006.

Logan BE. 2001. Assessing the outlook for perchlorate remediation. Environ Sci Technol 35:482A–487A.

Logan BE, Zhang HS, Mulvaney P, Milner MG, Head IM, Unz RF. 2001. Kinetics of perchlorate- and chlorate-respiring bacteria. Appl Environ Microbiol 67:2499–2506.

Long CM, Borden RC. 2006. Enhanced reductive dechlorination in columns treated with edible oil emulsion. J Contam Hydrol 87:54–72.

Mace RE, Fisher RS, Welch DM, Parra SP. 1997. Extent, mass, and duration of hydrocarbon plumes from leaking petroleum storage tank sites in Texas. Geological Circular 97-1. Bureau of Economic Geology University of Texas, Austin, TX, USA.

Morse JJ, Alleman BC, Gossett JM, Zinder SH, Fennell DE, Sewell GW, Vogel CM. 1998. Draft Technical Protocol: A Treatability Test for Evaluating the Potential Applicability of the Reductive Anaerobic Biological *In Situ* Treatment Technology (RABITT) to Remediate Chloroethenes. ESTCP, Arlington, VA, USA. http://www.estcp.org/viewfile.cfm?Doc=Rabitt%5FProtocol%2Epdf. Accessed June 11, 2008.

Newell CJ, Connor JA. 1997. Characteristics of Dissolved Petroleum Hydrocarbon Plumes: Results from Four Studies. Bulletin No. 8. American Petroleum Institute, Washington, DC, USA.

Rice DW, Grose RD, Michaelsen JC, Dooher BP, MacQueen DH, Cullen SJ, Kastenberg WE, Everett LG, Marino MA. 1995. California Leaking Underground Fuel Tank (LUFT) Historical Case Analyses. UCRL-AR-122207. California State Water Resources Control Board, Underground Storage Tank Program, Sacramento, CA, USA.

Sturchio NC, Hatzinger PB, Arkins MD, Suh C. Heraty LJ. 2003. Chlorine isotope fractionation during microbial reduction of perchlorate. Environ Sci Technol 37:3859–3863.

Tan K, Anderson TA, Jackson WA. 2004. Degradation kinetics of perchlorate in sediments and soils. Water Air Soil Pollut 151:245–259.

CHAPTER 9

PERMEABLE ORGANIC BIOWALLS FOR REMEDIATION OF PERCHLORATE IN GROUNDWATER

Bruce M. Henry,[1] Michael W. Perlmutter[2] and Douglas C. Downey[3]

[1]Parsons, Denver, CO 80290; [2]CH2M HILL, Atlanta, GA 30328; [3]CH2M HILL, Denver, CO 80439

9.1 INTRODUCTION

Enhanced *in situ* anaerobic bioremediation using permeable organic biowalls can be an effective method of degrading perchlorate in groundwater, and is anticipated to be a cost-effective and widely applicable treatment technology for shallow groundwater plumes. Biowalls constructed with organic substrates such as mulch and compost are intended to provide a long-term source of organic carbon to stimulate anaerobic biodegradation of contaminants over periods of several years, typically 3 to 5 years without modification based on experience at the former Naval Weapons Industrial Reserve Plant (NWIRP) in McGregor, Texas (CH2M HILL, 2006). Biowall systems may be replenished with fluid substrates (e.g., emulsified vegetable oil) to operate over longer periods of time.

Permeable biowalls are used in a biobarrier configuration to intercept and degrade perchlorate (and other contaminants subject to anaerobic degradation processes) dissolved in shallow groundwater (e.g., Perlmutter et al., 2000). Biowalls may be an effective strategy to treat perchlorate in groundwater for large plumes having poorly defined, widely distributed, or inaccessible source areas. For example, biowalls may be employed upgradient of a property boundary or point of regulatory compliance to prevent plume migration to potential receptors.

9.1.1 Applications to Date

The use of biowalls for perchlorate has been demonstrated by the Navy at NWIRP McGregor, McGregor, Texas. Several permeable organic biowalls were installed at NWIRP McGregor in 1999 as part of state-approved interim remediation measures for perchlorate in groundwater. The remedy was approved as operating properly and successfully in June 2006, allowing for transfer of the property to the City of McGregor (CH2M HILL, 2006). Several mulch biobarriers have been installed for remediation of perchlorate in groundwater for industrial clients in Virginia and Arkansas by Environmental Alliance, Inc., although data regarding these sites have not been published to date (personal communication, Kevin Morris).

Several other permeable mulch biowalls have been installed for remediation of chlorinated solvents in groundwater at Air Force Bases in Nebraska, Oklahoma, Delaware, Missouri, Wyoming, and South Dakota (AFCEE, 2008). The Army also has installed biowalls with state approval to treat chlorinated solvents at the Seneca Army Depot Activity in New York and at the Red River Army Depot in Texas (personal communication, Farrukh Ahmad). This experience should increase the ability to transfer this technology to remediation of perchlorate at other sites, and also facilitate acceptance at the state level of biowalls to treat perchlorate.

9.1.2 Technology Description

Remediation of groundwater using permeable organic biowalls relies on the flow of groundwater under a natural hydraulic gradient through a biowall trench to promote contact with organic matter. Microbial biodegradation of organic carbon within the biowall produces reducing conditions that support the anaerobic degradation of perchlorate. A "biowall" is one form of biobarrier, but the term is reserved for systems consisting of a trench filled with organic materials that serve as a permeable barrier creating reducing conditions within and downgradient of the trench.

Backfill materials suitable for use in biowalls (e.g., tree mulch, compost, sand and gravel) can be readily obtained at relatively low cost compared to other organic substrates commonly used for enhanced *in situ* anaerobic bioremediation (e.g., emulsified vegetable oil or hydrogen release compound [HRC®]), or reagents used for chemical reduction (e.g., zero-valent iron [ZVI]). The past performance of most biowalls indicates that organic backfill materials such as tree mulch and compost can sustain reducing activity for at least 3 to 5 years. Therefore, biowall systems can be expected to operate effectively over periods of 3 to 5 years with minimal operation, maintenance and monitoring (OM&M) costs other than periodic performance monitoring.

Periodic amendment of permeable biowalls with supplemental substrate may be required after a period of 3 years, or when the biowall can no longer meet its remediation goals. Biowall designs often include piping to facilitate substrate amendment. Usually the cost of an amendment event is low compared to installation costs. The low OM&M requirements of biowalls offer a substantial cost savings compared to other enhanced bioremediation systems that require frequent or periodic injection of fluid substrates and maintenance of injection systems.

Advantages related to use of permeable organic biowalls include:

- Biowalls are effective for shallow groundwater plumes in very low to moderate permeability or highly heterogeneous formations because the trench physically removes the formation and replaces it with a homogeneous mixture of mulch and sand. In this way, all of the contaminant mass passing through the biowall in groundwater is subject to contact with the anaerobic reaction zone.
- Mulch, compost, sand and gravel are relatively inexpensive when purchased in bulk quantities. Tree mulch can often be obtained for the cost of shipping and handling alone.
- Biowalls require little OM&M other than periodic performance monitoring. However, they may need to be periodically replenished with substrate to sustain anaerobic degradation processes when the remedy can no longer meet its remediation goals.
- Trenches can be readily modified to include permanent wells or perforated pipe for addition of fluid substrates to supplement substrate loading, as needed.

Limitations of permeable biowalls include:

- The depth that can be trenched in a practical and cost-effective manner is limited to approximately 11 to 12 meters (~35 to 40 feet). In addition, trenching may interfere with site infrastructure and utilities.
- The effectiveness of a biowall to treat high concentrations of perchlorate may be limited, particularly when the rate of groundwater flow and/or concentrations of native electron acceptors are also high. The contaminant retention time in the trench and the

rate at which organic carbon is added to the groundwater passing through the trench may be limited under these conditions. In some cases the use of wider trenches or multiple parallel trenches may be necessary to treat high concentrations of perchlorate or to deplete high concentrations of native electron acceptors.

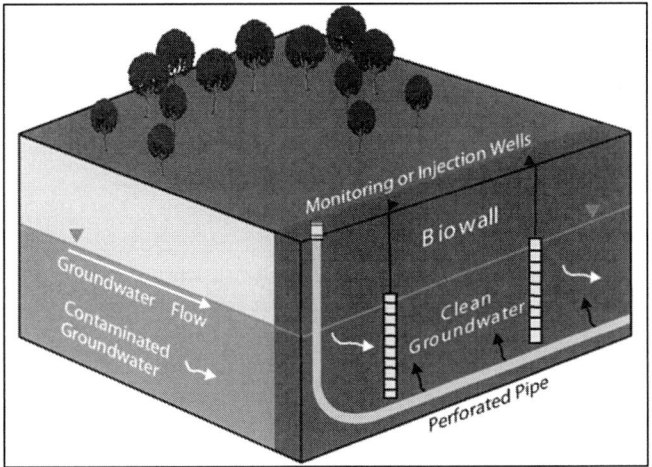

Figure 9.1 Biowall conceptual design (from AFCEE, 2008)

Figure 9.1 illustrates a conceptual design for a permeable organic biowall. The biowall trench should extend to an aquitard or past the total depth of contamination. In many cases, perforated pipes (vertical or horizontal) are installed within the biowall for future addition of slow release substrates such as emulsified vegetable oil. Biowalls may also be recharged with substrate without dedicated piping by using direct-push techniques.

9.2 SITE SUITABILITY

Not all sites are suitable for application of permeable organic biowalls. Site screening criteria that should be evaluated during the technology selection process include:
- Site infrastructure and use,
- Contaminant concentration and distribution (particularly depth),
- Hydrogeology, and
- Groundwater geochemistry.

Site selection for biowalls requires subsurface site characterization and development of an adequate conceptual site model. This conceptual model needs to include the depth and stratigraphic horizons of perchlorate contamination, the rate and direction of groundwater flow, and aquifer geochemistry. In general, microorganisms that facilitate degradation of perchlorate are ubiquitous in the shallow subsurface, and the microbiology of the subsurface can be assumed adequate unless adverse geochemical conditions exist. Site screening criteria are summarized in Table 9.1, and the following subsections further describe these screening criteria.

9.2.1 Land Use and Infrastructure

Trenching may interfere with utilities or roadways, and may not be practical in many situations. Trenching around utilities or other infrastructure may leave gaps in the biowall system allowing for contaminant bypass, although multiple and overlapping trench sections located up and downgradient of the obstruction may be effectively used. A flat and stable ground surface is typically required to accommodate heavy construction equipment for biowall installation. Finally, potential settling and compaction of biowall materials should be taken into consideration if roadways or buildings are to be built over a biowall trench.

Table 9.1 Suitability of Site Characteristics for Biowalls

Site Characteristic	Suitable for Biowalls	Suitability Uncertain	Suitability Unclear - Requires Further Evaluation
Infrastructure and Land Use	No infrastructure or utilities to interfere with trenching or excavation.	Some utilities (*e.g.*, sewer lines) or roadways may be moved or temporarily breached during construction.	Presence of buildings or utility lines that cannot be breached, leaving gaps in the biobarrier.
Perchlorate and Co-contaminants	Concentrations of Perchlorate <50 ppm	Concentrations of perchlorate > 100 ppm	Mixed contaminant plumes
Lithology	Low permeability, cohesive silts and clays or silty sands. Weathered or poorly consolidated bedrock.	-	Loose sands and gravels, well consolidated or hard bedrock.
Depth	<10 meters (<33 feet) to base of contaminant plume	11 to 14 meters (35 to 45 feet) to base of contaminant plume using benching	Contamination >15 meters (>50 feet deep), beyond practical depth of trenching or excavation.
Hydraulic Conductivity	$<3 \times 10^{-4}$ cm/sec (<1.0 ft/day)	3×10^{-6} to 3×10^{-4} cm/sec (1.0 to 10 ft/day)	$>3 \times 10^{-4}$ cm/sec (>10 ft/day)
Groundwater Velocity	<0.3 m/day (< 1.0 ft/day)	0.3 m/day to 3.0 m/day (1.0 ft/day to 10 ft/day)	>3.0 m/day (>10 ft/day)
pH	6.0 to 8.0	5.0 to 6.0	<5.0
Nitrate Concentration (expressed as nitrate)	< 10 mg/L	10 to 20 mg/L	>20 mg/L with caution

Modified from AFCEE, 2008.

Notes: m/day = meters per day; ft/day = feet per day; m/yr = meters per year; ft/yr = feet per year; cm/sec = centimeters per second; mg/L = milligrams per liter; ppm = parts per million.

9.2.2 Contaminant Concentration and Distribution

The distribution and mass of perchlorate in groundwater must be taken into account in designing an organic biowall. The ability to treat high concentrations of perchlorate in groundwater is inversely related to the rate of groundwater flow and directly related to the residence time in the treatment zone. Biowalls installed at NWIRP McGregor have treated concentrations of perchlorate as high as about 20 mg/L. However, for plumes with high rates of groundwater flow, it has not been demonstrated that a single biowall trench can treat perchlorate concentration over 100 mg/L. Even though the treatment zone may be extended in a downgradient direction by soluble organic carbon that is released from the biowall trench, in some cases multiple biowalls may be required to adequately contain the perchlorate plume. The lateral and vertical distribution of perchlorate also must be known, to ensure that the biowall trench will fully intercept the perchlorate groundwater plume and prevent contaminant bypass under or around the biowall.

9.2.3 Hydrogeology

Groundwater velocity is important as it is related to contaminant residence time in the reaction zone, including within the biowall trench and the area downgradient of the biowall where sufficient soluble organic carbon is released to sustain an anaerobic treatment zone. Groundwater seepage velocities greater than 0.3 to 3.0 m/day (1.0 to 10 ft/day) require careful consideration of the design to avoid perchlorate breakthrough. In lieu of such conditions, hydraulic controls could be implemented to encourage the movement of perchlorate in groundwater through the biowall. In addition, the permeability of the biowall must be higher than that of the surrounding formation to prevent flow around the biowall trench.

9.2.4 Geochemistry

Careful consideration should be given to the geochemical conditions, such as aquifer buffering capacity and the potential for pH excursions, as well as the levels of dissolved oxygen (DO) and nitrate, which can act as competing electron acceptors. Sufficient substrate must be available to deplete the likely influx of DO and nitrate into the reaction zone during the expected lifetime of the biowall. Conversely, excessive substrate and low aquifer buffering capacity may cause pH to decrease to adverse levels. Reduction of perchlorate has been reported to be inhibited by pH levels less than 5.7 in aquifer materials from a site in Maryland (Envirogen, 2002).

9.2.5 Co-Contaminants

The presence of co-contaminants should be considered in the design of a biowall system. Chlorinated solvents such as perchloroethene (also termed tetrachloroethene or PCE) and trichloroethene (TCE) may be readily degraded in a biowall system, but the degradation of sequential dechlorination products such as dichloroethene (DCE) and vinyl chloride (VC) may require a longer residence time in the reaction zone to prevent accumulation of these regulated compounds. Other co-contaminants may not be subject to anaerobic degradation processes (e.g., polycyclic aromatic hydrocarbons or PAHs), and it is uncertain whether other contaminants will inhibit perchlorate-reducing microorganisms.

Many potentially limiting site-specific conditions can be mitigated through proper design or design alternatives. After a site is screened and found suitable for application of biowall technology, the next step involves development of a biowall design and appropriate engineering specifications.

9.3 DESIGN OF PERMEABLE BIOWALLS

Key design considerations for biowall applications include:
- Site-specific hydrogeology and contaminant distribution
- Dimensions and configuration (in relation to residence times and degradation rates)
- Selection of fill materials

- Modifications and contingencies
- Regulatory concerns

Design considerations revolve around the site-specific conditions and the appropriate biowall dimensions and configuration to ensure interception and adequate treatment of the contaminant plume. Groundwater flow characteristics and the residence time of the contaminants in the biowall reaction zone are the key features of design. Materials must be selected and procured, and any modifications or contingencies identified to optimize the operation of the biowall system. In addition, there are regulatory considerations for biowalls that must be addressed.

9.3.1 Site-Specific Hydrogeology and Contaminant Distribution

Site-specific hydrogeology and contaminant distribution should be adequately characterized during site selection. If not, additional characterization may be warranted. Parameters to evaluate include the rate and direction of groundwater flow, the vertical and horizontal contaminant distribution, and any potential for preferential flow paths. Additional characterization may also be warranted for local areas within the site, where the biowall is planned to be emplaced, to learn about the bedrock topography (i.e., to key in the biowall) and to confirm the location of any utilities. Additionally, any temporal variations in depth to groundwater or flow direction and velocity should be considered in the design, because the biowall may be in place for several years and such hydrogeological variations can lead to overflow or bypassing of the biowall.

9.3.2 Dimensions, Configuration and Residence Time

The biowall design must ensure sufficient residence time for perchlorate (and any co-contaminants) to degrade to performance objectives within the treatment zone. Estimates of the maximum groundwater seepage velocity should be used for residence time calculations. A conservative estimate of the potential degradation rates should be used, particularly when these are based on past experience or reported rates in the literature instead of site-specific testing. The maximum contaminant concentration and the anticipated degradation rate can be used to determine the residence time required to meet remedial objectives. The groundwater flow velocity is then used to determine the biowall dimensions needed to create a reaction zone of sufficient residence time. Multiple biowalls may be required to achieve sufficient contaminant residence time in the reaction zone at sites with high perchlorate concentrations, rapid groundwater flow velocities, and/or high concentrations of competing electron acceptors (e.g., DO or nitrate).

The lateral and vertical dimensions of a biowall system must be sufficient to avoid contaminant bypass or breakthrough. If possible, the biowall should be keyed into a confining layer. This may not always be possible, as current trenching techniques suitable for installation of mulch and sand mixtures are limited to 8 to 11 meters (25 to 35 feet in depth). Deeper trenches may be constructed using biopolymer slurries, but it may be difficult to install the mulch mixture at depth in a thick slurry and the cost of biopolymer slurry installation is typically not competitive relative to enhanced anaerobic bioremediation using direct injection of fluid substrates. The biowall should at least extend below the depth of contamination to be treated. It is also important that the biowall have a greater permeability than the surrounding formation, or a portion of the impacted groundwater may flow around the biowall, over the biowall (if mounding occurs), or even underneath it (if the biowall is not keyed into bedrock or a confining horizon).

9.3.3 Biowall Materials

Selection of biowall materials is often influenced by the materials that are locally available. Tree mulch is a common source of organic material for biowalls. Table 9.2 summarizes important features of some biowall materials, but it is not inclusive and there are many other potential sources of organic material. The table also lists common bulking materials that are used to maintain biowall permeability and geotechnical integrity as the organic materials decompose over time.

Mulch is an excellent growth medium for microorganisms, and will maintain its structure under saturated, anaerobic conditions for many years. Wood mulch is composed primarily of cellulose, hemicellulose, and lignin. Cellulose is readily degraded under both aerobic and anaerobic conditions. However, lignin is not readily degraded under either aerobic or anaerobic conditions and much of the cellulose within the plant cell wall is bound in lignin. It is not unusual to find wood materials in landfills that are 20 years or more old. Because wood mulch is degraded very slowly under anaerobic conditions, biowalls are often amended with other sources of organic materials that provide a greater amount of bioavailable organic carbon. Examples include compost, cotton gin trash (leaves and hulls from removing the cotton bolls), and vegetable oil.

Analysis of biowall materials for organic carbon content and analysis of nutrients such as nitrogen, phosphorous, and potassium can be used to optimize selection of biowall materials. If the quality of the materials is in question, bench-scale tests (e.g., column studies) may be conducted to assess substrate options (e.g., Perlmutter et al., 2000; Ahmad et al., 2007).

Sand or pea gravel is typically used to maintain permeability and to prevent compaction, commonly comprising 50 percent or more of the total trench volume. Crushed limestone may also be used if needed to buffer the pH close to neutral.

9.3.4 Recharge Options and Alternative Configurations

Additional substrate will be required to sustain reducing conditions once the bioavailable organic carbon in the biowall materials is depleted. The most common and practical approach is to replenish the biowall with a long-lasting, fluid substrate such as emulsified vegetable oil. This approach has been used at the former NWIRP McGregor, Texas (see Section 9.7).

Alternative configurations include groundwater recirculation within biowalls and *in situ* bioreactors. While permeable biowalls are designed primarily as passive flow-through reactive barriers, they may be readily modified to capture and extract groundwater for recirculation through the treatment zone. *In situ* bioreactors are treatment cells constructed of mulch and sand that are installed in source area excavations (Parsons, 2006; Appendix E.3 in AFCEE, 2008), and may be coupled with recirculation to create a larger treatment zone.

Table 9.2 Characteristic of Materials Used for Permeable Biowalls

Material	Availability	Usefulness
Organic Materials		
Tree Mulch	Readily available	Bulk source of cellulose; easily obtained.
Cotton Gin Trash	Common in Mid-West United States	Biodegradation activator; high nitrogen content.
Cottonseed Meal	Common in Mid-West United States	Biodegradation activator; high nitrogen content. Availability may be limited due to use of cottonseed meal cake as a high protein feed source.
Alfalfa Hay	Common	High nitrogen content.
Mushroom Compost	Common	Dark, rich, moist mixture of wheat straw, peat moss, cottonseed meal, cottonseed hulls, corncobs, cocoa bean shells, gypsum, lime, chicken litter, horse stable bedding, and/or other organic materials used in commercial mushroom farms.
Chitin	Common near Coastal Areas	Supports anaerobic degradation; longevity may be limited.
Rice Hulls	Common on Gulf and West Coasts of the United States	Longevity may be limited; may plug pore spaces.
Bulking Materials (to Maintain Trench Permeability)		
Sand	Readily available	Maintain permeability, reduce compaction.
Gravel	Readily available	Maintains permeability and reduces compaction; also useful as a weighting material.
Limestone	Varies locally	Crushed gravel for pH control.

9.3.5 Regulatory Compliance

Two regulatory concerns must often be addressed when considering the use of a biowall. One is the potential for contaminants in the mulch or compost material, such as herbicides or pesticides. Fortunately, most commercial pesticides, herbicides, and defoliants are biodegradable and typically degrade within a period of two to four weeks. Stockpiling of mulch usually provides sufficient time for degradation of any residual pesticides or herbicides to occur. However, it may be necessary to sample and analyze organic materials to satisfy regulatory concerns.

Perhaps of greater concern is the management of the materials excavated during trenching. Costs associated with off-site disposal of contaminated materials may make the technology more costly than a biobarrier created by injected substrates such as emulsified vegetable oil. Since biowalls are typically installed across contaminated groundwater plumes, and not in source areas, the potential for trench spoils to present a hazard is proportional to the concentration of the contaminant present, its sorption potential, and the amount of organic carbon in the aquifer sediments. Perchlorate has a low potential for sorption, and should not accumulate within the aquifer matrix. In most cases the materials will not pose a risk, but a management plan with confirmation sampling may be required. As a contingency, the materials may be managed on site using land treatment or composting.

9.4 BIOWALL INSTALLATION AND CONSTRUCTION

9.4.1 Construction Methods

There are three commonly used methods for installation of biowall trenches. The easiest and most cost effective is use of a *conventional backhoe* in soils that are not prone to cave-in or sloughing. For small trenches less than a few hundred feet in length and less than about 6 meters (20 feet) in depth, conventional trenching with trench boxes and shoring (if needed) may be the most economical alternative. Needed equipment is readily available, and these methods are less expensive relative to the high cost of mobilizing specialized chain trenchers and long-arm excavators. This approach was used to install the biowalls at Seneca Army Depot Activity, New York (Appendix E.1 of AFCEE, 2008), Whiteman Air Force Base (AFB), Missouri (CH2M HILL, 2004), and Air Force Plant 4, Texas (Wice et al., 2006). But in general, the number of sites where conventional trenching can be used will be few.

More commonly, biowall trenches are installed using a *continuous one-pass trencher*. Continuous chain trenchers are capable of rapidly installing permeable biowalls in a one-pass operation where the trench is cut and the biowall material emplaced in one continuous operation (Figure 9.2). Continuous chain trenchers are currently capable of trenching to depths of 11 meters (35 feet) in unconsolidated sediments. An additional 1.5 to 3.0 meters (5 to 10 feet) of depth may be gained by excavating a bench for the trenching rig, but the extent of benching is dependent on the difference in elevation between the ground surface and water table. Therefore, benching may not be an option at sites with a shallow water table.

Figure 9.2. Continuous one-pass trencher in operation at Ellsworth Air Force Base, South Dakota (from AFCEE, 2008)

The trencher is a track-mounted vehicle that has a cutting boom resembling a large chain saw (i.e., a linked chain belt with cutting teeth). A steel box with a hopper assembly is fitted atop the cutting boom. The cutting boom excavates a trench by simultaneously rotating the cutting chain and advancing the boom until the desired depth of excavation relative to the ground surface has been achieved. The steel box and hopper assembly provide for stabilization of the trench sidewalls during excavation and subsequent placement of the sand and mulch mixture, which is introduced through the feed hopper. Simultaneous excavation and placement of backfill materials eliminates concerns associated with open excavations.

Large backhoes equipped with long-arm booms (called *long-arm excavators*) are also capable of trenching to depths of 9 to 12 meters (30 to 40 feet) or more. Because the trench walls are not supported during excavation, use of a biopolymer slurry is typically required to maintain an open trench for installing the biowall materials. However, cost estimates to use this technique have not been competitive compared to continuous one-pass trenchers or to using injectable substrates.

9.4.2 Quality Assurance/Quality Control

The overall goal of a construction Quality Assurance/Quality Control (QA/QC) program is to ensure that proper construction techniques and procedures are used and to verify that the biowall installation meets project specifications. Additionally, the QA/QC program is used to identify and define problems that may occur during construction, and ensure that these problems are corrected before construction is complete.

The key elements of a QA/QC program include a project QC manager, who is responsible for implementing the QC Program, QC meetings (pre-construction, progress, and deficiency meetings), submittals, inspections and testing, and documentation. The inspection and testing component of the QA/QC program for the installation of a biowall often focuses on the following:

- Weights, volumes, quantity, and quality of delivered materials (e.g., mulch, gravel, or geotextile fabric).
- The width, depth, and location of the trench relative to surveyed benchmarks. In addition, critical geological features (e.g., observation of clay lenses and/or approximate elevation of the groundwater table) are often recorded as part of the trench inspection program.
- Characteristics of the organic trench backfill and compacted surface materials. QC inspections are frequently conducted to ensure that the mixture is being prepared with suitable materials and with the required ratios, that the maximum amount of delivered materials are placed into the trench, and minimal loss occurs due to mixing, transport, and installation.
- Fabrication and installation of a piping system for injection of supplemental organic substrate.

At completion of the work, a construction completion report is commonly prepared to document that construction has been performed in accordance with the design standards and specifications. The main emphasis of the QC program is careful documentation during the entire process, from the selection of materials through the installation of equipment.

9.4.3 Waste Management Plan

A Waste Management Plan (WMP) provides general information for the characterization, handling, transportation and disposal of wastes, which includes discarded materials and

waste-like materials, but also includes environmental media such as groundwater and soil when managed as waste. At a minimum, the waste streams associated with the construction of a mulch biowall may include:
- Uncontaminated and/or contaminated soil from trench installation,
- Uncontaminated and/or contaminated groundwater collected during trench installation,
- Personal protective equipment (PPE) and spent sampling equipment, and
- Uncontaminated general construction debris (such as caution tapes, barricades, signs, packing materials, trees and shrubs).

The following elements should be included in the WMP to address the wastes generated during the mulch biowall construction:
- Waste characterization/profile.
- Management of soil and groundwater. Previously collected investigation data may be used to assess the final disposal location of the excavated soil or groundwater. As discussed above, because mulch biowalls are commonly installed at the distal end of plumes, waste soil and groundwater are often disposed onsite. Alternatively, offsite disposal at a regulated landfill may be required.
- Stored waste management. Hazardous wastes may be accumulated for less than 90 days from the date of generation, while other wastes should be removed from the site as soon as possible.
- Transportation and disposal.

9.5 PERFORMANCE MONITORING

Methods commonly used to assess the effectiveness of *in situ* enhanced anaerobic bioremediation using permeable organic biowalls include evaluations of changing contaminant concentration/mass over time or distance, changes in hydrogeology and groundwater geochemistry, and an evaluation of the efficiency (rate) and extent of biodegradation. Intermediate degradation products of perchlorate include chlorate and chlorite. These ions have detection limits that are much higher than those for perchlorate, and generally do not accumulate during perchlorate biodegradation. However, monitoring for chlorate and chlorite is often conducted for enhanced anaerobic bioremediation of perchlorate.

Groundwater contaminant and geochemical data are collected during system monitoring to document that appropriate aquifer redox and geochemical conditions have been attained, and to identify any adverse conditions that may reduce the efficacy of the biowall system in degrading perchlorate. Evaluation of field data as it applies to bioremediation of chlorinated solvents is described in further detail in USEPA (1998), AFCEE et al. (2004), and AFCEE (2008).

9.5.1 Biogeochemistry

Various lines of evidence are used to determine if a biowall has stimulated anaerobic conditions conducive to the degradation of perchlorate. Changes in biogeochemical conditions that are commonly evaluated for biowall systems include native electron acceptors, general indicators of redox conditions, and availability of organic substrate.

Native Electron Acceptors. Native electron acceptors that may be preferred over perchlorate during anaerobic biodegradation include DO and nitrate (often measured as nitrate nitrogen). For example, nitrate may be preferred over perchlorate by microorganisms capable of utilizing both nitrate and perchlorate as electron acceptors (Herman and Frankenberger, 1999). Iron reducing and sulfate reducing microorganisms, and fermentative methanogens, may compete for and consume an adverse amount of available substrate but are not likely to inhibit perchlorate reduction. In general, concentrations of DO less than 0.5 mg/L and concentrations of nitrate less than 0.5 mg/L are desirable to stimulate anaerobic biodegradation of perchlorate (Chaudhuri et al., 2002; Coates and Achenbach, 2004).

Oxidation-Reduction Potential. The oxidation-reduction potential (ORP) of groundwater (often referenced to a hydrogen electrode [Eh] in the laboratory, but typically referenced to a silver/silver chloride [Ag/AgCl] electrode when using field meters) is a measure of electron activity and an indicator of the relative tendency of a solution to accept or transfer electrons. Redox reactions in groundwater containing organic substrates are usually biologically mediated, and therefore the ORP of a groundwater system reflects the biodegradation processes that are occurring. Therefore, ORP is a common measurement to document that appropriate reducing conditions have been achieved. In general, a negative ORP value is necessary to stimulate anaerobic perchlorate biodegradation.

Substrate Availability. The amount of substrate available for microbial activity is typically determined by measuring soluble organic carbon in groundwater. Measurement of total organic carbon (TOC) or dissolved organic carbon (DOC, filtered samples) is most common. Volatile fatty acids (VFAs) may be measured as an indication of the amount of metabolic acids present in groundwater. In general, concentrations of TOC greater than 10 mg/L are desired to stimulate perchlorate biodegradation (EnSafe, 2005).

9.5.2 Perchlorate Degradation

The effectiveness of a permeable organic biowall is measured by reductions in perchlorate concentration or mass. Reductions in post-installation concentrations of perchlorate relative to pre-installation baseline conditions, or relative to concentrations upgradient of the biowall, can be used to show that biodegradation is occurring. Plots of perchlorate concentration over time or along a path of groundwater flow through the biowall reaction zone can be useful in evaluating the effectiveness of the anaerobic treatment zone.

If biodegradation has been stimulated in a biowall, an increase in the rate of biodegradation of perchlorate should be observed. Calculation of a biodegradation rate constant prior to and after system construction may help demonstrate the effectiveness of the application. Degradation rate constant estimates can be calculated by many methods; USEPA (1998) and Newell et al. (2002) provide examples and discussion for estimating biodegradation rate constants.

A simple method for measurement of perchlorate degradation rate is to calculate an average degradation rate based on the concentration of perchlorate entering and leaving the treatment zone (usually determined by a monitoring location within or immediately downgradient of the biowall), and using an average residence time for perchlorate in the reaction zone. This requires that the hydraulics of the system be well characterized. Perchlorate has a low potential for sorption to the aquifer matrix and the effect of retardation due to sorption is negligible. Methods to calculate residence time may be found in AFCEE (2008).

9.5.3 Sustaining the Reaction Zone

Without some form of source reduction, biowalls may need to be maintained over periods of many years. Mulch and other organic substrates used in biowalls are depleted over time due to consumption by biological processes; therefore, the substrate may need to be periodically replenished over time. The minimum or threshold concentration of substrate required to stimulate and sustain anaerobic degradation of perchlorate may vary from site to site depending on the prevailing groundwater geochemistry. Concentrations of TOC are typically used as an indication of substrate availability, with concentrations of TOC greater than 10 to 20 mg/L commonly thought to be necessary to support anaerobic degradation processes (USEPA, 1998).

However, because mulch and compost are solid substrates and provide an excellent growth medium, an arbitrary TOC threshold alone may not be a good indication that anaerobic degradation is being sustained. For example, the level of TOC necessary to sustain reduction of perchlorate at sites with low nitrate concentrations (less than a few mg/L) is likely to be much lower than a site with high nitrate concentrations (perhaps greater than 10 mg/L).

Therefore, multiple lines of evidence should be evaluated to determine when groundwater conditions within the biowall reaction zone are no longer able to sustain effective degradation of perchlorate. Based on site-specific observations, a number of key indicator parameters may be identified that indicate when the supply of organic substrate within a biowall system should be replenished. This evaluation can be incorporated into an OM&M plan, along with contingencies for replenishing the substrate.

9.6 BIOWALL SYSTEM COSTS

9.6.1 Installation and Trenching Costs

The primary cost for installation of permeable biowalls is for the trenching, which may account for up to 70 percent of the total cost for construction. Mobilization of specialized equipment is a large portion of the trenching cost. For example, mobilization for continuous one-pass trenchers ranges from $20,000 to $60,000, based on trenching bids obtained for various U.S. Air Force and Army projects. The cost per foot of trenching is highly scale-dependent, both in terms of the depth and length of the trench to be installed. Cost per linear foot ranges from $150 to $300, depending on the length of trench and capabilities of the trenching machinery used.

Materials are a relatively low percentage of installation costs, on the order of 10 to 15 percent of the total. The cost of mulch is dependent upon handling and delivery, which may range up to $10 to $15 per cubic yard. Biowall trenches less than 150 linear meters (500 linear feet) can typically be installed for less than $200,000 using a one-pass trencher, including the cost of trenching, materials, and installation of monitoring wells.

Other factors that impact capital construction costs include permitting requirements, installation of piping or recirculation systems, additional amendments (e.g., compost or vegetable oil), surveying, installation of the monitoring network (number and depth of wells), and site restoration.

9.6.2 Operations and Monitoring Costs

OM&M over the first few years after biowall construction consists primarily of performance monitoring. Monitoring on a semi-annual basis is usually sufficient for passive biowall systems. Monitoring costs are proportional to the size of the biowall system and the monitoring network. Annual monitoring and reporting costs may range from $20,000 per year for semi-annual sampling of a small biowall system with only one or two well transects, up to perhaps $200,000 for quarterly monitoring of large scale applications with multiple biowall sections and monitoring transects.

After an initial performance evaluation, the monitoring protocol should be optimized to include only those monitoring locations and sample protocols necessary to document that performance objectives are being achieved and to determine when system optimization (e.g., recharge) is required. In addition to monitoring costs, long-term maintenance may require addition of supplemental organic substrate in the biowall system. An estimated cost to recharge a 90-meter length (300-foot length) of biowall to a depth of 8 meters (25 feet) with emulsified vegetable oil is on the order of $30,000 for a one-time application.

9.6.3 Summary of Life Cycle Costs

The life-cycle cost of a biowall system can be broken down into capital construction and the cost to operate and maintain the system, including performance monitoring. To illustrate the cost of a typical biowall application, costs are presented for the BG05 biowall at Ellsworth AFB, South Dakota (Table 9.3). A 177-meter long by 10-meter deep (580-foot long by 32-foot deep) biowall was installed using a continuous one-pass trencher in June 2005. Total capital costs for system installation were less than $300,000, with the trenching subcontract accounting for over half of that amount. A total of $30,000 was spent on biowall materials.

Table 9.3 Biowall Technology Costs, Site BG05, Ellsworth AFB, South Dakota (from AFCEE, 2008)

Element	Cost ($)
Capital Cost	
Work Plan and Procurement	$19,300
Mobilization/Demobilization/Permitting	$9,600
Site Labor	$38,000
Equipment and Appurtenances	
- Monitoring Wells	$16,800
- Biowall Materials	$30,100
- Monitoring Equipment and Supplies	$3,200
Trenching Subcontractor	$154,600
Baseline Laboratory Analyses	$7,800
Surveying	$1,200
Reporting	$12,600
Total Capital Costs	**$293,200**
Operating Costs (Performance Monitoring)	
Mobilization/Demobilization	$3,000
Site Labor (sampling)	$15,000
Sampling Equipment and Supplies	$4,000
Laboratory Analyses	$14,000
Project Management/Reporting	$6,000
Total Annual Operating Costs (per year, semi-annual sampling events)	**$42,000**

The capital cost also includes work plan development, permitting, mobilization, installation of the monitoring network, baseline sampling, site restoration (grading and seeding), and a construction completion report. The capital construction cost may be compared to the cost of a permeable reactive barrier (PRB) ZVI wall. A 180-meter (600-foot) ZVI wall to a depth of 10 meters (32 feet) would cost over $1,000,000, three times the cost of a comparable size biowall (ITRC, 2005; USEPA, 1999).

The initial annual monitoring (two semi-annual events) and reporting costs of about $42,000 for this demonstration includes mobilization of a field crew, sampling three well transects, and an extensive analyte list. Annual monitoring by a base contractor using an optimized and more streamlined monitoring approach would be closer to $30,000/year for two semi-annual sampling events. In any event, the cost of monitoring is of consequence and may exceed the capital construction cost over a period of 10 years or more.

9.7 CASE STUDY: FORMER NWIRP MCGREGOR, MCGREGOR, TEXAS

NWIRP McGregor was a 39 square kilometer (9,700-acre) government-owned, contractor-operated facility in McGregor, Texas approximately 20 miles southwest of Waco. The U.S. Army Ordnance Corps originally established it in 1942 as the Bluebonnet Ordnance Plant. Over the facility's 50-year history, owners included the U.S. Army, U.S. Air Force, and the U.S. Navy's Naval Air Systems Command (NAVAIR). The NWIRP McGregor mission focused on the research, testing, and manufacturing of solid propellant rocket motors used in missiles such as the Shrike, Sparrow, Phoenix and Sidewinder.

At its operational peak, the facility employed nearly 1,400 people, making it the largest employer in the area. Revenues from the facility supported the economies of many local communities including McGregor, Gatesville, Oglesby, and Valley Mills. Because, the facility closure adversely impacted the local economy, Congress enacted special legislation that would help turn the property over to the City of McGregor for economic redevelopment.

Prior to the facility closure, a multi-phased Resource Conservation and Recovery Act (RCRA) Facility Investigation (RFI) was initiated in 1992. The RFI responded to environmental issues raised in the RCRA Facility Assessment completed by the Texas Commission on Environmental Quality (TCEQ). However, a potential setback for economic redevelopment occurred in March 1998 when perchlorate was identified as a contaminant of concern in both soil and groundwater, with maximum groundwater concentrations up to 91 mg/L.

The TCEQ notified the Navy that the United States Environmental Protection Agency (USEPA) had provided documentation stating that NWIRP McGregor used, stored and disposed of ammonium perchlorate, a contaminant posing an environmental issue throughout the United States. This revelation had the potential of slowing down the goal of cleaning up the site and property transfer.

Perchlorate had migrated from the NWIRP McGregor site onto offsite properties, potentially impacting the drinking water of over 500,000 citizens of central Texas. The subsequent perchlorate-focused environmental investigation and remediation was expedited as a result of the cooperation between the U.S. Navy, the TCEQ, and USEPA. With effective remedies in place, all 39 square kilometers (9,700 acres) of the facility was transferred to the City of McGregor for redevelopment. The new tenants, including Dell Computer Corporation, Ferguson Plumbing Company, General Micrographics, In Situ Forms, SpaceX, and McLennan County Electrical Cooperative, have helped to generate 1,000 new jobs for the City of nearly 6,000 residents.

9.7.1 Fast Track Cleanup and Innovative Technology Implementation

Three technologies have been used to remediate groundwater at NWIRP McGregor: 1) *ex situ* treatment using a biological fluidized-bed reactor (FBR); 2) *in situ* treatment using biowalls and bioborings; and 3) natural attenuation. These technologies, as well as some limited soil treatment, are described below.

9.7.2 *Ex Situ* Groundwater Treatment

As part of interim and full-scale remediation measures in Area M, the Navy constructed more than approximately 1,500 meters (~5,000 feet) of cutoff trenches and surface water holding ponds along the facility's southwest boundary to intercept perchlorate-contaminated groundwater and surface water before it migrated offsite. Subsequently, three 3,800 cubic meter (1,000,000 gallon) soil cells and one 38,000 cubic meter (10,000,000 gallon) pond were constructed in the same area to enhance storage capacity, promote groundwater cleanup, and provide a mixing basin to polish treated water as needed.

Figure 9.3 *Ex situ* **bioreactor at NWIRP McGregor, McGregor, Texas** *(Photo courtesy of NAVFAC Southeast)*

The intercepted groundwater was pumped through a biological FBR that degrades perchlorate to concentrations below laboratory detection limits (<0.43 micrograms per liter [µg/L]) (Figure 9.3). Since January, 2002, more than 2,040 kilograms (4,500 pounds) of perchlorate have been removed from more than 1.5 million cubic meters (400,000,000 gallons) of recovered groundwater. Because of the groundwater collection and treatment system, the perchlorate plume in Area M has significantly retracted from flowing toward Lake Belton, which is the major drinking water resource for central Texas.

9.7.3 *In Situ* Groundwater Treatment

Biowalls and bioborings (large diameter soil borings backfilled with an organic mixture) were used as the primary groundwater remedy at the site except for onsite Area M, where the *ex situ* treatment system was installed. These innovative and cost-effective biological permeable biobarriers quickly evolved from rudimentary bench-scale studies conducted in 1999 to full-scale status, such that nearly three miles of biowalls were installed from 2002 to 2005 to remediate onsite and offsite groundwater. During the initial bench-scale studies, the Navy determined that:

- Site groundwater contained indigenous bacteria that could degrade perchlorate and volatile organic compounds (VOCs) under amended geochemical conditions.
- Biobarriers could be installed in the shallow contaminated aquifer at a fraction of the cost of a conventional pump and treat system. Biobarriers would also be effective at treating groundwater flowing in a weathered and fractured limestone matrix in which secondary porosity features dominated.
- No discharge permits had to be obtained as long as remediation occurred in the ground.
- Biobarriers would not impact site operations (i.e., farming and cattle grazing).

As a result, the Navy developed an aggressive biobarrier remediation construction schedule based on the following key factors and priorities:

- Intercept the groundwater-to-surface water pathway.
- Prevent further groundwater migration to offsite property.
- Remediate offsite contamination on a timetable that would eliminate the need to place institutional controls (deed restrictions) on the groundwater of offsite property owners.

More than 4,000 meters (13,000 feet) of biowall and 1,300 bioborings were installed by the summer of 2005. The biowalls were excavated with a hydraulic excavator (shallow trenches) or rock-trencher (deep trenches) and backfilled with a mixture of mushroom compost, pine wood chips, soybean oil, and limestone aggregate.

As shown in Figure 9.4, construction of the biowalls at NWIRP McGregor is shown in Photographs 1 to 5. In general, after the biowall alignment was cleared and the overburden removed (Photo 1), the biowalls were trenched (Photo 2), and the backfill was prepared (Photo 3) and added to the trench (Photo 4). The preparation of the injection piping is shown in Photo 5; depending on the design, the injection piping is placed on the bottom of the trench before the backfill is applied or near the top after it is added.

Typical construction specifications for the biowall segments are illustrated in Figure 9.5. Diffuser pipes were installed on the bottom of each trench to allow for future injections of soybean oil or other carbon substrates as needed. Rows of bioborings, which

Photo 1 Biowall site preparation

Photo 2 Biowall trenching

Photo 3 Biowall media before placement

Photo 4 Media backfilling

Photo 5 Injection pipe construction

Figure 9.4. Construction of biowalls at NWIRP McGregor *(Photo courtesy of NAVFAC Southeast)*

194 B.M. Henry et al.

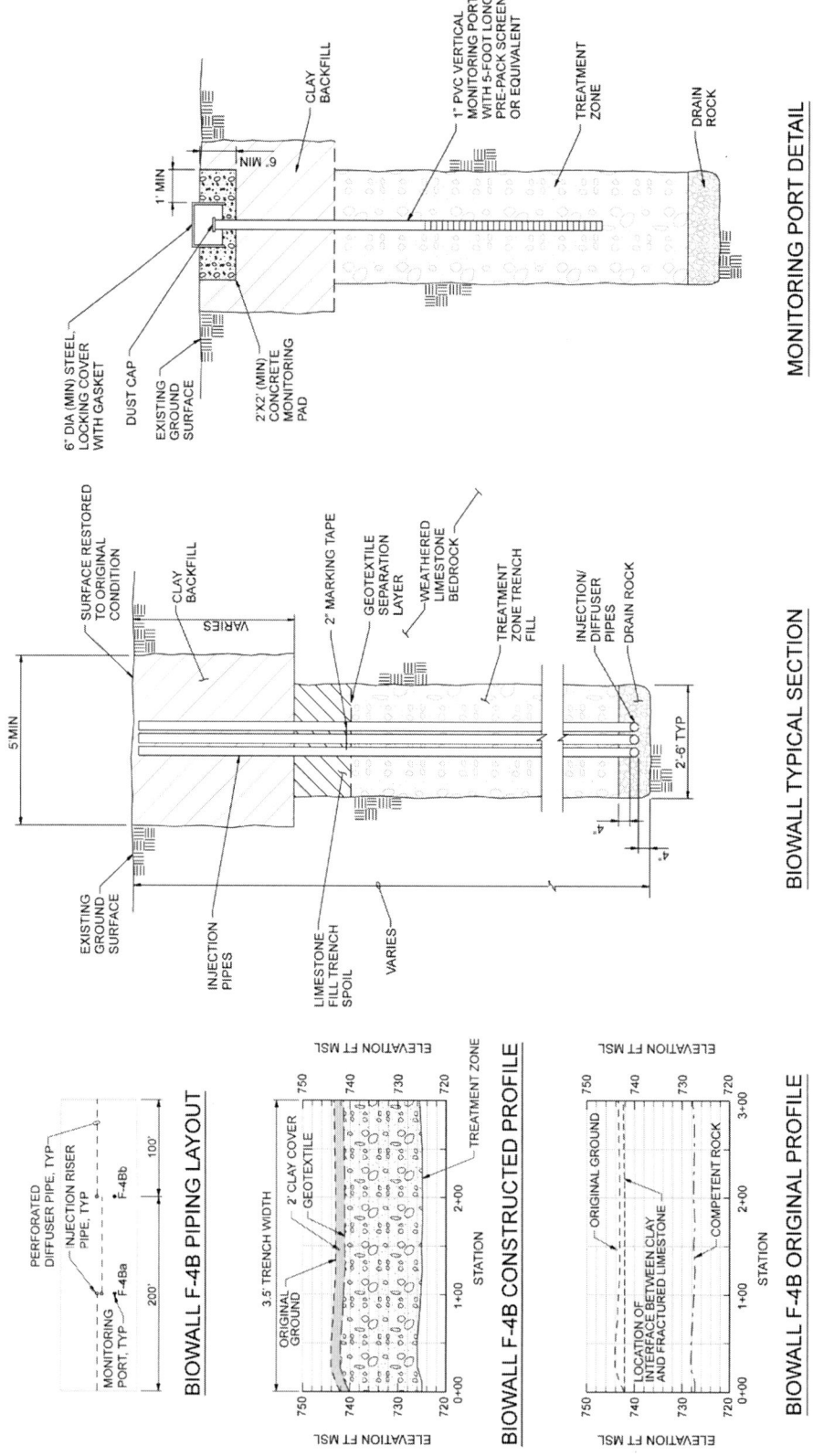

Figure 9.5. Typical biowall construction specifications at NWIRP McGregor, McGregor, Texas (from CH2M HILL, 2007)

consisted of multiple, closely-spaced 30 centimeter (12-inch) diameter soil borings backfilled with the biobarrier media, were installed where biowall trench construction was difficult.

The trenches were capped with a compacted clay layer to limit seeps and surface infiltration. Reducing conditions were quickly established within the biowalls, leading to perchlorate removal. Monitoring results indicated that the biowalls had reduced perchlorate concentrations from approximately 1,000 µg/L to below the laboratory detection limit. The perchlorate protective concentration limit (PCL) at NWIRP McGregor is 51 µg/L.

In August, 2006, after four years of passive operation, nearly 10,000 kilograms (22,000 pounds) of emulsified vegetable oil and dilution water were injected into 15 onsite and offsite biowall segments in Area S, where the initial sections of the full-scale system were installed in 2002. The addition of supplemental organic carbon was triggered by routine monitoring results conducted as part of long-term OM&M. Groundwater collected from monitoring ports within several of the Area S biowalls, the oldest on the site, had TOC and perchlorate concentrations and ORP levels that indicated that the original source of organic substrate was no longer sustaining reducing conditions required to meet perchlorate treatment goals. Only one organic carbon rejuvenation event has been completed to date. Therefore, the duration of emulsified vegetable oil effectiveness is not currently known. The impact of the supplemental substrate injection on long-term biowall effectiveness will continue to be evaluated with ongoing long-term groundwater monitoring at the site. Re-injection is expected to occur annually or as needed to maintain reducing conditions. Because the groundwater flow rates are highly sensitive to precipitation at NWIRP McGregor, the re-injection frequency may be variable as the perchlorate mass discharge varies from year to year.

9.7.4 Natural Attenuation in Groundwater

Perchlorate concentrations throughout the groundwater plumes at NWIRP McGregor continue to decrease via natural attenuation (dilution) at a rate of 5 to 7 percent every year, which complements engineered perchlorate remediation. This natural decrease can be attributed to the following factors:
- The facility is closed; no additional ammonium perchlorate has been delivered to the site since 1998.
- All significant soil sources have been identified and removed.
- Perchlorate is highly mobile and moves with groundwater.
- Due to site geology and hydrogeology, rain easily infiltrates into the shallow aquifer further diluting the perchlorate plumes.

The use of strategically placed biobarriers and natural attenuation has proven to be a highly cost-effective approach, saving the U.S. Navy millions of dollars while achieving the offsite cleanup goals.

9.7.5 *Ex Situ* Soil Treatment

Three multipurpose soil cells were constructed in 2001 to address perchlorate-contaminated soil and source area groundwater in Area M. These soil cells were designed to provide the following functions:
- Soil treatment units using anaerobic land farming.
- Amendment infiltration basins for source area groundwater treatment.

- Groundwater treatment lagoons for *ex situ* treatment system polishing.
- Supplemental water storage units during the wet season.

The soil cells enabled anaerobic soil treatment while allowing carbon- and nutrient-rich water to infiltrate into the perchlorate-contaminated aquifer. Subsequent water, soil, and sediment analyses indicate that the soil cells have successfully reduced perchlorate concentrations below laboratory detection limits.

9.7.6 Operations and Maintenance

Monitoring results from early biowalls suggest that the effective duration of treatment without replenishing with additional substrate may range from 3 to 5 years or more. An OM&M plan is useful to help maintain biowall performance over time. A formal OM&M plan was developed by EnSafe (2005) for the former NWIRP McGregor biowall system.

The NWRIP McGregor monitoring protocol was based on an optimization study of the site (EnSafe, 2005). Parameters evaluated during the optimization study included perchlorate, DO, nitrate, methane, ORP, pH, TOC, VFAs, humic and fulvic acids, and dissolved hydrogen. TOC was deemed to be the most useful parameter that indicated effective biodegradation of perchlorate. Depletion of TOC followed a first order rate, and the minimum range at which breakthrough occurred appeared to be between 5 and 10 mg/L. Native microbial populations that utilize perchlorate as an electron acceptor may preferentially use nitrate for metabolism. Perchlorate degradation was observed to be sensitive to the presence of nitrate at low concentrations of nitrate ranging from 0.1 to 0.5 mg/L (i.e., perchlorate reduction diminished when nitrate reduction diminished).

ORP was also a useful indicator. Increases in ORP to greater than -50 millivolts (mV) may be the first evidence of impending perchlorate breakthrough, although this did not occur at all locations. Another parameter that appeared to be useful was the concentration of methane. Methane indicates highly reducing conditions, much more reducing than required to sustain perchlorate degradation. However, a decrease in methane to less than 2.0 mg/L appeared to indicate depletion of the biowall substrate, and a good correlation was observed between a reduction in methanogenesis and a reduction in the rate of perchlorate degradation.

Given these observations, the parameters chosen for quarterly monitoring at NWRIP McGregor included perchlorate, VOCs (where present), TOC, ORP, nitrate, methane, DO, and pH. DO and pH were retained primarily as stabilization parameters for well purging. A scoring matrix was established to determine when to recharge the biowalls. The scoring matrix included perchlorate, TOC, ORP, nitrate, and methane. Perchlorate and TOC were weighted higher than the other parameters; methane was weighted the least. Other considerations included the number of sample locations indicating recharge was required. For example, recharge is initiated when two or more of four total sample locations in a biowall section indicate recharge is needed.

In summary, multiple lines of evidence should be used to determine when recharge is necessary. The recharge protocol should take into account any temporal divergences from ideal conditions; confirmation of groundwater conditions over two consecutive monitoring events may prevent unnecessary or excessive recharge activities. The parameters most useful to determine when to recharge will be highly site-specific.

Many organic substrates can be injected into a biowall or bioreactor, including soluble substrates (e.g., lactate, molasses, or fructose) and slow-release substrates (e.g., emulsified vegetable oil and HRC®). Emulsified vegetable oil was selected for recharge of biowalls at the

NWIRP McGregor site, and is the substrate most commonly considered for biowall recharge (AFCEE, 2008). Emulsified vegetable oil is a suitable substrate for recharge of biowall systems based on (1) ability to distribute the substrate throughout the biowall matrix, (2) its longevity, and therefore the frequency of injections that will be required, and (3) relatively lower cost compared to other slow-release substrates.

9.8 SUMMARY

Biowalls can be effective for treating shallow groundwater plumes in both homogeneous and highly heterogeneous formations having low to moderate permeability. It is yet to be determined whether the achievable retention times and substrate loading rates in permeable biowalls can be sufficient for degrading concentrations of perchlorate in excess of 100 mg/L, or for treating sites with groundwater velocities greater than 0.3 to 3.0 m/day (1 to 10 ft/day). The use of wider trenches (greater than 0.6 m [2 ft] in width) or multiple parallel trenches may be necessary to treat higher perchlorate concentrations at sites with high rates of groundwater flow or with high concentrations of native electron acceptors (i.e., DO or nitrate).

REFERENCES

AFCEE (Air Force Center for Environmental Excellence). 2008. Technical Protocol for Enhanced Anaerobic Bioremediation using Permeable Mulch Biowalls. Prepared by Parsons Infrastructure and Technology Group, Inc., Denver, CO, USA.

AFCEE, Naval Facilities Engineering Service Center (NFESC), and the Environmental Security Technology Certification Program (ESTCP). 2004. Principles and Practices of Enhanced Anaerobic Bioremediation of Chlorinated Solvents. Prepared by Parsons Infrastructure & Technology Group, Inc., Denver, CO, USA. http://www.afcee.af.mil/shared/media/document/AFD-071130-020.pdf. Accessed June 12, 2008.

Ahmad F. 2007. Personal communication with Bruce Henry, Parsons, Denver, CO, USA.

Ahmad F, Schnitker SP, Newell CJ. 2007. Remediation of RDX- and HMX-contaminated groundwater using organic mulch biowalls. J Contam Hydrol 90:1-20.

Chaudhuri SK, O'Connor SM, Gustavson RL, Achenbach LA, Coates JD. 2002. Environmental factors that control microbial perchlorate reduction. Appl Environ Microbiol 68:4425-4430.

Coates JD, Achenbach LA. 2004. Microbial perchlorate reduction: Rocket-fuelled metabolism. Nat Rev Microbiol 2:569-580.

CH2M HILL. 2004. Treatability Study Environmental Cleanup Plan for Whiteman Air Force Base, Site DP-32, Disposal Pit; Old Hospital Incinerator. Prepared for Whiteman Air Force Base, Missouri. March.

CH2M HILL. 2006. Draft Remedial Action Effectiveness Report. Prepared for the Naval Weapons Industrial Reserve Plant (NWIRP) McGregor, McGregor TX, USA.

CH2M HILL. 2007. Response Action Effectiveness Report. Naval Weapons Industrial Reserve Plant, McGregor, TX. Revision No. 02. Prepared for Naval Facilities Engineering Command, Southeast Division. November.

EnSafe, Inc. 2005. Operation and Maintenance Manual for Biowalls, NWIRP McGregor, McGregor, Texas. Prepared for the Naval Facilities Engineering Command, North Charleston, SC, USA.

Envirogen, Inc. 2002. In Situ Bioremediation of Perchlorate. Project CU-1163 Final Report. Prepared for the Strategic Environmental Research and Development Program (SERDP), Arlington, VA, USA. http://www.estcp.org/viewfile.cfm?Doc=CU%2D1163%2DFR%2D01%2Epdf. Accessed June 12, 2008.

Herman DC, Frankenberger WT. 1999. Bacterial reduction of perchlorate and nitrate in water. J Environ Qual 28:1018-1024.

ITRC (Interstate Technology & Regulatory Council). 2005. Technical and Regulatory Guidance Permeable Reactive Barriers: Lessons Learned/New Directions. PRB-4, ITRC Permeable Reactive Barriers Team. February. http://www.itrcweb.org/Documents/PRB-4.pdf. Accessed June 12, 2008.

Morris K. 2007. Personal communication with Erica Becvar, AFCEE, Brooks City-Base, San Antonio, TX, USA.

Newell CJ, Rafai HS, Wilson JT, Connor JA, Aziz JA, Suarez MP. 2002. Calculation and use of first-order rate constants for monitored natural attenuation studies. EPA/540/S-02/500. USEPA National Risk Management Research Laboratory, Cincinnati, OH, USA. http://www.epa.gov/ada/download/issue/540S02500.pdf. Accessed February 25, 2008.

Parsons. 2006. Final Technical Report, Bioreactor Demonstration at Landfill 3, Altus Air Force Base, Oklahoma. Prepared for ESTCP and Altus AFB, OK, USA. November.

Perlmutter MW, Britto R, Cowan JD, Patel M, Craig M. 2000. Innovative technology: In situ biotreatment of perchlorate-contaminated groundwater. Proceedings 93rd Annual Conference, Air Waste Management Association, Salt Lake City, UT, USA, June 18-22, 2000. CD 00-283 Session #WR-2G, Paper #00-214.

USEPA (U.S. Environmental Protection Agency). 1998. Technical Protocol for Evaluating Natural Attenuation of Chlorinated Solvents in Groundwater. EPA/600/R-98/128. National Risk Management Research Laboratory, Office of Research and Development, USEPA, Cincinnati, OH, USA.

USEPA. 1999. Field Applications of In Situ Remediation Technologies: Permeable Reactive Barriers. EPA 542-R-99-002. USEPA Office of Solid Waste and Emergency Response, Washington, DC, USA. April. http://www.epa.gov/swertio1/download/remed/field-prb.pdf. Accessed June 12, 2008.

Wice RB, Rogers R, Walters G. 2006. Biowall Design, Installation, and Performance Monitoring at Landfill No. 3, Air Force Plant 4, Fort Worth, TX. Presentation at the 2006 Air Force Symposium, Session #24, February 27.

CHAPTER 10

COST ANALYSIS OF *IN SITU* PERCHLORATE BIOREMEDIATION TECHNOLOGIES

Thomas A. Krug,[1] Christopher Wolfe,[2] Robert D. Norris[3] and Carrolette J. Winstead[4]

[1]Geosyntec Consultants, Inc., Guelph, ON Canada; [2]HydroGeologic, Philadelphia, PA 19103; [3]Brown & Caldwell, Longmont, CO 80501; [4]Brown & Caldwell, Phoenix, AZ 85004

10.1 BACKGROUND

Once the effectiveness and reliability of different remedial approaches have been established, cost becomes a significant factor in selecting a remedial alternative. Evaluation of costs for various remedial approaches requires estimates of not just the initial capital costs, but also the costs of operations and maintenance (O&M), and the necessary monitoring over the life of the project. Such a life-cycle cost analysis can help responsible parties to evaluate not just the total costs of different options, but also the rate and timing of spending. Some remedies will require larger up front investments but have lower life-cycle costs, while other remedies may have relatively low up front costs, but higher O&M costs when compared to other options.

The costs of site remediation begin with the initial or Phase I investigations and continue through closure. The early investigations including preparation of remedial investigation work plans and evaluation of data are relatively independent of the final remedy. As work proceeds, the range of potential remedies may become smaller and investigation activities may begin to include some sampling and testing that is remedy-specific and the investigation and evaluation costs amongst various remedies may begin to diverge.

If a site is not sufficiently delineated, an inadequate or inappropriate remedy might be selected based on an insufficient understanding of the mass distribution of perchlorate and/or the site geology/hydrogeology. Insufficient delineation can also lead to treating portions of the aquifer not requiring treatment or missing areas that must be addressed later, often requiring additional mobilizations and longer times to achieve closure. In the case of an *in situ* treatment approach, it is critical to develop an accurate conceptual model of the geology and hydrogeology to understand how (and whether) the appropriate reagents can be delivered in sufficient quantities to reach the contaminants of concern (COCs) throughout the treatment area. If the appropriate technology is selected but the limitations on delivery are not understood, costs overruns may ensue as modifications to the system are required during implementation.

The Remedial Action Objectives (RAOs) selected for the site can also have a significant impact on remediation costs. The cost of remediation and possibly the remedy selection/design will differ depending on whether federal maximum contaminant levels (MCLs), risk based concentrations, background levels, state cleanup levels or some other site-specific objectives are required. More achievable RAOs may result from site-specific risk assessments or in cases where monitored natural attenuation (MNA) is allowed as a polishing step. Open communication with regulators and the public can often improve the likelihood of identifying and implementing appropriate site-specific cleanup standards, as well as obtaining approval of novel and potentially more cost-effective remedies, including MNA as a polishing step or to address the distal portions of the plume.

To ensure that the most appropriate remedy is selected for a specific site, some level of costing must be conducted for alternatives that can reliably achieve the RAOs. The accuracy of these cost estimates will depend in large part on the completeness of the designs for each alternative with early stage designs imparting more uncertainty than more developed ones. Unfortunately, because the design process has costs of its own, detailed engineering designs are not often prepared for all of the alternatives being considered, and some costing uncertainty must therefore be expected. This uncertainty is the reason that screening level estimates used in a feasibility study are generally expected to have an accuracy range of −50% to +100% (USACE/USEPA, 2000).

It is also important to realize that different remedial approaches will have different capabilities, limitations, and performance risks. As a result, it may be difficult to obtain a straight "apples to apples" cost comparison when evaluating different approaches.

This chapter provides some insight into the comparative costs of the four *in situ* perchlorate bioremediation approaches covered previously and highlights how some changes in site conditions may impact the costs associated with each approach. Because no two sites are identical, the reader should only use the information contained in this chapter as a guide from which to develop his/her own cost estimates. This chapter discusses the costs for the four alternatives to treat perchlorate-impacted groundwater but does not consider other key factors in selecting a remedial approach such as effectiveness, reliability, potential impacts on secondary water quality, and the potential for the specific approach to reduce the permeability of the treatment zone and cause diversion of groundwater around the treatment zone. Site owners may select a more expensive remedial alternative if there is greater certainty that the alternative can achieve RAOs or if the alternative avoids unwanted side effects, such as negative impacts on secondary water quality characteristics or the potential to divert impacts on groundwater around the treatment zone.

10.2 COSTING METHODOLOGY

Approaches used to compare costs of alternative technologies often include a comparison of various remedies historically applied to different sites (McDade et al., 2005) or the use of a single template site (Quinton et al., 1997). The former approach makes use of published data based on actual experience, but does not allow for a direct comparison under identical conditions. If enough sites are included in the study data base, average costs and factors impacting costs can be compared. However this would not take into account the possibility that combinations of site conditions at the historical sites might have deemphasized (or overemphasized) impacts of certain cost drivers. On the other hand, a comparison based on a single template site might be inappropriate if the template site is substantially different than the site of interest.

The approach presented here is a refinement of the template approach. Initially, we developed a template site that is representative of shallow relatively permeable sites where each of the four *in situ* perchlorate treatment technologies (active biobarrier, semi-passive biobarrier, passive injection biobarrier, and passive trench biowall) that were presented in the preceding chapters would be appropriate. The four authors of the perchlorate treatment technology chapters were asked to generate cost estimates for their respective technologies based on the assumed site characteristics of the hypothetical template site representing a typical site with perchlorate-impacted groundwater. The cost estimates were assembled and evaluated to provide insight into the comparative capital, O&M, and monitoring costs of four separate perchlorate remediation approaches and to identify the cost drivers for each of these approaches. A cost estimate was also prepared for a conventional pump and treat system to provide a point of comparison with this approach.

The costs developed for comparison included: capital costs, operations and maintenance, and long-term monitoring for treatment of plumes of groundwater containing perchlorate. For this exercise, capital costs included design and permitting activities, mobilization, site preparation, well installation, chemical reagents, and management and derived waste disposal. Labor associated with the planning and implementation of the above-mentioned categories is also included.

Specifically excluded from consideration in this exercise are the costs of pre-remedial investigations (e.g., plume delineation, risk determination, and related needs) and treatability studies on the basis that all of the remedial technologies involve bioremediation and the costs for these activities will be similar for each. The remedial alternatives evaluated herein are focused on treatment of a contaminated plume of groundwater and costs for possible source zone treatment are not included. In reality, it may be appropriate to treat source areas which may contain a significant mass of perchlorate and contribute slowly to elevated concentrations in groundwater. A perchlorate "source" may take a variety of forms including:

- Perchlorate in the geological media above the water table (the "vadose zone") which is carried into the groundwater by water infiltrating from the surface and flushing the perchlorate into the groundwater.

- Perchlorate in the vadose zone which dissolves into the groundwater as groundwater elevations increase (possibly on an intermittent basis) and saturate the vadose zone containing the perchlorate.

- Perchlorate disposed below the water table in a manner that allows the perchlorate to be released into the groundwater over an extended period of time.

- Perchlorate which was released into the groundwater at high concentrations and diffused into low hydraulic conductivity (K) units in the geological media and which continue to diffuse out of the low K units as the upgradient source of perchlorate is depleted.

If the "source" material is not treated, it may continue to feed the plume for an extended period of time and it may be necessary to operate the remedial action to treat the plume for a longer period of time until the source zone is sufficiently depleted to allow water quality objectives to be achieved. The active, semi-passive, and passive remedial approaches described earlier in this document could be used in a modified configuration to treat certain source areas below the water table, but estimating the costs for these applications in treating source areas is beyond the scope of this costing exercise. Sources of perchlorate above the water table may be treated using other approaches such as enhanced flushing of the vadose zone.

The O&M costs considered in this analysis include mobilization to conduct O&M activities, labor costs for O&M related activities, replacement equipment and supplies, such as electron donor required for ongoing system operation. Long-term O&M costs for the different alternatives have a high degree of uncertainty because none of the technical approaches considered have actually been operated for anywhere near 30 years and the O&M requirements beyond 5 to 6 years are based on extrapolations from short-term operating experience.

Long-term monitoring cost estimates were developed to include costs for field sampling, analysis and regulatory reporting.

To obtain a clearer picture of life-cycle costs for the base case examples, we developed a spreadsheet to calculate future costs and the Net Present Value (NPV) of future costs. The spreadsheet provides cash flow analysis for up to 30 years, showing the costs by category for each year. The future costs are only carried forward for 30 years because the NPV of future costs beyond the 30-year timeframe are small, and the future costs beyond the 30-year period of time are difficult to predict. O&M and monitoring costs are discounted, using a 3% discount rate, to develop the NPV estimates of future costs (DoD, 1995). The discount rate of 3% was chosen based on the U.S. Federal Government Office of Management and Budget (OMB) "Real Interest Rates on Treasury Notes and Bonds" for 20-year and 30-year notes and bonds, which is currently set at 2.8% (OMB, 2008).

Post remediation and decommissioning costs were not included. Furthermore, we have not made provisions for replacing permanent wells, if this were necessary.

We also assessed the impacts of changes in individual site characteristics and design parameters on the costs of each of the alternative treatment technologies. Using the template site's properties as a base case, design parameters (depth to groundwater, contaminant plume width, groundwater velocity, etc.) were varied one at a time. The four authors were then asked to identify how capital and O&M costs would be affected as a result of each parameter modification. This combination of the template site cost comparison and the parameter specific impact analyses was designed to help the reader improve the understanding of the comparative analysis and predict to some extent how design modifications will impact costs.

10.3 TEMPLATE SITE CHARACTERISTICS AND VARIATIONS CONSIDERED

The template base case site represents a situation where perchlorate is present in a shallow groundwater aquifer consisting of homogeneous silty sands and each of the four technologies being considered would likely be effective. The specific site characteristics for the base case are presented in Table 10.1, along with twelve variations on the base case characteristics that were considered in the evaluation of the sensitivity of costs to the various parameters. The physical configuration of the plume and biobarriers or biowall for the base case is shown in Figure 10.1. The costing for the template site and other cases considered assumes that source zone treatment is complete or at least that there is no continuing source of groundwater contamination. If the source is not treated, operation of the biobarrier or biowall beyond the anticipated time period required to achieve clean up objectives would almost certainly be required.

The base case assumes a homogenous silty sand aquifer from a depth of approximately 3 meters (m) (~10 feet [ft]) below ground surface to 12 m (40 ft) below ground surface with a hydraulic conductivity of 0.001 cm/sec, a horizontal gradient of 0.008 m/m and a porosity of 0.25. These aquifer characteristics result in a groundwater seepage velocity of approximately 10 m/year (yr) (~33 ft/yr). The plume of perchlorate-impacted groundwater extends along the direction of groundwater flow for 240 m (800 ft) and is 120 m (400 ft) in width. The concentration of perchlorate at the upgradient side of the plume is 2 mg/L and the concentration on the downgradient side is 1.1 mg/L. Oxygen and nitrate will contribute demand for electron donor and the assumed concentrations of dissolved oxygen and nitrate are 5 mg/L and 15 mg/L, respectively.

Cost Analysis of *In Situ* Perchlorate Bioremediation Technologies

Table 10.1. Summary of Site Characteristics and Design Parameters for Biological Treatment of Perchlorate-Impacted Groundwater

Design Parameter	units	Base Case	Accelerated Clean Up Case	Low Perchlorate Conc. Case	High Perchlorate Conc. Case	Low Donor Demand Case	High Donor Demand Case	Low GW Velocity Case	High GW Velocity Case	Deep GW Case	Thin Interval Case	Thick Interval Case	Narrow Plume Case	Wide Plume Case
		Case 1	Case 2	Case 3	Case 4	Case 5	Case 6	Case 7	Case 8	Case 9	Case 10	Case 11	Case 12	Case 13
Width of Plume	meters	120	120	120	120	120	120	120	120	120	120	120	30	240
	feet	400	400	400	400	400	400	400	400	400	400	400	100	800
Length of Plume	meters	240	240	240	240	240	240	240	240	240	240	240	240	240
	feet	800	800	800	800	800	800	800	800	800	800	800	800	800
Porosity		0.25	0.25	0.25	0.25	0.25	0.25	0.25	0.25	0.25	0.25	0.25	0.25	0.25
Gradient		0.008	0.008	0.008	0.008	0.008	0.008	0.0008	0.016	0.008	0.008	0.008	0.008	0.008
Hydraulic Conductivity*	cm/sec	0.001	0.001	0.001	0.001	0.001	0.001	0.001	0.001	0.001	0.001	0.001	0.001	0.001
Upgradient Perchlorate Concentration	mg/L	2	2	0.4	10	2	2	2	2	2	2	2	2	2
Downgradient Perchlorate Concentration	mg/L	1.1	1.1	0.22	5.5	1.1	1.1	1.1	1.1	1.1	1.1	1.1	1.1	1.1
Nitrate Concentration	mg/L	15	15	15	15	5	30	15	15	15	15	15	15	15
Dissolved Oxygen Concentration	mg/L	5	5	5	5	2	8	5	5	5	5	5	5	5
Depth to Water	m bgs	3	3	3	3	3	3	3	3	30	3	3	3	3
	ft bgs	10	10	10	10	10	10	10	10	100	10	10	10	10
Vertical Saturated Thickness	m	9	9	9	9	9	9	9	9	9	3	15	9	9
	ft	30	30	30	30	30	30	30	30	30	10	50	30	30
GW Seepage Velocity	m/year	10	10	10	10	10	10	1	20	10	10	10	10	10
	ft/year	33	33	33	33	33	33	3.3	66	33	33	33	33	33
Perchlorate Treatment Objective	mg/L	0.0245	0.0245	0.0245	0.0245	0.0245	0.0245	0.0245	0.0245	0.0245	0.0245	0.0245	0.0245	0.0245
Assumed Number of Pore Volumes to Flush Plume		2	2	2	2	2	2	2	2	2	2	2	2	2
Number of Barriers Perpendicular to GW Flow		1	5	1	1	1	1	1	1	1	1	1	1	1
GW Travel Time to Barrier(s)	years	24	5	24	24	24	24	240	12	24	24	24	24	24
Years to Clean Up GW	years	48	10	48	48	48	48	480	24	48	48	48	48	48

Notes: * - Hydraulic conductivity based on uniform silty sand aquifer
▒ - Input parameters changed from base case

bgs – below ground surface; Conc. – Concentration; cm/sec – centimeters per second; ft - feet; GW – groundwater; kg – kilograms; L – liters; m – meters; mg/L – milligrams per liter

Site/Project Parameters: sand with some silt; clay aquitard below aquifer; no source treatment required; no buildings on foot print; no access problems; 100 miles from office; electric power available 100 feet from plume; agencies are accommodating; and no significant stakeholder issues.

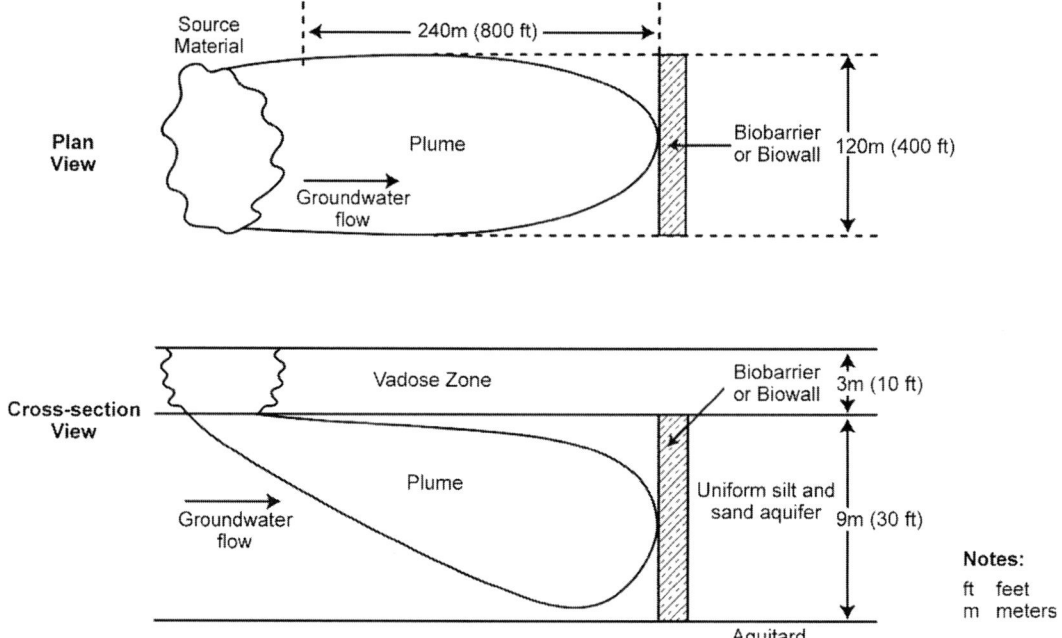

Figure 10.1. Base case plume and biobarrier or biowall configuration

The base case also assumes that two pore volumes of clean water will need to flush through the impacted areas to achieve the cleanup objectives. In reality, the number of pore volumes of clean water required to flush through the subsurface to achieve target treatment objectives will be determined by a number of factors, such as the degree of heterogeneity of the geological media. Variations in the K of the aquifer material can allow a significant fraction of the total mass of perchlorate to diffuse into low K layers and then act as an ongoing source of perchlorate to the higher K zone as the perchlorate is flushed from these more transmissive zones. In most geological settings, more than two pore volumes will be required to achieve treatment objectives, and longer-term operation of the remedial measures will be required. The assumption that two pore volumes of flushing are required to achieve treatment objectives could only be valid for situations where there is very uniform K of the geological media and is likely an optimistic assumption for most real world situations.

The base case design incorporates one biobarrier or biowall on the downgradient edge of the plume to treat water as it flows across the line of the biobarrier or biowall. Based on a groundwater seepage velocity of 10 m/yr (33 ft/yr), a plume that extends for 240 m (800 ft) along the direction of groundwater flow, and the assumed need to flush two pore volumes of clean water through the impacted aquifer to achieve clean up standards, it would be expected to take approximately 48 years for the plume to be treated in the base case. If more than two pore volumes of flushing are actually required to achieve treatment objectives, the biobarrier or biowall would need to be operated well beyond the 30-year timeframe considered in this costing exercise, but the concentrations to be treated would likely be reduced significantly and operating requirements also reduced. The costs of this potential future operation would be incurred more than 30 years into the future and the NPV of these costs would not be as significant as the costs incurred for operation in the near and medium term (i.e., less than 30 years).

The perchlorate treatment objective that was used for the template site was based on the chronic exposure reference dose (and the resulting drinking water equivalent concentration) selected by the U.S. Environmental Protection Agency (USEPA) in 2005 (http://www.epa.gov/iris/subst/1007.htm). It should be noted that various states have since implemented (or are in the process of implementing) MCLs (in the cases of Massachusetts, New Jersey and California), cleanup/action levels, or health-based goals that are lower than this concentration. Because of the uncertainty surrounding the future of perchlorate regulation and due to the relatively small difference (from a concentration perspective) between the 2005 number and the projected state numbers, it was decided that the 2005 USEPA number would be used. Obviously, a lower treatment objective would increase the costs associated with the implementation of any of the treatment approaches presented here.

Each of the bioremediation alternatives considered can achieve low treatment criteria (i.e., below 0.004 mg/L). To achieve lower target treatment criteria, a higher safety factor will be required in the design and operation of each of the remedies such that pockets or layers of low K geological material containing untreated groundwater do not remain or transmit perchlorate in groundwater following treatment. Also, the system may need to be operated for a longer period of time. If a very low target treatment objective is required, even small pockets or layers of untreated groundwater could result in groundwater samples exceeding the target criteria. Layers of low K geological material exist at many sites where inter-bedded clay, silts and sands are present and can serve as longer-term repositories for perchlorate from which diffusion is the dominant transport mechanism. These pockets or layers may release perchlorate to flowing groundwater after treatment of perchlorate in the higher K units has been completed.

As discussed above, the presence of significant low K repositories of perchlorate and low target treatment concentrations would affect the assumption used in the base case that two pore volumes of groundwater need to be flushed through the plume to achieve the target treatment objectives. If additional clean groundwater needs to be flushed through the plume area to achieve lower concentration RAOs then the treatment system will need to be operated for a longer period of time and incur additional long-term O&M and monitoring costs. The additional safety factor in design and possibly longer-term operation will increase costs to achieve lower target treatment objectives but the impact of a specific change in the target treatment concentration is difficult to predict without extensive and very detailed site characterization and contaminant transport modeling.

The first variation of the base case, Case 2: Accelerated Clean Up Case, utilizes five biobarriers or biowalls aligned perpendicular to the direction of groundwater flow and distributed every 48 m (160 ft) within the 240 m (800 ft) long plume. This will provide treatment of the plume at one downgradient and four intermediate locations rather than just at the downgradient edge of the plume. Based on the seepage velocity of 10 m/yr (33 ft/yr) and the assumption that two pore volumes of clean water need to flow through the plume area to achieve clean up, this case will require approximately 10 years to treat the groundwater rather than the 48 years of the base case.

Cases 3 and 4 incorporate reduced and elevated concentrations of perchlorate in groundwater as shown in Table 10.1. Cases 5 and 6 assume lower and higher concentrations of nitrate and dissolved oxygen which will result in a higher and lower demand for electron donor. Cases 7 and 8 incorporate lower and higher groundwater seepage velocities resulting from changes in the hydraulic gradient from the base case. Case 9 assumes that the depth to groundwater is 30 m (100 ft) rather than the 3 m (10 ft) in the base case. Cases 10 and 11 assume thin and thick vertical intervals of 3 m (10 ft) and 15 m (50 ft) rather than the 9 m

(30 ft) of the base case. Cases 12 and 13 assume a narrow plume (30 m [100 ft] in width) and a wide plume (240 m [800 ft] in width) rather than the 120 m (400 ft) width of the base case.

The costs of the four alternative approaches for the base case are presented in Section 10.4, and the cost impacts of the 12 different cases are discussed in Section 10.5.

10.4 COST ESTIMATES FOR BASE CASE SITE CHARACTERISTICS

Tables 10.2 through 10.6 show the estimated capital costs, O&M costs and long-term monitoring costs for implementation of the base case for each of the four technical approaches considered and for a pump and treat alternative to *in situ* biological treatment. The tables also show the NPV and total of the O&M and monitoring costs. The capital costs and NPV of other costs provide the respective life-cycle costs adjusted to take into account the time value of money. Table 10.7 presents a summary of the capital costs, NPV of O&M and monitoring costs and the total NPV of lifetime costs for base case conditions for each of the four alternative approaches to *in situ* bioremediation of perchlorate and the pump and treat alternative considered in this evaluation.

The active biobarrier alternative assumes that a series of four extraction and five injection wells will be installed along the alignment of the biobarrier and a groundwater recirculation system will be constructed to recirculate groundwater and distribute electron donor across the biobarrier. Groundwater will be recirculated between injection and extraction wells and a soluble electron donor will be added to the water being recirculated to distribute the electron donor across the plume of perchlorate-impacted groundwater. The electron donor will promote the biodegradation of the perchlorate in the groundwater in the vicinity of the injection wells. The costing has been developed based on circulating groundwater and adding electron donor on a continuous basis. The capital cost including design, installation of wells, installation of the groundwater recirculation and amendment system, and system start up and testing is approximately $430K and the NPV of the O&M represents an additional $1,200K of costs over a 30-year life. The O&M costs include costs for labor for system O&M, costs for equipment repair and replacement, and cost for electron donor. The NPV of the long-term monitoring costs is estimated to be $350K, to give a total current value lifetime cost for the alternative of $1,980K. The total cost of the remedy over 30 years is estimated to be $2,700K.

The semi-passive biobarrier alternative also assumes that a series of injection and extraction wells will be installed along the alignment of the biobarrier and a groundwater recirculation system will be constructed to recirculate groundwater and distribute electron donor across the biobarrier. Groundwater will be recirculated between injection and extraction wells and a soluble electron donor will be added to the water being recirculated to distribute the electron donor across the plume of perchlorate-impacted groundwater. This initial system installation is identical to the active alternative. The costing has been developed based on circulating groundwater and adding electron over a period of 3 weeks, after which the recirculation system will be shut down for a period of 9 months. Operation will continue on a cycle of 3 weeks of groundwater recirculation and addition of electron donor every 9 months. The capitals costs for the installation would be similar to that of the active system, but the operating costs would be reduced as a result of the reduced operating requirements and reduced potential for biofouling of injection wells. The capital cost, including design, installation of wells, installation of the groundwater recirculation and amendment system, and system start up and testing is approximately $430K, and the NPV of the operation and maintenance represents an additional $780K of costs over a 30-year life. The O&M costs for this alternative are considerably less than for the active alternative because the groundwater recirculation system is being operated on an intermittent basis which results in reduced labor and equipment maintenance costs and also lower costs to deal with biofouling in the injection wells. The NPV of the long-term monitoring costs is estimated to be $350K, to give a total current value cost for the alternative of $1,560K. The total cost of the remedy over 30 years is estimated to be $2,060K.

Table 10.2. Cost Components for Active Biobarrier Treatment of Perchlorate-Impacted Groundwater

	Year Cost is Incurred							NPV of Costs*	Total Costs
	1	2	3	4	5	6	7 to 30		
CAPITAL COSTS									
System Design	90,611	-	-	-	-	-		90,611	90,611
Well Installation	86,292	-	-	-	-	-		86,292	86,292
System Installation	235,862	-	-	-	-	-		235,862	235,862
Start-up and Testing	17,122	-	-	-	-	-		17,122	17,122
SUBCOST ($)	429,887	-	-	-	-	-		429,887	429,887
OPERATION AND MAINTENANCE COSTS									
System Operation and Maintenance	35,759	60,759	60,759	60,759	60,759	60,759	60,759 every year	1,201,630	1,797,770
SUBCOST ($)	35,759	60,759	60,759	60,759	60,759	60,759		1,201,630	1,797,770
LONG TERM MONITORING COSTS									
Sampling/Analysis/Reporting	35,240	35,240	35,240	35,240	35,240	11,780	11,780 every year	348,483	470,700
(Quarterly through 5 years then Annually)									
SUBCOST ($)	35,240	35,240	35,240	35,240	35,240	11,780		348,483	470,700
TOTAL COST ($)	500,886	95,999	95,999	95,999	95,999	72,539		1,980,000	2,698,357

Notes:
NPV - Net Present Value
* NPV calculated based on a 3% discount rate

Table 10.3. Cost Components for Semi-Passive Biobarrier Treatment of Perchlorate-Impacted Groundwater

	Year Cost is Incurred							NPV of Costs*	Total Costs
	1	2	3	4	5	6	7 to 30		
CAPITAL COSTS									
System Design	90,611	-	-	-	-	-		90,611	90,611
Well Installation	86,292	-	-	-	-	-		86,292	86,292
System Installation	235,862	-	-	-	-	-		235,862	235,862
Start-up and Testing	17,122	-	-	-	-	-		17,122	17,122
SUBCOST ($)	429,887	-	-	-	-	-		429,887	429,887
OPERATION AND MAINTENANCE COSTS									
System Operation and Maintenance	24,269	39,269	39,269	39,269	39,269	39,269	39,269 every year	777,780	1,163,070
SUBCOST ($)	24,269	39,269	39,269	39,269	39,269	39,269		777,780	1,163,070
LONG TERM MONITORING COSTS									
Sampling/Analysis/Reporting	35,240	35,240	35,240	35,240	35,240	11,780	11,780 every year	348,483	470,700
(Quarterly through 5 years then Annually)									
SUBCOST ($)	35,240	35,240	35,240	35,240	35,240	11,780		348,483	470,700
TOTAL COST ($)	489,396	74,509	74,509	74,509	74,509	51,049		1,556,151	2,063,657

Notes:
NPV - Net Present Value
* NPV calculated based on a 3% discount rate

Table 10.4. Cost Components for Passive Injection Biobarrier Treatment of Perchlorate-Impacted Groundwater

	Year Cost is Incurred								NPV of Costs*	Total Costs
	1	2	3	4	5	6	7	8 to 30		
CAPITAL COSTS										
System Design	68,100	-	-	-	-	-	-	-	68,100	68,100
Well Installation (30 1" PVC Wells)	32,713	-	-	-	-	-	-	-	32,713	32,713
Substrate Injection	175,784	-	-	-	-	-	-	-	175,784	175,784
Start-up and Testing**		-	-	-	-	-	-	-	0	0
SUBCOST ($)	276,597	-	-	-	-	-	-	-	276,597	276,597
OPERATION AND MAINTENANCE COSTS										
Substrate Injection	-	-	-	166,284	-	-	166,284	166,284 every 3 years	985,956	1,496,556
SUBCOST ($)	0	0	0	166,284	0	0	166,284		985,956	1,496,556
LONG TERM MONITORING COSTS										
Sampling/Analysis/Reporting	35,240	35,240	35,240	35,240	35,240	11,780	11,780	11,780 every year	348,483	470,700
(Quarterly through 5 years then Annually)										
SUBCOST ($)	35,240	35,240	35,240	35,240	35,240	11,780	11,780		348,483	470,700
TOTAL COST ($)	311,837	35,240	35,240	201,524	35,240	11,780	178,064		1,611,036	2,243,853

Notes:
NPV - Net Present Value
* NPV calculated based on a 3% discount rate
** No "Start-up and Testing" costs are included because no operating equipment is left behind following substrate injection

Table 10.5. Cost Components for Passive Trench Biowall Treatment of Perchlorate-Impacted Groundwater

	Year Cost is Incurred										NPV of Costs*	Total Costs
	1	2	3	4	5	6	7	8	9	10 to 30		
CAPITAL COSTS												
System Design	62,100	-	-	-	-	-	-	-	-	-	62,100	62,100
Well Installation	23,217	-	-	-	-	-	-	-	-	-	23,217	23,217
Trench Installation	181,917	-	-	-	-	-	-	-	-	-	181,917	181,917
Substrate Injection	50,000	-	-	-	-	-	-	-	-	-	50,000	50,000
Start-up and Testing**		-	-	-	-	-	-	-	-	-	0	0
SUBCOST ($)	317,234	-	-	-	-	-	-	-	-	-	317,234	317,234
OPERATION AND MAINTENANCE/REAPPLICATION COSTS												
Substrate Injections	-	-	-	-	139,017	-	-	-	139,017	123,017 every 3 years	780,561	1,251,153
SUBCOST ($)	-	-	-	-	139,017	-	-	-	139,017		780,561	1,251,153
LONG TERM MONITORING COSTS												
Sampling/Analysis/Reporting	35,240	35,240	35,240	35,240	35,240	11,780	11,780	11,780	11,780	11,780 every year	348,483	470,700
(Quarterly through 5 years then Annually)												
SUBCOST ($)	35,240	35,240	35,240	35,240	35,240	11,780	11,780	11,780	11,780		348,483	470,700
TOTAL COST ($)	352,474	35,240	35,240	35,240	174,257	11,780	11,780	11,780	150,797		1,446,278	2,039,087

Notes:
NPV - Net Present Value
* NPV calculated based on a 3% discount rate
** No "Start-up and Testing" costs are included because no operating equipment is left behind following substrate injection

Table 10.6. Cost Components for Extraction and Treatment of Perchlorate-Impacted Groundwater

	Year Cost is Incurred							NPV of Costs*	Total Costs
	1	2	3	4	5	6	7 to 30		
CAPITAL COSTS									
System Design	90,611	-	-	-	-	-	-	90,611	90,611
Well Installation	86,292	-	-	-	-	-	-	86,292	86,292
System Installation	292,362	-	-	-	-	-	-	292,362	292,362
Start-up and Testing	25,000	-	-	-	-	-	-	25,000	25,000
SUBCOST ($)	494,265	-	-	-	-	-	-	494,265	494,265
OPERATION AND MAINTENANCE COSTS									
System Operation and Maintenance	49,009	74,009	74,009	74,009	74,009	74,009	74,009 every year	1,469,127	2,195,270
SUBCOST ($)	49,009	74,009	74,009	74,009	74,009	74,009		1,469,127	2,195,270
LONG TERM MONITORING COSTS									
Sampling/Analysis/Reporting	35,240	35,240	35,240	35,240	35,240	11,780	11,780 every year	348,483	470,700
(Quarterly through 5 years then Annually)									
SUBCOST ($)	35,240	35,240	35,240	35,240	35,240	11,780		348,483	470,700
TOTAL COST ($)	578,514	109,249	109,249	109,249	109,249	85,789		2,311,875	3,160,235

Notes:
NPV - Net Present Value
* NPV calculated based on a 3% discount rate

Table 10.7. Summary of Capital Costs and NPV of Costs for Operation, Maintenance and Monitoring for Biological Treatment of Perchlorate-Impacted Groundwater

Alternative	Capital Costs	NPV of 30 Years of O&M Costs	NPV of 30 Years of Monitoring Costs	NPV of 30 Years of Total Remedy Costs	Total 30-Year Remedy Costs
Active Biobarrier	$430	$1,200	$350	$1,980	$2,700
Semi-Passive Biobarrier	$430	$780	$350	$1,560	$2,060
Passive Injection Biobarrier	$280	$990	$350	$1,610	$2,240
Passive Trench Biowall	$320	$780	$350	$1,450	$2,040
Pump and Treat	$490	$1,470	$350	$2,310	$3,160

Notes: All costs are in thousands of dollars
NPV - Net Present Value; current value of future costs based on a 3% annual discount rate
O&M - Operation and Maintenance

The passive injection biobarrier alternative assumes an initial injection of emulsified vegetable oil (EVO) to create a passive biobarrier and reinjection of EVO every 3 years. The EVO will promote biological activity to degrade the perchlorate as groundwater flows through the area where the EVO has been injected. The capital cost for the initial injection is approximately $280K and the NPV of the operation and maintenance costs represents an additional $990K of costs. The capital costs for the passive injection biobarrier are significantly less than for the active and semi-passive alternatives because of the limited infrastructure required to support this alternative. The NPV of the O&M costs for the passive injection approach under the assumptions of the base case are lower than for the active biobarrier but higher than for the semi-passive biobarrier. The NPV of the long term monitoring costs for the passive injection biobarrier alternative is estimated to be $350K, to give a total current value of lifetime costs for the alternative of $1,610K. This total NPV cost is more than for the semi-passive approach but less than for the active approach.

The passive trench biowall alternative assumes an initial installation of a mulch and EVO biowall in a trench aligned perpendicular to the direction of the groundwater flow in order to intercept the plume of impacted groundwater. The biowall is installed using a trenching method and the biowall is rejuvenated 4 and 8 years after installation and then every 3 years thereafter. The organics in the mulch and in the EVO will promote biological activity to degrade the perchlorate as groundwater flows through the biowall. Because the organics in the mulch will eventually be depleted, additional EVO will be added to the biowall on a periodic basis.

The capital cost for the initial installation is approximately $320K and the NPV of the O&M represents an additional $780K of costs. The capital costs for the initial injection are more than the initial costs for the passive injection biobarrier because of the higher costs for construction of a trench biowall relative to the costs for the initial injection of EVO. However, the ongoing O&M costs for the passive trench biowall alternative are lower than for the passive injection biobarrier, because of higher costs for addition of EVO on an ongoing basis for the passive injection biobarrier relative to the trench biowall. The NPV of the long-term monitoring costs is estimated to be $350K to give a total current value lifetime cost for the alternative of $1,450K, giving this the lowest total NPV costs of the four alternatives considered.

The groundwater extraction and treatment or pump and treat system included for comparison would be similar to the active biobarrier system in that a row of extraction and injection wells would be used to bring groundwater to the surface and to re-inject the groundwater. But, rather than just amending the groundwater with electron donor the groundwater would be treated to remove perchlorate prior to reinjection. The groundwater treatment component of this system would be a small-scale bioreactor to degrade perchlorate. The capital cost for this alternative is $490K; somewhat higher than for the active and semi-passive biobarriers at $430K each and significantly higher than for the passive injection biobarriers and passive trench biowalls at $280K and $320K, respectively. The NPV of the O&M costs for the pump and treat approach are estimated to be $1,470K, higher than any of the other *in situ* alternatives considered in the evaluation.

Using the base case site conditions and design parameters, the two passive approaches (passive trench biowall and passive injection biobarrier) have a lower initial capital investment ($320K and $280K, respectively) than both the semi-passive and active biobarrier alternatives ($430K each). The NPV of the O&M costs over the 30-year period considered in this evaluation of the four alternatives, however, changes the overall costs for the different approaches. The NPV of O&M costs for the passive trench biowall and passive injection biobarrier ($780K and $990K, respectively) are significant as a result of the costs to amend the biowall or biobarrier with additional electron donor on a regular basis (every 3 to 4 years). The NPV of O&M costs for the active system are also high ($1,200K) as a result of the labor, equipment, and maintenance costs associated with operating a continuous groundwater recirculation system on an ongoing basis. The NPV of O&M costs for the semi-passive alternative are lower than for the active alternative because the groundwater recirculation system is only operated on an intermittent basis, thus reducing labor costs, equipment maintenance costs and maintenance activities required to control biofouling of injection well. The numbers in Table 10.7 only consider the first 30 years of operation of each of the remedies but it is likely that operation will be required beyond this time period. If operation beyond 30 years is taken into consideration, approaches which have lower O&M costs such as the semi-passive biobarrier and passive trench biowall will look more favorable.

The NPV of long-term monitoring costs are equivalent for each of the four alternatives at $350K. The monitoring costs in these estimates do not account for potential savings in monitoring methods and procedures which may be developed and accepted during the 30-year operating period considered in the estimates.

It must be emphasized that this analysis only holds for the specific template case utilized here and any use of these findings in determining projected costs at a separate, "real-world", site must be done with extreme caution. Consideration must be given to the potential costs associated with issues such as:

- The likelihood that operation beyond the 30 years considered in this evaluation will be required to achieve treatment objectives.

- The potential with some approaches that additional costs may be incurred to address an expanded plume which may result from loss of hydraulic conductivity in the biobarrier or biowall treatment zone diverting groundwater around or beneath the treatment zone.

- The potential with some approaches that excessive amounts of electron donor may have negative impacts on groundwater quality which may need to be mitigated.

10.5 IMPACTS OF CHANGES IN SITE CHARACTERISTICS ON COSTS

As discussed earlier, an assessment was conducted to predict the impact of varying certain site characteristics and design parameters from the base case conditions on the projected costs of the different remedial approaches. A detailed re-costing of each of the 12 alternative cases was not conducted but an assessment of the impact of each of site characteristics/design parameters on the capital, O&M and monitoring costs was conducted and is presented below and summarized in Table 10.8. The impact of each of the changes in site characteristics on the design, O&M and monitoring requirements are discussed below and based on the impacts of the changes in site conditions, a factor change in each of the capital, the NPV of O&M and the NPV of monitoring costs has been estimated relative to the base case. These estimated factor changes relative to the base case are show in the Table 10.8 and are used in Table 10.8 to estimate the capital, NPV of O&M and NPV of monitoring costs for each approach and for each of the alternative cases.

10.5.1 Case 2: Accelerated Clean Up

The accelerated clean up case assumes that five parallel biobarriers will be installed at 48 m (160 ft) intervals within the plume so that perchlorate impacted groundwater will pass through one of the treatment biobarriers within a shorter period of time than in the base case. Based on the seepage velocity of 10 m/yr (33 ft/yr) and an assumption that the aquifer impacted by perchlorate will require flushing with two pore volumes of clean water, the additional biobarriers would be expected to reduce the clean up time from about 50 years with the base case, to about 10 years. Each of the four alternatives could still be used to treat groundwater under this scenario but the capital costs would be expected to increase by a factor of about 4.5. The remedial systems would be five times larger but there would be some economy of scale in the design and implementation of each of the alternatives and the factor increase in costs would be similar for each alternative.

The annual O&M costs for the 10 years of operation would be significantly higher than for the base case (approximately 4 times) but the remedies would be operated for a shorter period of time (10 years versus 30 years). The result is that the NPV of the O&M costs would be approximately 1.75 times the NPV of O&M costs for the base case. Monitoring costs would likely increase for all treatment alternatives. Assuming that groundwater monitoring would be focused on the area downgradient of the most downgradient barrier rather than at each of the individual treatment barriers, the annual monitoring costs for the larger system may be about double the costs for the base case but the monitoring would be required for only 10 years. The NPV of the monitoring costs for 10 years at two times the base case annual monitoring costs is about 25% higher than the NPV of monitoring costs for the base case for 30 years. The monitoring costs for this case would therefore increase by a factor of 1.25.

10.5.2 Cases 3 and 4: Reduced and Elevated Concentrations of Perchlorate

The costs for each of the treatment alternatives would decrease very slightly in the case of reduced perchlorate concentrations and increase very slightly in the case of elevated

Table 10.8. Summary of Impact of Site Characteristics and Design Parameters on Costs for Biological Treatment of Perchlorate-Impacted Groundwater

Design Alternative / Cost Component	Base Case		Accelerated Clean Up Case		Low Perchlorate Concentration Case		High Perchlorate Concentration Case		Low Donor Demand Case		High Donor Demand Case		Low GW Velocity Case		High GW Velocity Case		Deep GW Case		Thin Interval Case		Thick Interval Case		Narrow Plume Case		Wide Plume Case	
	Case 1		Case 2		Case 3		Case 4		Case 5		Case 6		Case 7		Case 8		Case 9		Case 10		Case 11		Case 12		Case 13	
	Cost		Factor	Cost	Factor	Cost	Factor	Cost	Factor	Cost	Factor	Cost	Factor	Cost	Factor	Cost	Factor	Cost	Factor	Cost	Factor	Cost	Factor	Cost	Factor	Cost
Active Biobarrier																										
Capital Cost	$430		4.50	$1,935	0.98	$421	1.05	$452	0.95	$409	1.15	$495	0.90	$387	1.15	$495	1.25	$538	0.90	$387	1.15	$495	0.35	$151	1.85	$796
NPV of O&M Costs	$1,200		1.75	$2,100	0.95	$1,140	1.05	$1,260	0.90	$1,080	1.15	$1,380	0.90	$1,080	1.10	$1,320	1.00	$1,200	0.90	$1,080	1.15	$1,380	0.45	$540	1.75	$2,100
NPV of Monitoring Costs	$350		1.25	$438	1.00	$350	1.00	$350	1.00	$350	1.00	$350	1.00	$350	0.90	$315	1.00	$350	1.00	$350	1.00	$350	0.50	$175	1.50	$525
NPV of Total Costs	$1,980		2.26	$4,473	0.97	$1,911	1.04	$2,062	0.93	$1,839	1.12	$2,225	0.92	$1,817	1.08	$2,130	1.05	$2,088	0.92	$1,817	1.12	$2,225	0.44	$866	1.73	$3,421
Semi-Passive Biobarrier																										
Capital Cost	$430		4.50	$1,935	0.98	$421	1.05	$452	0.95	$409	1.15	$495	0.90	$387	1.15	$495	1.25	$538	0.90	$387	1.15	$495	0.35	$151	1.85	$796
NPV of O&M Costs	$780		1.75	$1,365	0.95	$741	1.05	$819	0.90	$702	1.20	$936	0.90	$702	1.10	$858	1.00	$780	0.90	$702	1.15	$897	0.45	$351	1.75	$1,365
NPV of Monitoring Costs	$350		1.25	$438	1.00	$350	1.00	$350	1.00	$350	1.00	$350	1.00	$350	0.90	$315	1.00	$350	1.00	$350	1.00	$350	0.50	$175	1.50	$525
NPV of Total Costs	$1,560		2.40	$3,738	0.97	$1,512	1.04	$1,621	0.94	$1,461	1.14	$1,781	0.92	$1,439	1.07	$1,668	1.07	$1,668	0.92	$1,439	1.12	$1,742	0.43	$677	1.72	$2,686
Passive Injection Biobarrier																										
Capital Cost	$280		4.50	$1,260	0.98	$274	1.05	$294	0.90	$252	1.30	$364	0.80	$224	1.20	$336	1.35	$378	0.50	$140	1.50	$420	0.35	$98	1.85	$518
NPV of O&M Costs	$990		1.75	$1,733	0.95	$941	1.1	$1,089	0.90	$891	1.40	$1,386	0.80	$792	1.20	$1,188	1.00	$990	0.65	$644	1.40	$1,386	0.45	$446	1.75	$1,733
NPV of Monitoring Costs	$350		1.25	$438	1.00	$350	1.00	$350	1.00	$350	1.00	$350	1.00	$350	0.90	$315	1.00	$350	1.00	$350	1.00	$350	0.50	$175	1.50	$525
NPV of Total Costs	$1,620		2.12	$3,430	0.97	$1,565	1.07	$1,733	0.92	$1,493	1.30	$2,100	0.84	$1,366	1.14	$1,839	1.06	$1,718	0.70	$1,134	1.33	$2,156	0.44	$719	1.71	$2,776
Passive Trench Biowall																										
Capital Cost	$320		4.50	$1,440	0.98	$314	1.05	$336	0.95	$304	1.15	$368	0.85	$272	1.15	$368	NF	NF	0.50	$160	1.60	$512	0.35	$112	1.85	$592
NPV of O&M Costs	$780		1.75	$1,365	0.95	$741	1.1	$858	0.90	$702	1.40	$1,092	0.80	$624	1.15	$897	NF	NF	0.65	$507	1.40	$1,092	0.45	$351	1.75	$1,365
NPV of Monitoring Costs	$350		1.25	$438	1.00	$350	1.00	$350	1.00	$350	1.00	$350	1.00	$350	0.90	$315	NF	NF	1.00	$350	1.00	$350	0.50	$175	1.50	$525
NPV of Total Costs	$1,450		2.24	$3,243	0.97	$1,405	1.06	$1,544	0.94	$1,356	1.25	$1,810	0.86	$1,246	1.09	$1,580	NF	NF	0.70	$1,017	1.35	$1,954	0.44	$638	1.71	$2,482

notes: All costs are in thousands of dollars
Factor - factor increase or decrease in costs relative to the Base Case
NF - not feasible, costs not estimated

NPV - Net Present Value
O&M - Operation and Maintenance

perchlorate concentrations. In the base case, the electron donor demand of the perchlorate makes up a very small percentage (less than 10%) of the total electron donor demand (dissolved oxygen and nitrate make up the bulk of the demand for electron donor) and a change in perchlorate concentrations would have a small impact on the costs for electron donor. The capital costs for the alternatives would change very little with a decrease or increase in perchlorate concentrations. The NPV of O&M costs would decrease or increase somewhat more than capital costs because a slightly lower or higher dose of electron donor would be required on an ongoing basis. The cost of electron donor represents a small fraction of the O&M costs for the active, semi-passive alternatives and the impact of a change in concentration on costs would be small. The O&M costs for electron donor represent a higher fraction of costs for the passive injection and passive trench biowall alternatives and the cost impacts would be somewhat more significant than with the other alternatives.

The greatest impact of an increase in perchlorate concentration would likely be on the total time required to achieve target treatment concentrations. If the initial concentration of perchlorate is higher, it is likely that a greater mass of perchlorate will diffuse into low K layers and the time required for this increased mass of perchlorate to diffuse out of the low K units to be treated will be longer. Since this evaluation only considers the O&M costs for the first 30 years of operation, the impact of this requirement for longer-term operation is not reflected in the numbers in Table 10.8.

10.5.3 Cases 5 and 6: Lower and Higher Electron Acceptor Concentrations

Cases 5 and 6 assume lower and higher concentrations of nitrate and dissolved oxygen in the groundwater to be treated. These changes will result in changes in demand for electron donor that are greater than the changes in electron donor demand due to the changes in Case 3 and Case 4 because the electron donor demand of the nitrate and dissolved oxygen are significantly greater than the electron donor demand of the perchlorate. The decrease in nitrate and dissolved oxygen would drop the electron donor demand by a factor of about 2 and the increase in concentrations would result in an increase in electron donor demand by a factor of about 1.75.

The costs for each of the treatment alternatives would decrease somewhat in the case of reduced concentrations of electron acceptors and increase somewhat in the case of elevated electron acceptor concentrations. The capital costs for the active, semi-passive and passive trench alternatives would change only a small amount with a decrease or increase in nitrate and dissolved oxygen concentrations but capital costs for the passive injection biobarrier would change somewhat more because the initial injection of electron donor represents a greater portion of the capital costs for this alternative. As with changes in perchlorate concentrations, the NPV of O&M costs would decrease or increase somewhat more than for capital costs because a lower or higher dose of electron donor would be required on an ongoing basis. The cost of electron donor represents a small fraction of the O&M costs for the active and semi-passive alternatives but the impact of a change in concentration on costs would be larger than in Case 3 and Case 4. The O&M costs for electron donor represent a higher fraction of costs for the passive injection alternative and the cost impact would be somewhat more significant than with the other alternatives.

10.5.4 Cases 7 and 8: Low and High Groundwater Seepage Velocities

Under these scenarios, the groundwater seepage velocities decrease or increase as a result of changes in the hydraulic gradient of the water table relative to the base case. The capital costs for the active and semi-passive systems would both decrease a small amount with lower velocities and increase with higher velocities. With these systems, the capital cost savings would result from the lower capacity pumps, piping and other equipment required to circulate groundwater at lower rates for the slower groundwater velocity. The capital cost would increase somewhat for the case with the higher groundwater velocity. The annual O&M costs would also decrease or increase with reduced or increased groundwater seepage velocities primarily associated with the reduced or increased amount of electron donor required and lower and higher costs associated with lower or higher groundwater recirculation rates used during the operating phase with these alternatives.

The NPV of the O&M costs for the higher groundwater velocity case decrease from the base case because the higher groundwater velocity will reduce the time to achieve clean up standards. If the assumption that two pore volumes are required to achieve clean up standards holds true, the clean up time would be reduced from 48 years in the base case to 24 years for the high groundwater velocity case. This scenario would reduce the number of years during which O&M and monitoring costs are incurred. The time to achieve clean up with the slower groundwater velocity case extends well beyond the 30 years considered in this cost evaluation and these very long-term costs do not impact the cost estimates presented here.

The capital costs for the passive injection alternative would decrease for lower groundwater seepage velocities because a lower dose of electron donor could be used during the initial implementation relative to the base case. Capital costs for the passive injection alternative would increase for a higher seepage velocity because more electron donor would be used. The NPV of O&M costs would also decrease and increase with lower and higher groundwater velocities as a result of the decrease or increase in the amount of electron donor required. As with the active and semi-passive systems, clean up would be achieved faster with the high groundwater velocity case and the number of years during which O&M and monitoring costs would be incurred would be reduced. This would counteract the increase in the annual O&M costs so the increase in NPV would not be as high as the increase in the annual costs.

The lower or higher seepage velocities would have a small impact on the capital costs for the passive trench biowall because the design of the biowall is not likely to change significantly with the modest changes in groundwater velocity in these two cases. The NPV of the O&M costs are likely to be impacted somewhat as rejuvenation of the mulch biowall would be less frequent for the low groundwater velocity and more frequent for the high groundwater velocity. The increase in O&M costs with the high velocity groundwater will be offset somewhat by the shorter operating time for the remedy with this case.

10.5.5 Case 9: Deep Groundwater

This case assumes that the depth to groundwater is 30 m (100 ft) rather than 3 m (10 ft) as in the base case. This change will have a significant impact on the capital costs for all of the remedial alternatives and eliminates the passive trench biowall alternative from consideration as it would not be practical to install a trench biowall for an aquifer 30 m to 40 m (100 to 130 ft) below the ground surface. The capital costs for the active, semi-passive and passive alternatives will increase as the costs to install injection points and groundwater recirculation wells goes up with the increased depth to water. The capital costs for all three of these alternatives would likely increase by similar amounts as the deeper groundwater would require installation of injection points or wells to a greater depth.

The NPV of the O&M costs for the active and semi-passive systems would not change significantly with the deep groundwater because the additional costs to install deeper wells would be covered in the increased capital costs. The NPV of the O&M costs for the passive alternative will also not increase significantly if permanent wells are installed initially and can be used for subsequent injection of electron donor.

10.5.6 Cases 10 and 11: Thin and Thick Saturated Vertical Intervals

Cases 10 and 11 assume thin and thick saturated vertical intervals of 3 m (10 ft) and 15 m (50 ft) rather than the 9 m (30 ft) of the base case. The capital cost for the active and semi-passive systems would decrease somewhat for the thin vertical interval and increase for the thick saturated interval. The decreased and increased capital costs would be associated with the reduced or increased drilling costs for injection wells and the decreased or increased electron donor because a smaller or larger volume of groundwater would require treatment. The capital costs for the passive injection and the passive trench biowall would likely change by a larger factor because the costs are more directly related to the total thickness of the interval to be treated which determines the amount of electron donor purchased and the depth of the trench biowall.

The NPV of the O&M costs for all of the alternatives would decrease or increase as the groundwater to be treated is smaller or larger with the thin or thick vertical interval. The impact (lower or higher) would be greatest on the NPV of O&M costs for the passive alternatives because the electron donor costs represent a higher percentage of the O&M costs because a greater excess of electron donor is used in the passive alternatives than in the semi-passive or active alternatives.

10.5.7 Cases 12 and 13: Narrow and Wide Plumes

Cases 12 and 13 assume narrow and wide plumes of 30 m (100 ft) and 240 m (800 ft) rather than the 120 m (400 ft) of the base case. The capital costs and NPV of O&M cost for a narrow and wide plume would change nearly in direct proportion to the width of the plume. Capital and NPV of O&M costs will be made up of cost components that are fixed regardless of the size of application, such as some design elements, permitting, reporting and mobilization and cost components that are proportional to the magnitude of the system to be installed. For the treatment alternative considered in this evaluation, capital costs that are fixed regardless of the size of the system may represent 10 to 20 % of the capital costs and 20 to 30% of the O&M costs of the base case costs. The capital costs and O&M costs for the case where the plume is 30 m (100 ft) wide relative to a base case of 120 m (400 ft) would be expected to decrease costs by a factor of about 0.35 and 0.45, respectively. The capital costs and O&M costs for the case where the plume is 240 m (800 ft) wide relative to a base case of 120 m (400 ft) would be expected to decrease costs by a factor of about 1.85 and 1.75.

10.6 SUMMARY

This chapter is not intended to provide detailed cost estimates or site-specific cost comparisons for different approaches. Rather, this chapter provides a point of reference and an analysis of the key cost drivers for different approaches, to assist those who need to develop cost estimates or to assess different alternatives for a particular set of site conditions. This analysis should allow managers and cost estimators to understand the key cost drivers and the factors that are most critical for a given site, and also to ground-truth specific cost estimates.

It is also important to realize that different remedial approaches will have different capabilities, limitations, and performance risks. As a result, it may be difficult to obtain a straight "apples to apples" cost comparison when evaluating different approaches.

The cost comparison approach used in evaluating variations from the base case varied only one parameter at a time and did not take into account the possibility that multiple parameter changes might have a synergistic impact on costs. Additionally, we have not taken into account items such as the fact that dosage requirements might change over time based on observations of performance. An attempt has been made to address changes in monitoring frequency over time in the costing of the base case, but for purposes of comparison, monitoring costs are considered to be constant across the technologies and assumed that the methods and procedures will not be optimized and change over time.

Some general observations regarding the relative costs of the different approaches and the sensitivity of the costs to different factors are discussed below.

- The active biobarrier is generally the most expensive approach. This technique entails both high capital costs, because of the equipment required, and relatively high continuing O&M costs. In particular, the costs for managing biofouling are generally higher for this option than for other alternatives. However, this approach does provide the most consistent concentration of electron donor, which can be an important consideration in at least two ways. First, it minimizes the impacts of electron donor addition on secondary water quality characteristics by minimizing the shift in the oxidation reduction potential (ORP) of the groundwater being treated. This more consistent geochemical environment may be particularly important at sites used for drinking water supply. In addition, the consistent donor supply during active treatment minimizes the total amount of electron donor required. This feature can become significant in several situations, such as Case 4 (a site with a relatively high electron donor demand) and Case 11 (a site with a thick vertical interval) where the electron donor demand is high and the costs for electron donor make up a significant portion of the costs. If the electron donor demand if very high, the active approach may be more cost effective than other approaches because it allows for lower quantities of electron donor to be used than for the passive approaches where the strategy is to provide a significant excess of electron donor.

- The semi-passive approach requires lower O&M costs than the active approach. Although the capital costs for the initial equipment and installation may be similar for the two techniques, the semi-passive approach has significantly lower O&M costs because the system is operated on an intermittent basis, which reduces the labor and maintenance required to operate the system, and the injection wells are significantly less susceptible to biofouling. The semi-passive approach will have more fluctuations in the electron donor concentration than the active approach but careful monitoring can be conducted to minimize the impacts on secondary water quality and operate only slightly above the minimum amount of electron donor required to achieve degradation of perchlorate. In many situations, the semi-passive approach can provide an appropriate balance between the very high level of control on electron donor concentrations but high O&M of the active approach and the excess dosing of electron donor and lower costs of the passive approaches.

- The passive injection biobarrier approach generally has the lowest capital costs. Permanent wells or trenches are not required, reducing capital costs. However, the O&M costs associated with ongoing reinjection of significant excess quantities of electron donor can make the NPV of this approach higher than for the semi-passive approach in most situations. However, it should be recognized that the O&M costs for this approach are very sensitive to the costs for the electron donor used for any ongoing injections. In many field situations, groundwater-monitoring results may show that less electron donor is required during subsequent re-injections, and the costs for electron donor may decrease significantly over time, reducing the O&M costs in later years.

- The passive trench biobarrier approach will often be the least costly technique. It was the most economical option in the base case considered here and in many but not all of the variety of alternate cases considered. The capital costs for initial application of this approach are slightly higher than for the passive injection biobarrier approach, because of the need to excavate the trench, but the O&M costs are generally lower, because the presence of the mulch can reduce the amount of electron donor required on an ongoing basis. As with the passive injection approach, monitoring may show that less electron donor is required during subsequent re-injections and the costs for electron donor may decrease over time. The passive trench biobarrier becomes less favorable and eventually infeasible as the depth of the groundwater interval to be treated increases.

- The most significant cost driver in the cases tested in this chapter was the decision to accelerate the clean up of the entire zone of perchlorate impacted groundwater rather than treating groundwater only at the downgradient limit. Given the size and groundwater flow velocity of the plumes considered in this exercise, several separate biobarrier systems would be required to provide enough coverage of the impacted groundwater to significantly accelerate site restoration. The costs for each of the approaches increases as the number of barriers increases, although there is a relative cost advantage for the two passive approaches as an increasing number of barriers are required, because these approaches have lower capital costs but higher O&M costs relative to the active and semi-passive systems.

- There are several potential risks that are not addressed in this analysis, but should be considered when evaluating costs for a specific site. In particular, there are potential costs associated with the following issues: (1) the likelihood that operation beyond the 30 years considered in this evaluation will be required to achieve treatment objectives; (2) the potential with some approaches that additional costs may be incurred to address an expanded plume which may result from loss of hydraulic conductivity in the biobarrier or biowall treatment zone diverting groundwater around or beneath the treatment zone; and (3) the potential with some approaches that excessive amounts of electron donor may have negative impacts on groundwater quality which may need to be mitigated.

REFERENCES

DoD (Department of Defense). 1995. DoD Instruction 7041.3: Economic Analysis for Decision-Making. November 7, 1995. http://www.dtic.mil/whs/directives/corres/pdf/704103p.pdf. Accessed June 12, 2008.

McDade JM, McGuire TM, Newell CJ. 2005. Analysis of DNAPL source-depletion costs at 36 field sites. Remediation (Spring 2005):9–18.

OMB (Office of Management and Budget). 2008. Discount Rates for Cost-Effectiveness, Lease, Purchase and Related Analysis. http://www.whitehouse.gov/omb/circulars/a094/a94_appx-c.html. Accessed June 12, 2008.

Quinton GE, Buchanan Jr. RJ, Ellis DE, Shoemaker SH. 1997. A method to compare groundwater cleanup technologies. Remediation (Autumn 1997):7–16.

USACE/USEPA (U.S. Army Corps of Engineers/U.S. Environmental Protection Agency). 2000. A Guide to Developing and Documenting Cost Estimates during the Feasibility Study. EPA-540-R-00-002/OSWER 9355.0-75. July. http://www.epa.gov/superfund/policy/remedy/pdfs/finaldoc.pdf. Accessed June 12, 2008.

CHAPTER 11

EMERGING TECHNOLOGIES FOR PERCHLORATE BIOREMEDIATION

Valentine A. Nzengung[1], M. Tony Lieberman[2], Hans F. Stroo[3] and Patrick J. Evans[4]

[1]University of Georgia, Athens, GA 30602; [2]Solutions-IES, Inc. Raleigh, NC 27607; [3]HydroGeoLogic, Inc., Ashland, OR 97520; [4]CDM, Bellevue, WA 98005

11.1 INTRODUCTION

Remediation of perchlorate in groundwater by monitored natural attenuation (MNA), phytoremediation to treat perchlorate in soils and groundwaters, and *in situ* bioremediation of perchlorate sources in the vadose zone, are three promising innovative approaches. These approaches have solid technical underpinnings, but there are uncertainties that need to be addressed to allow them to become as mature and as widely applied as the technologies described in the preceding chapters.

As shown in earlier chapters, perchlorate degraders are widely distributed and are easily stimulated by the addition of electron donors to degrade perchlorate sequentially via chlorate (ClO_3^-) and chlorite (ClO_2^-) to chloride (Cl^-) (Coates et al., 1999; van Ginkel et al., 1996; Herman and Frankenberger, 1999; Logan, 2001; Logan et al., 2001a; Losi et al., 2002; Urbansky, 1998). The more mature technologies discussed earlier involve the addition of electron donors into contaminated groundwater to stimulate the indigenous perchlorate reducers (e.g., Rikken et al., 1996; Hatzinger et al., 2002). However, the source of organic carbon utilized for electron donors could also be natural organic matter or exudates of plant roots (such as acetate, organic acids, sugars, and dead root biomass) for MNA and phytoremediation, respectively. Also, it may be possible to add electron donors into the vadose zone to stimulate biodegradation of the long-term sources of groundwater contamination. More detailed descriptions of these promising *in situ* bioremediation technologies, along with a discussion of their scientific basis, current status and specific advantages and limitations are provided in the following sections.

11.2 MONITORED NATURAL ATTENUATION

11.2.1 Basis

Natural attenuation has been defined as the "biodegradation, diffusion, dilution, sorption, volatilization, and/or chemical and biochemical stabilization of contaminants to effectively reduce contaminant toxicity, mobility or volume to levels that are protective of human health and the environment" (USEPA, 1998). The term monitored natural attenuation refers to a remedial approach that relies on these natural attenuation processes, with careful monitoring and contingency plans, to achieve site-specific remedial goals.

MNA is an important and common component of remedial strategies for groundwater contaminated with several pollutants including gasoline and its water soluble constituents (ASTM, 1998), chlorinated solvents (USEPA, 1998), metals and radionuclides (Brady et al., 2003), methyl tertiary butyl ether (MTBE) (Wilson et al., 2005), and wood preservatives (Stroo et al., 1997), and it has been proposed for nitroaromatic explosives (Pennington et al., 1999). Although it has become a widely accepted approach for remediation of many contaminants, MNA of perchlorate is still in an immature stage of development.

MNA relies on several physical non-destructive processes (i.e., dilution, dispersion, volatilization, sorption) as well as destructive processes such as abiotic and biotic degradation. In the case of perchlorate, biodegradation is critical for MNA because perchlorate is not readily sorbed, volatilized or abiotically degraded. However, as discussed in prior chapters, the technical basis for natural biodegradation of perchlorate is sound. The biodegradation pathways are well understood and the microorganisms involved in perchlorate biodegradation are known. They can use a variety of different organic substrates as electron donors, are relatively ubiquitous in soil and groundwater environments, and function as strict or facultative anaerobes. These findings suggest that natural attenuation of perchlorate should occur at sites with favorable conditions (Cooley et al., 2005) and that MNA may be effective in managing the risks posed by perchlorate contamination of groundwater.

However, it is not clear how often environmental conditions that are favorable for natural attenuation of perchlorate will occur. Many aquifers are naturally aerobic, so that little perchlorate reduction would be expected. In many other aquifers, the electron donor demand due to the perchlorate flux alone can overwhelm the natural reducing capacity of the subsurface, so MNA may not be a sustainable long-term remedial option.

Site-specific evaluations will therefore be essential to the application of MNA. Such evaluations will be similar in many respects to those for other reductive processes, although the favorable oxidation-reduction (redox) range for perchlorate reduction is relatively broad, compared for example to the dehalogenation of chlorinated solvents. Site-specific evaluations of MNA typically rely on three lines of evidence: (1) plume stability; (2) geochemical indicators; and (3) biological indicators (e.g., USEPA, 1999). These are discussed separately below.

11.2.1.1 Plume Stability

Acceptance of MNA generally requires that the historical trends in the concentrations of the contaminants of concern (and any toxic byproducts) show that the contaminant plume is stable or retreating. A stable or retreating perchlorate plume indicates that biodegradation is removing perchlorate from the groundwater at least as fast as the source is releasing it to the plume.

Regardless of the contaminant, assessing plume stability is a challenge at most sites due to the inherent temporal and spatial variabilities, and it may therefore require extensive and expensive monitoring, over long periods of time, before stability can be concluded with confidence. There are a considerable number of publications on how to monitor and assess plumes and their stability (e.g., USEPA, 1999). As perchlorate is not sorbed to a significant degree, the perchlorate plume will migrate at a rate that is close to the average groundwater flow velocity. Therefore, a simple method to estimate whether the plume is stable or not is to compare its overall and expected length, given an estimate of when it first impacted the groundwater. If the plume is substantially shorter than expected, it is likely that perchlorate is being attenuated. This assessment requires factoring in the effects of dilution and dispersion, and source area width and concentration on plume lengths over time. In many cases, there are too few monitoring points to reliably evaluate the plume status, or the monitoring points are not properly located for the needed analyses.

11.2.1.2 Geochemical Indicators

Our understanding of perchlorate biodegradation indicates that anoxic and reducing conditions (i.e., no or low dissolved oxygen and negative redox potentials), as well as appropriate electron donors (e.g., reduced organic compounds) are the key conditions necessary for perchlorate biodegradation. These are the same conditions that are required for denitrification (the conversion of nitrate to nitrogen gas), which is not surprising as perchlorate reduction is similar in many ways to nitrate reduction. In fact, the nitrate reductase enzyme can also catalyze perchlorate reduction (Romanenko et al., 1996), and some denitrifying bacteria are capable of reducing perchlorate (Okeke et al., 2002). Similarly, many perchlorate degraders can reduce nitrate as well (Herman and Frankenberger, 1999).

Therefore, evidence for the reduction of nitrate (e.g., decrease of nitrate), or observance of nitrite production (a step in the denitrification of nitrate), along with a decrease in perchlorate concentration along the flow path, may be good indicators of the natural attenuation of perchlorate. However, if the concentrations of nitrate (and other electron acceptors such as sulfate) are too high, it can inhibit perchlorate reduction (Chaudhuri et al., 2002; Krauter et al., 2005), which may be a problem because nitrate levels in groundwater can be orders of magnitude higher than the perchlorate levels. Also, pH values outside the range of roughly 5.0 to 8.0 may inhibit perchlorate degradation even if all other conditions are favorable.

Increased levels of chloride also may be directly indicative of perchlorate biodegradation, although background chloride levels are often relatively high. Total salinity may also be relatively high in some perchlorate-impacted groundwaters. However, perchlorate reducers can be extremely salt-tolerant (Logan et al., 2001b).

11.2.1.3 Biological Activity Indicators

The most commonly used estimates of biological activity and contaminant degradation typically come from indirect analyses such as laboratory microcosm tests and/or measurements of the key organisms in field samples. It is possible to identify that organisms capable of perchlorate reduction are present in soil and groundwater samples (Waller et al., 2004), although presence alone does not indicate biodegradation *in situ*. Microcosm tests using site samples can both demonstrate biodegradation potential and provide rate estimates, but laboratory conditions inevitably differ from the field.

Direct microbial or biochemical evidence for natural biodegradation of perchlorate is more difficult to obtain. Molecular probes are available for some perchlorate reducers (Zhang et al., 2005), and probes specific for the chlorite dismutase (CD) enzyme may be useful as an indicator of perchlorate biodegradation (Bender et al., 2002; O'Connor and Coates, 2002). The CD enzyme: (1) is present in organisms grown anaerobically with perchlorate; (2) is not present in these same organisms grown aerobically; and (3) is not present in closely related organisms that cannot grow by perchlorate reduction. DNA and messenger RNA enzyme assays are commercially available to evaluate the genetic capability of the bacterial population to produce the CD enzyme and the on-going activity of the enzyme, respectively.

Microorganisms preferentially use lighter isotopes in their metabolic processes (Mariotti et al. 1981; Heaton, 1986) and as a contaminant is degraded, the isotopic composition of the remaining material becomes progressively heavier. Compound-specific isotope analysis may be a sensitive technique for evaluating and monitoring natural attenuation, because of the relatively large chlorine isotope fractionation effect that occurs during perchlorate biodegradation (Sturchio et al., 2003). Bohlke et al. (2005) have shown that stable isotope analyses ($^{37}Cl/^{35}Cl$ and $^{18}O/^{17}O/^{16}O$) of perchlorate can distinguish between synthetic and natural sources of the material and Lieberman et al. (2006b) demonstrated actual *in situ* perchlorate isotopic fractionation to definitively illustrate biodegradation in an electron-donor enhanced *in situ* permeable reactive biobarrier.

11.2.1.4 Status

There are few examples of MNA at perchlorate-contaminated sites. However, it is likely that MNA use will increase as the technical understanding and regulatory comfort increase. In one recent case, the U.S. Environmental Protection Agency (USEPA) amended the Record of Decision (ROD) for a nitrate- and perchlorate-contaminated site to include MNA (USEPA, 2005a). The ROD stated that "there is sufficient evidence that the contaminants are naturally biodegrading to the extent that continued monitoring of this MNA activity will be as effective as any other technology." The evidence included the presence of perchlorate-reducing bacteria and modeling of the natural restoration rate in a plume that posed little risk.

Research is underway to identify sites undergoing natural attenuation of perchlorate and assess the potential lines of evidence that may be used to better predict and monitor perchlorate biodegradation. Researchers are testing various methods for monitoring natural biodegradation of perchlorate, and intend to develop a protocol for evaluating the suitability of MNA for perchlorate-contaminated sites (Lieberman, 2004).

11.2.2 Advantages and Limitations

The primary advantages of MNA of perchlorate are similar to MNA of all other contaminants of concern. These include the low capital and maintenance costs, the lack of remediation equipment or other impacts to the site activities, and the lack of artificial impacts to the groundwater geochemistry and biology. Although often considered a low-cost alternative, it is not a "no action" alternative, default option or presumptive remedy. In fact, MNA often requires a significant life-cycle cost because monitoring may be needed for decades in order to ensure that the natural processes remain protective of human health and the environment.

The use of MNA for perchlorate may be limited by the hydrogeology, groundwater geochemistry, high contaminant concentrations, microbiology, and location of receptors. Perchlorate is highly soluble and can therefore migrate quickly, particularly at sites with rapid groundwater flows. Contaminated plumes can be very large and the perchlorate concentrations can be high enough to overwhelm the assimilative capacity of the natural attenuation processes.

11.2.3 Case Studies

At seven Department of Defense (DoD) sites where perchlorate concentrations appeared to decline as groundwater migrated downgradient from the presumed source area, data showed a wide spread of conditions across each of the sites and a variety of co-contaminants typical of many DoD installations (Lieberman et al., 2006a). In general, perchlorate concentrations decreased away from the source. The screening results indicated that low oxidation-reduction potential and low nitrate concentrations are important conditions, but that higher oxygen concentrations and low natural organic carbon did not absolutely negate the possibility of natural attenuation of perchlorate. The DNA assay generally showed presence of perchlorate reductase in the native populations. This study also found that the CD enzyme assay was a promising indicator of perchlorate reductase bioactivity.

Microcosm studies were performed on groundwater from the Naval Surface Warfare Center in Indian Head, Maryland. Groundwater from a well at the distal end of the plume contained approximately 79 micrograms per liter (µg/L) perchlorate. The results (Figure 11.1) showed that natural biodegradation does occur without added electron donors, particularly in the groundwater with relatively low perchlorate concentrations, and that natural biodegradation can reduce concentrations from approximately 100 µg/L to non-detectable levels within 6 months. When groundwater was incubated in the presence of an electron donor, the perchlorate concentration dropped to non-detect in less than 34 days. But the perchlorate concentration also declined to non-detect when the site groundwater was incubated at room temperature without any amendment, although removal required a longer time (188 days). The natural degradation rate was calculated to be 0.4 µg/day. An elevated starting concentration of 5200 µg/L was created by spiking the natural groundwater with ammonium perchlorate. In this treatment, by Day 188, one replicate went to non-detect, while the other two showed lesser degradation. The average rate of degradation for the three replicates was 16.6 µg/day (Lieberman et al., 2006a).

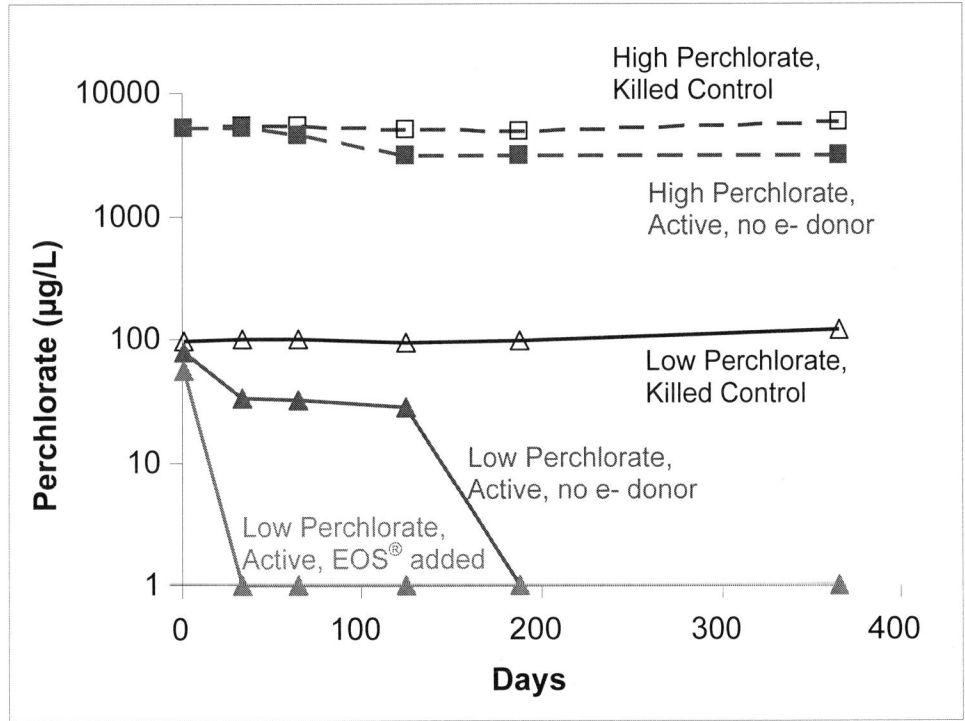

Figure 11.1. Perchlorate concentrations in microcosms containing groundwater collected from Naval Surface Warfare Center, Indian Head, Maryland. Microcosms included: (1) site groundwater amended with Edible Oil Substrate (EOS®) as a positive control; (2) groundwater with either relatively high or low perchlorate levels, with no electron donor added; and (3) groundwater with no added electron donor, but killed by autoclaving to provide negative (sterile) controls. Results shown are averages for triplicate active and duplicate killed microcosms. Unpublished results; data from Lieberman et al., 2007.

11.3 PHYTOREMEDIATION

Phytoremediation is the use of vegetation and in some cases the associated root zone microorganisms for *in situ* treatment of hazardous wastes. It is a demonstrated low-cost technology that has effectively treated a wide range of contaminants, involving many different plant-mediated mechanisms. The technology has been commercially applied for treatment of soils, groundwaters and wetlands impacted by several contaminants. Reported field-scale demonstrations of phytoremediation of perchlorate include the use of dynamic flow-through wetlands (Krauter et al., 2005) and poplar trees to treat contaminated groundwater (Schnoor et al., 2004). Phytoremediation of perchlorate by stimulating rhizodegradation has been reported recently in bench-scale tests (Yifru, 2006; Yifru and Nzengung, 2008a, b) but is yet to be demonstrated at the field scale.

11.3.1 Basis

Phytoremediation can be an effective bioremediation technology for several inorganic and organic contaminants (USEPA, 2006; Schnoor et al., 1995). Its use to date has been mostly for remediation of soils, and shallow groundwaters, contaminated with metals, chlorinated solvents, pesticides, and petrochemicals (Kramer, 2005). Many recent bench-scale studies confirm that phytoremediation is a potentially promising technology for the clean up of perchlorate-contaminated surface water, groundwater and soils (Nzengung et al., 1999; Susarla et al., 1999; Nzengung et al., 2004; Nzengung and McCutcheon, 2003; Schnoor et al., 2002; Yifru and Nzengung, 2008a, b).

It is important to realize that phytoremediation is not one technology, but it is actually a suite of technologies (Figure 11.2). It can involve the use of many different plant species, and several plant-mediated processes may be responsible for the contaminant removal and destruction as well (Pilon-Smits, 2005). Shrubs, trees, grasses, or wetland plants may be used, and the plants (and/or their associated microorganisms) may extract, sequester, transform, degrade or transpire the contaminants (McCutcheon and Schnoor, 2003).

Many species of plants can remove perchlorate from contaminated water and soils (Susarla et al., 1999; Nzengung and McCutcheon, 2003). Terrestrial plants shown to be effective in perchlorate removal include black willow (*Salix nigra, Salix caroliniana*), eastern cottonwood (*Populus deltoides*), eucalyptus (*Eucalyptus cinerea*), and loblolly pine (*Pinus taeda*). Aquatic plants that have been tested successfully include water weed (*Elodea cadadensis*), parrot-feather (*Myriophyllum aquaticum*), duckweed (*Spirodela polyrhiza*) and cattails (*Typha spp*). In addition, microbial mats and green algae have also shown promise (Nzengung and McCutcheon, 2003).

Two predominant mechanisms of phytoremediation of perchlorate (Figure 11.3) have been identified: (1) uptake and phytodegradation; and (2) rhizodegradation (Nzengung et al., 1999; Nzengung et al., 2004). Uptake and phytodegradation may pose ecotoxicological risks because the slow phytodegradation steps result in the accumulation (phytoaccumulation) of a fraction of the extracted perchlorate, primarily in the leaf tissue (Nzengung et al., 1999). However, many bacteria found in the plant rhizosphere can biodegrade perchlorate, at relatively rapid rates (Shrout et al., 2006). Perchlorate half-lives in the rhizosphere range from minutes to a few hours, depending on the availability of dissolved organic carbon (DOC) or other electron donors and favorable biodegradation conditions in the rhizosphere (Yifru and Nzengung, 2008a). Rhizodegradation also results in the complete destruction of perchlorate to innocuous chloride, so that current research is focused on enhancing rhizodegradation and minimizing the fraction taken up into the plants (Yifru and Nzengung, 2008b).

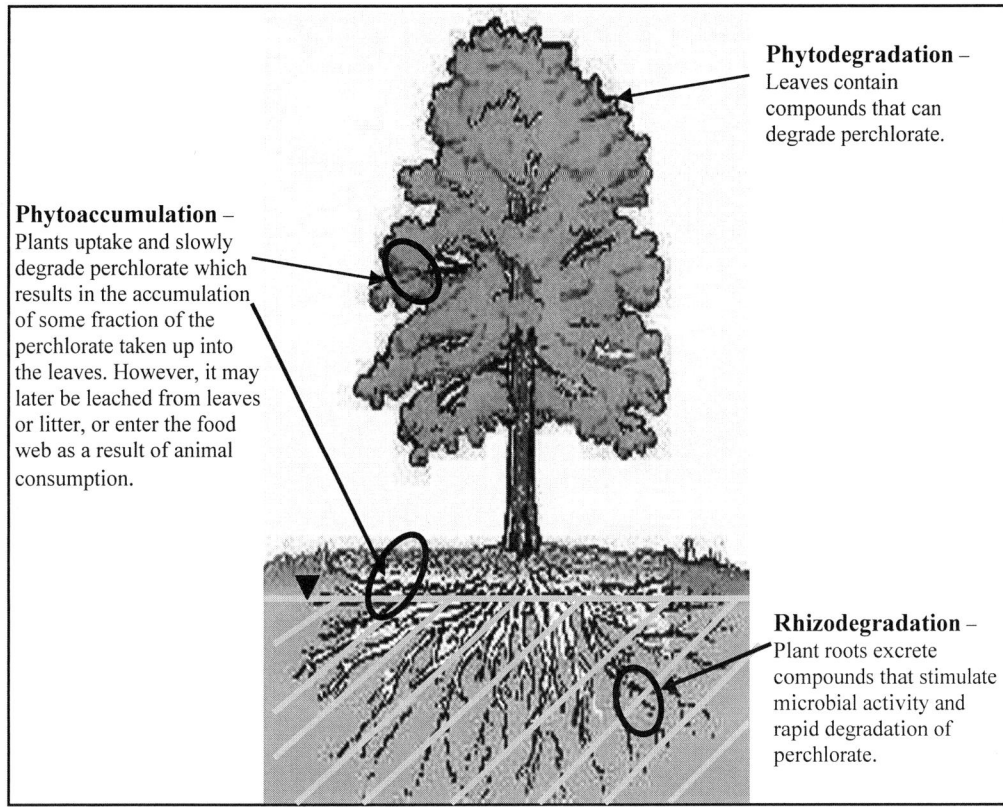

Figure 11.2. Processes that may occur during phytoremediation of perchlorate.

Bench-scale tests have demonstrated that perchlorate can be completely removed from root-zone bioreactors in one week, and that rhizodegradation of perchlorate can account for 90 to 99% of the total removal, given low dissolved oxygen levels, sufficient DOC and little or no nitrate-N (Dondero, 2001). However, at higher concentrations of nitrate (a competing terminal electron acceptor), removal rates decreased by approximately one order of magnitude, and the contribution of rhizodegradation decreased to roughly 75% (Nzengung et al., 2004).

The importance of limiting plant uptake has been highlighted by recent studies that have shown that several plants are capable of extracting perchlorate from contaminated waters and translocating it to the foliage (Sundberg et al., 2003; Tan et al., 2006; Urbansky et al., 2000). Plants from contaminated sites that have been demonstrated to phytoaccumulate perchlorate include tobacco (Ellington et al., 2001; EWG, 2005; Smith et al., 2004), lettuce (Hogue, 2003; EWG, 2005), grasses (Smith et al., 2004), terrestrial and aquatic field plants (Smith et al., 2004; Yifru, 2006) and many food crops (Hogue, 2003). These findings have led to concerns about potential human and animal exposures to perchlorate in vegetation, or subsequent releases back to the environment after leaching or exudation (Tan et al., 2006). Recent detections of perchlorate in dairy milk and processed dairy products and breast milk (Kirk et al. 2005; USEPA 2006) and human urine (Blount et al. 2007) have been attributed in part to the consumption of perchlorate-contaminated produce.

In a detailed study of the fate of perchlorate in field plants, Yifru (2006) collected and analyzed various species of trees and grasses from Longhorn Army Ammunition Plant (LHAAP) in Karnack, Texas, and the Las Vegas Wash (LVW), Nevada, for two years. All species of vegetation collected from LHAAP and LVW contained detectable amounts of

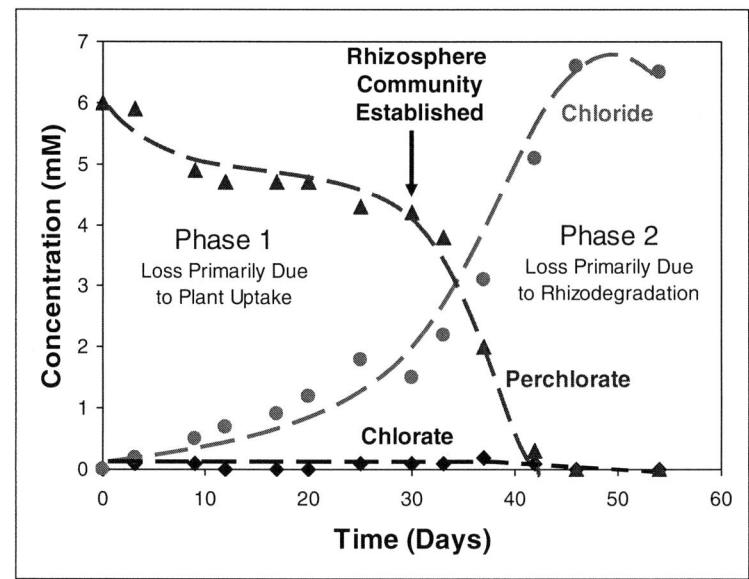

Figure 11.3. Concentrations of perchlorate, chlorate and chloride in growth solutions from hydroponics bioreactors. The relatively slow perchlorate removal over the first 30 days was primarily due to plant uptake. However, after the rhizosphere community became established in the bioreactors, perchlorate removal was more rapid because rhizodegradation was the dominant mechanism. Modified from Yifru, 2006.

perchlorate. The phytoaccumulation of perchlorate in terrestrial vegetation varied with plant species, perchlorate concentration in the rhizosphere and stage of plant maturity. The highest perchlorate accumulation was obtained in willows and grasses at LHAAP and in salt cedar (*Tamarix ramosissima*) trees at LVW. Litter fall collected from both LHAAP and LVW contained significant amounts of perchlorate, implying that plants recycle perchlorate and in fact serve as both source and sink of perchlorate contamination. Since degradation of perchlorate in plant tissue is a slower process, most of the perchlorate taken up gets accumulated in plants for long periods of time. As a result, leaves collected late in the growing season contained more perchlorate than those collected in spring.

To overcome the limitation of rhizodegradation resulting from an insufficient supply of electron donors, external sources of DOC derived from agricultural waste or synthetic sources can be provided to stimulate rapid biodegradation and rhizodegradation of perchlorate (Krauter 2001; Krauter et al., 2005; Nzengung et al., 1999; Nzengung and Wang, 2000; Nzengung and McCutcheon, 2003; Yifru and Nzengung, 2008a, b). In bench-scale hydroponics and soil bioreactor studies, Yifru (2006) successfully stimulated and sustained rapid rhizodegradation of perchlorate by providing electron donors such as acetate, chicken litter extract and a mushroom compost "tea" (Figure 11.4). This treatment resulted in the removal of greater than 99% of perchlorate by rhizodegradation in under one week, while reducing the concentration of perchlorate taken up in the leaf tissue by one order of magnitude compared to controls. Krauter et al. (2005) measured perchlorate degradation in simulated wetlands amended with electron donors and found that virtually all of the perchlorate could be removed by rhizodegradation, with minimal uptake and phytoaccumulation.

Effective treatment of perchlorate in groundwater and soil using rhizodegradation is potentially promising, based on the results of laboratory research done to date (AFCEE, 2002; Nzegung and McCutcheon, 2003). Current research focuses on gaining a better understanding of the mechanisms active during phytoremediation and on identifying both plants and rhizosphere bacteria that could be useful for treating perchlorate. Based on the available research data, phytoremediation of perchlorate can be engineered to biostimulate and/or enhance rhizoremediation by providing a sustained supply of electron donors to the root zone. The ecological benefits of enhancement of rhizodegradation include rapid attainment of site cleanup goals, minimization of the undesired uptake and phytoaccumulation of perchlorate and avoidance of the potential recycling of perchlorate during phytoremediation (Yifru and Nzengung, 2008a, b).

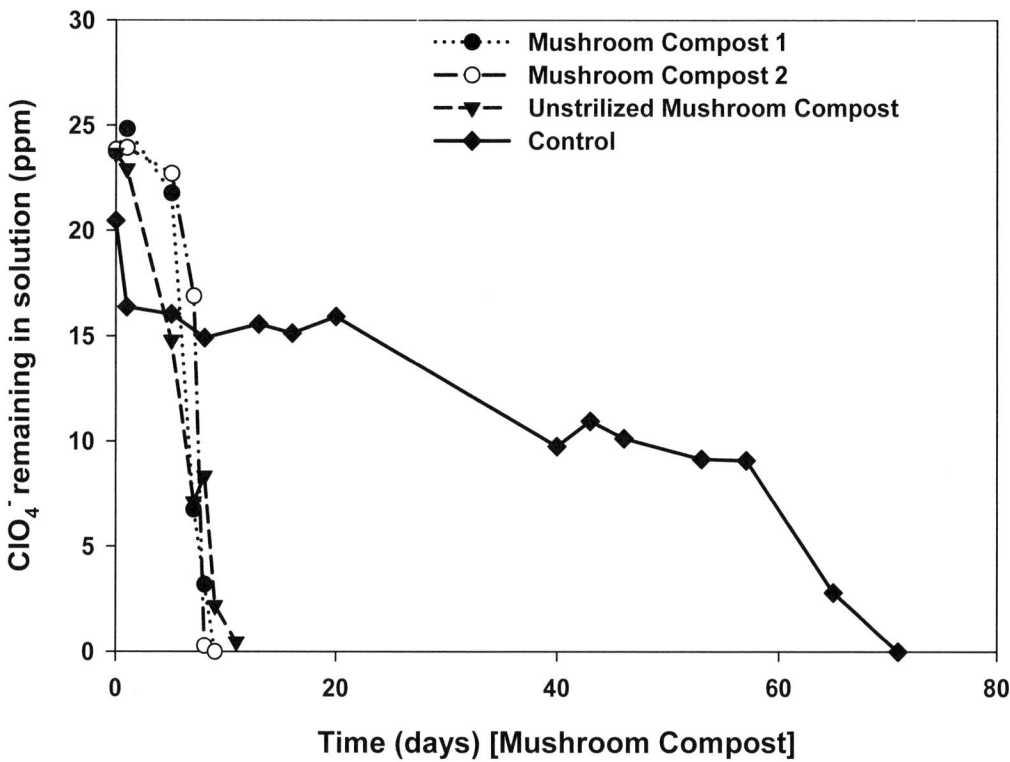

Figure 11.4. Effectiveness of stimulation and enhancement of rhizodegradation of perchlorate during phytoremediation using DOC supplied to the root zone as sterilized and unsterilized mushroom compost "tea". The complete removal of perchlorate in willow and poplar planted bioreactors was achieved in less than 10 days compared to about 70 days for controls (controls were not provided an external source of DOC). Reprinted from Yifru, 2006.

11.3.2 Status

Although phytoremediation of perchlorate appears promising, no known full-scale engineered perchlorate phytoremediation systems are in operation. However, field studies have demonstrated that plant uptake and transformation can be important components of natural attenuation in wetlands impacted by perchlorate (Tan et al., 2004, Krauter et al., 2005). The analysis of field plants growing on perchlorate-contaminated soils at Kerr McGee's site in the LVW, Las Vegas, Nevada and at the LHAAP in Karnack, Texas have shown that under natural field conditions some plant species, such as salt-tolerant trees, may accumulate significant quantities of perchlorate, thereby contributing to natural attenuation (Urbansky et al., 2000; Yifru, 2006).

At some sites, plant uptake of water and contaminants (phytoextraction) can be used for hydraulic control of impacted plumes (Eberts et al., 1999) or to reduce the flux of contaminants moving downgradient. There is, however, a need for effective engineered phytoremediation of perchlorate, with a focus on enhancing the rhizodegradation process to ensure destruction. Phytoremediation is most promising for sites with relatively shallow groundwater or where deep groundwater can be pumped to the surface and treated.

11.3.3 Advantages and Limitations

Phytoremediation has the advantages of low cost, high public acceptance and little secondary waste production. With respect to cost, phytotechnologies generally compare favorably with costs for aboveground treatment technologies (see cost comparisons at the Federal Remediation Technologies Roundtable website: http://www.frtr.gov/matrix2/section3/table3_2.pdf.

Phytoremediation can also treat other common co-contaminants such as chlorinated solvents and explosives including N-nitrosodimethylamine (Yifru and Nzengung, 2006). It can help meet other land-use goals as well (such as open land quotas, wetland acreage, or canopy coverage), and it can improve a site's aesthetic appeal. Finally, phytoremediation generally causes minimum disturbance to the environment or any on-site operations.

The limitations of phytoremediation of perchlorate include depth and climate restrictions as well as the potential for transfer of contaminants from soil and groundwater into the food chain. Phytoremediation of perchlorate is also a relatively slow process and subject to seasonal variations. Often, many plants spread over a large area are needed to adequately capture and treat perchlorate. Although rhizodegradation is the desirable process for phytoremediation of perchlorate, the natural supply of plant exudates alone is not sufficient to sustain rapid rhizodegradation of perchlorate at some sites, unless phytoremediation is engineered to enhance the rhizodegradation activity.

Perchlorate may exert a toxic effect on certain plant species. For example, concentrations of perchlorate above 1,000 mg/L are phytotoxic to young rooted cuttings of phreatophytes such as willows and poplars (Mbuya and Nzengung, 2006). Therefore, selecting a species suitable for achieving treatment goals is important. In addition, climatic changes can significantly impact plant growth, thus requiring variation in the treatment period. At this point, the relatively immature stage of the technology is also a limitation. Without cost and performance data, and greater regulatory comfort, it is difficult for remedial project managers to select and obtain approval for a phytoremediation alternative.

11.3.4 Case Studies

11.3.4.1 Groundwater Remediation

To date, one pilot-scale demonstration of phytoremediation of perchlorate-contaminated groundwater has been conducted (Schnoor et al., 2004). This field demonstration involved planting 425 hybrid poplars on a 0.3-hectare (0.7-acre) site at the LHAAP, Karnack, Texas in March 2003. The initial perchlorate concentration in the groundwater was 34 milligrams per liter (mg/L). After a year, the concentration of perchlorate in the treated groundwater had decreased to 23 mg/L. The mass of perchlorate taken up by the poplar trees and/or degraded within the rhizosphere was estimated as 0.114 ± 0.016 kilograms per day (kg/d). Between April 2003 and September 2004, it was estimated that 52 kg of perchlorate was removed from the groundwater by the hybrid poplar trees and/or the microbes present in the root zone (Schnoor et al., 2004). The report does not provide any evidence that phytoremediation at this site involved enhancement of rhizodegradation of perchlorate. The pilot-scale demonstration is ongoing. There is still a need for field-scale demonstration of phytoremediation of perchlorate with deliberate enhancement of rhizodegradation.

11.3.4.2 Constructed Treatment Wetlands

In November, 2000, the U.S. Department of Energy and Lawrence Livermore Laboratory (LLNL) designed and constructed an innovative containerized wetland system that biologically degrades perchlorate and nitrate under relatively low-flow conditions at Site 300 (Building 854) at LLNL (Krauter et al., 2005; Dibley and Krauter, 2005). Design of the wetland bioreactors was based on earlier studies showing that indigenous chlorate-respiring bacteria could effectively degrade perchlorate into nontoxic concentrations of chlorate, chlorite, oxygen, and chloride (Krauter, 2001). Treatability studies showed that addition of organic carbon (provided as acetate) enhanced microbial denitrification, without inoculation.

Prior to testing, groundwater was allowed to circulate through the bioreactor for three weeks to acclimate the wetland plants and to build a biofilm from indigenous microflora. Groundwater was pumped through granular activated carbon canisters to remove volatile organic compounds (VOCs). This VOC-treated groundwater, containing approximately 46 mg/L of nitrate and 13 µg/L of perchlorate, was gravity-fed continuously into a parallel series of 1,900-L (502 gallons [gal]) constructed wetland tank bioreactors. Each bioreactor contained coarse, aquarium-grade gravel and locally-obtained plant species such as cattails *(Typha* spp.), sedges (*Cyperus* spp.) and indigenous denitrifying microorganisms. An active flow rate of 3.8 liters per minute (L/min) (1 gal/min) was set to provide a minimum reactor hydraulic retention time (HRT) of 17 to 20 hours, which decreased as the plants matured and organic matter and rootlets accumulated in the bioreactors. The results showed that degradation of perchlorate and nitrate, without an added carbon source, required an HRT of four days and 20 hours, respectively. However, in the presence of a 0.25 grams per liter (g/L) solution of sodium acetate, the HRT decreased to 0.5 days. In about two years, the system processed over 3,463,000 L (914,232 gal) of groundwater and treated over 38 g of perchlorate and 148 kilograms of nitrate.

11.4 VADOSE ZONE BIOREMEDIATION

All of the other technologies described in this book have focused on *in situ* treatment within the saturated zone. However, perchlorate remaining in the vadose zone may represent a major continuing source of perchlorate to groundwater (Newman et al., 2005). Perchlorate contamination within vadose zone soils in source areas such as hogout operations, burn areas, live fire ranges, and ammonium perchlorate production and fine grinding facilities is common. Such residual contamination can increase the operating timeframe and associated costs for hydraulic containment (pump and treatment) and *in situ* groundwater treatment systems.

Near-surface contamination can generally be excavated and treated on site by composting, *ex situ* bioremediation, or by intrinsic bioremediation (Cox et al., 2000; Cox and Scott, 2003; Evans et al., 2008, Kastner et al., 2001, Nzengung et al., 2001; O'Niell and Nzengung, 2003; USEPA, 2005b). Composting of excavated perchlorate-contaminated soil involves mixing with bulking agents and organic amendments such as wood chips, hay, manure, and vegetative wastes. Selection of proper amendments is necessary to ensure adequate porosity and provide a balance of carbon and nitrogen to promote thermophilic microbial activity. Monitoring of moisture content and temperature are important for achieving maximum degradation efficiency (USEPA, 2005b; Cox et al., 2000).

Ex situ bioremediation of soil can also be conducted by adding water, an electron donor, and sometimes additional nutrients. For example, at the former Bermite site north of Los Angeles, California, soil is being treated at a rate of about 1,000 metric tons per day (Evans et al., 2008). Soil is amended with water to achieve a final moisture content of 14 to 17 percent, 450 mg/kg of glycerin, and 50 to 100 mg-N/kg of di-ammonium phosphate (DAP). The amended soil is then placed in an Ag-Bag® containment system or in concrete block containment cells and covered with plastic tarps (Figure 11.5). Perchlorate is biodegraded from 590 to 8,400 µg/kg to nondetectable concentrations (< 20 µg/kg) in about two weeks.

Figure 11.5. Full-scale *ex situ* soil bioremediation in (a) Ag-Bag® and (b) concrete containment cells (reprinted from Evans et al., 2008).

However, cost-effective treatment of deeper contamination represents an important challenge because there are many perchlorate sites with deep contamination over relatively large areas (AWMA, 2003; O'Niell and Nzengung, 2003). The technical challenge is to cost-effectively deliver electron donors to the areas where perchlorate may remain in the vadose zone and to maintain sufficient control of the environmental conditions within the vadose zone to enhance the activities of the perchlorate-reducing bacteria.

11.4.1 Basis

Biodegradation of perchlorate in soils by ubiquitous natural bacteria will occur rapidly under suitable field conditions, which include pH, redox conditions, moisture content and electron donors. Vadose zones are often aerobic, because of gaseous diffusion from the surface through unsaturated pore space. Even in deeper soils and geologic strata, reductive processes like perchlorate degradation will often be severely limited by the low supply and availability of electron donors. As a result, perchlorate in the vadose zone can be a persistent source for long-term groundwater contamination. However, if electron donors can be supplied, bioremediation of vadose zone contamination could be very effective in reducing the longevity of the groundwater contamination (Nzengung et al., 2001).

Perchlorate degraders are widespread in soils as well as groundwaters, and the process appears to occur under a wide range of environmental conditions (Coates et al., 1999). The challenge at most sites, therefore, is primarily one of cost effectively delivering electron donors, in a usable form, to deeper vadose zone materials. Liquid and gaseous delivery systems offer the two general approaches of supplying donors to the vadose zone. These are discussed separately in the following sections.

11.4.1.1 Liquid Delivery

Electron donors are often injected into the groundwater, and similar solutions can be directly applied to the vadose zone. Application methods could include sprinkler irrigation (O'Niell and Nzengung 2003; AWMA 2003), direct injection or periodic flooding via infiltration galleries. One method, known as Surface Application and Mobilization of Nutrient Amendments (SAMNAS), has been studied at the bench, pilot and field scale (Kastner et al., 2001; Nzengung et al., 2001; O'Niell and Nzengung, 2003; AWMA 2003). In this approach, liquid and solid amendments are mixed in with surface soils (0-1 meters [m] or 0–3 feet [ft]) and mobilized or leached with water to greater depths to stimulate perchlorate biodegradation. The liquid and solid amendments evaluated included ethanol, acetate, molasses, mushroom compost "tea", mushroom compost, cow and horse manure, and chicken manure. The liquid nutrient amendments generally perform better at sites where the contamination is deep or the clay content is high. The type and amount of soil amendments, clay content, field capacity, and water application rate determine the biodegradation rate of perchlorate (O'Niell and Nzengung, 2003).

Dilute aqueous solutions of ethanol applied as irrigation water to the surface of perchlorate-contaminated clay rich soils and mobilized to greater depths were used to stimulate vadose zone biodegradation of perchlorate at depths of up to 0.9-1.2 m (3–4 ft) (Kastner et al., 2001). To design infiltration systems for different soils types, the partition coefficient of the electron/carbon donor offers a valuable tool to directly estimate the amount of organic carbon that could be transported to defined depths based on application rates. In bench-scale soil column tests conducted with silty clay soils from LHAAP that were contaminated with perchlorate, Nzengung et al. (2001) observed that even at very low infiltration rates, complete breakthrough of ethanol was achieved in 3–4 days (Figure 11.6). This indicated that the soil had a very low capacity to hold organic carbon, as confirmed by the experimentally determined partition coefficient (K_d) of 3.1×10^{-5} L/kg (0.03 mL carbon/kg soil). These data should serve as useful design parameters when developing full-scale remediation strategies.

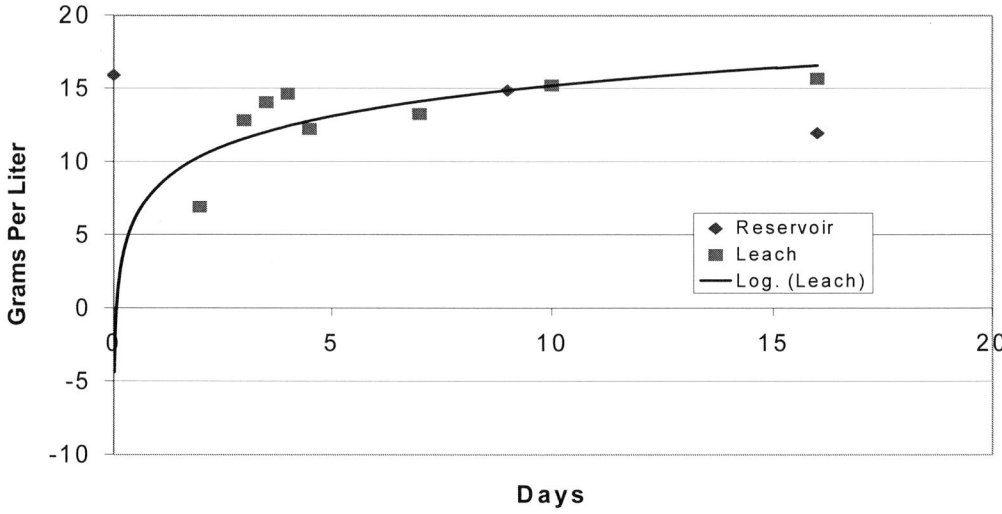

Figure 11.6. Breakthrough curve of ethanol transport through soil column. Inlet concentration of ethanol was 16 g/L. Reprinted from Nzengung et al., 2001.

11.4.1.2 Gaseous Delivery

Some donors (such as hydrogen, low molecular weight organic acids, esters, alkenes, and alkanes) are sufficiently volatile that they can be supplied as gases, similar to bioventing with oxygen (Figure 11.7). This "anaerobic bioventing" has been used for several other contaminants resistant to aerobic biodegradation including 1,1,1-trichloro-2,2-bis(p-chlorophenyl)ethane

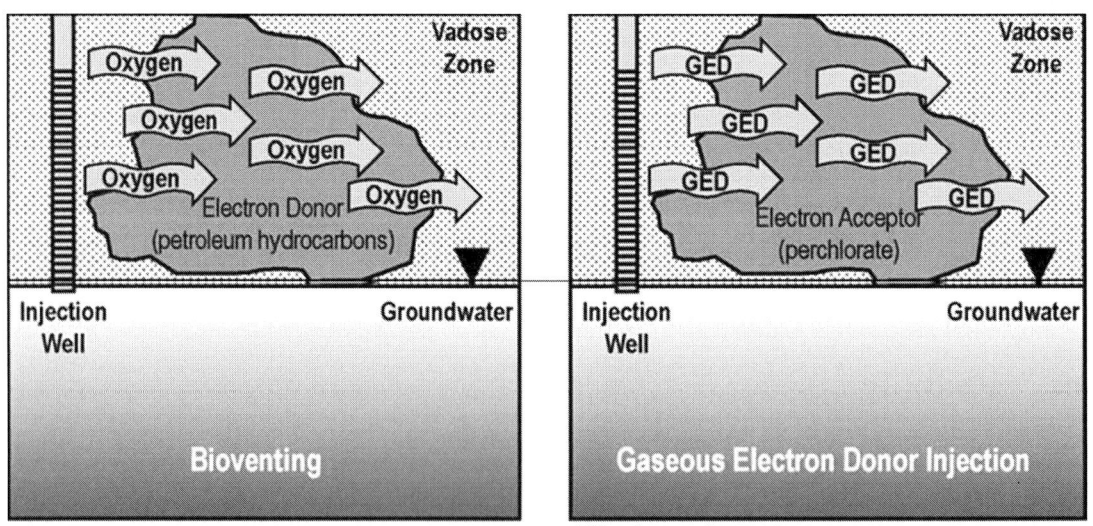

Figure 11.7. Schematics comparing gaseous electron donor injection and aerobic bioventing (reprinted from Evans and Trute, 2006).

(DDT), hexahydro-1,3,5-trinitro-1,3,5- triazine (RDX or **R**oyal **D**emolition e**X**plosive) and perchloroethene (PCE) (Gibbs et al., 1999; Milopoulos et al., 2002; Shah et al., 2001; USEPA, 2006).

One approach to anaerobic bioventing, known as Gaseous Electron Donor Injection Technology or GEDIT (Evans, 2007, Evans and Trute, 2006; Evans and Weaver, 2005), is currently being tested for perchlorate bioremediation in the DoD's Environmental Security Technology Certification Program (ESTCP). This first field demonstration of GEDIT involves injection of a constant low flow rate (about 50 liters per minute) of a gas mixture into soil. While different gas mixtures are possible, the mixture being used in this demonstration is comprised of 10 percent hydrogen, 10 percent liquefied petroleum gas (LPG), 1 percent carbon dioxide, and 79 percent nitrogen. While the test is ongoing, preliminary results indicate that nitrate concentrations were reduced by an order of magnitude in about two months (Figure 11.8) and nitrate is typically reduced prior to perchlorate. At the time of publication, perchlorate biodegradation may be starting but definitive data still need to be collected.

Anaerobic bioventing is an attractive option for vadose zone remediation because gases can disperse further into the unsaturated materials than liquids. Gases can also potentially diffuse more thoroughly through the unsaturated subsurface, to some extent minimizing the problems of preferential flow pathways that are more common with liquid flow (Milopoulos et al., 2002). Additionally, gaseous electron donor technology does not require the capture and treatment of infiltrated liquids that could otherwise adversely impact groundwater. While it is sometimes thought that gaseous injection may cause soils to lose moisture to levels that do not support biodegradation, previous work on bioventing of petroleum hydrocarbons has demonstrated that significant drying is likely to occur only in the immediate vicinity of the injection well (Leeson and Hinchee, 1995).

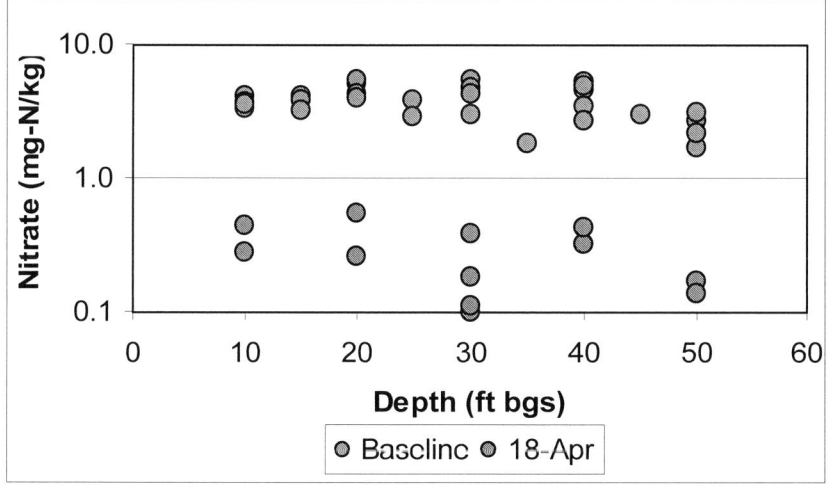

Figure 11.8. Preliminary results for *in situ* reduction of nitrate concentration in vadose zone soil. Data represent samples collected on April 18, 2008 after about two months of gaseous electron donor injection technology (GEDIT) application.

11.4.2 Status

The infiltration of electron donors and nutrient amendments to biostimulate the degradation of perchlorate in vadose zone soils has been demonstrated at the field scale and is a proprietary technology called SAMNAS (patent pending #11/280,760, PLANTECO Environmental Consultants, LLC. 337 South Milledge Avenue, Suite 202, Athens, GA, USA 30605). A current

ESTCP project is testing the addition of electron donors via irrigation and infiltration galleries to develop cost and performance data under field conditions (ESTCP, 2008a).

Gaseous delivery using GEDIT is also being tested under field conditions (ESTCP, 2008b). Preliminary pilot-scale results were discussed above and are promising. Project completion is scheduled for late 2008. Laboratory testing has shown that several gaseous electron donors (such as hydrogen, propane, ethanol, and 1-hexene) can stimulate perchlorate biodegradation (Brennan et al., 2006; Evans and Trute, 2006).

11.4.3 Advantages and Limitations

The primary advantage of vadose zone bioremediation is that it can remove much of the source of continuing groundwater contamination in a far more cost-effective manner than excavation or thermal treatment, for example. Bioremediation can completely degrade perchlorate to innocuous chloride. There is no secondary waste generation, and little infrastructure or disruption of ongoing site activities is required. It is also applicable to other reducible contaminants that may co-occur with perchlorate, notably many of the chlorinated solvents.

Any method of vadose zone bioremediation will be limited by the ability to distribute the electron donors within the subsurface. Delivery limitations can be particularly difficult at sites with deep vadose zone contamination (i.e., in the case of liquid electron donors), low permeabilities and/or highly heterogeneous geological conditions. Establishing and maintaining sufficiently reducing conditions can be difficult, especially in shallower zones, due to the migration of oxygen from the surface. Some of the donor sources, particularly the gaseous ones, may pose health and safety concerns due to flammability. Injection of aqueous donor sources may require capture and treatment of liquids to prevent adverse impacts to groundwater. Finally, geochemical limitations on perchlorate reducers, such as acidic or highly alkaline pH, presence of nitrate or other competitive ions, or toxicity due to metal concentrations, may be very difficult to overcome within the vadose zone.

11.4.4 Case Studies

Two full-scale and three pilot-scale demonstrations of anaerobic composting or *ex situ* bioremediation for treatment of perchlorate in soil have been identified. Three demonstrations of *in situ* bioremediation of perchlorate-contaminated vadose zone soils have been performed. These case studies are summarized in Table 11.1.

Table 11.1. Summary of Vadose Zone Bioremediation Case Studies by Location

Scale and Technology	Design and Operation	Performance Summary	Source
Longhorn Army Ammunition Plant (LHAAP), TX			
Pilot *In Situ* Bioremediation Surface Application and Mobilization of Nutrient Amendments (SAMNAS)	Based on the results of batch and laboratory column tests that evaluated the ability of different nutrient amendments to stimulate perchlorate degradation by naturally occurring bacteria. The tested nutrient amendments included cow manure, chicken manure, methanol, ethanol, acetate, molasses and cotton gin waste.	Period of performance: October 7, 2000 to August 27, 2001. Follow-up monitoring after 31 months. Perchlorate concentrations at all depths in soils treated with ethanol decreased from 300	Kastner et al., 2001 Nzengung et al., 2001 AWMA, 2003

Scale and Technology	Design and Operation	Performance Summary	Source
	Six 4.6-meter (m) by 2.7-m (15-feet [ft] by 9-ft) treatment plots were treated in duplicate with ethanol, horse manure and chicken litter, respectively, with one 4.6-m by 2.7-m (15-ft by 19-ft) plot used as an untreated control. The plots were hydraulically isolated with 30-cm (12-in) deep plastic-lined trenches. For the plots treated with solid nutrient amendments, the soil inside each plot was mixed with the amendment and tilled to approximately 30 cm (12 in) below ground surface. Water was added to achieve saturation within the top 30 cm (12 in), and subsequently down to 61 cm and 91 cm (24 and 36 in), respectively. Ethanol was added as a dilute solution in the applied water. Soil moisture content at depth was monitored with tensiometers, and redox potentials were measured at a number of locations and depths in each plot. The plots were covered periodically to prevent growth of vegetation. The targeted treatment depth was 0.9 m (3 ft) below ground surface (bgs) and the soil type was silty clay.	milligram/kilogram (mg/kg) to below the treatment level of 40 µg/kg in 10 months. The solid amendments stimulated perchlorate biodegradation in the clay-rich soils, with highest removal within the top 30 cm (12 in), but the treatment goal was not achieved in 10 months. Removal of perchlorate continued after 31 months in the treated plot, but not in the control plot. The concentration of perchlorate in plant tissues after treatment confirmed the reduction of perchlorate from soils and a significant reduction of ecological risk. The decrease of perchlorate observed in the control plot occurred mostly during the 10 months active treatment period and was attributed to the redistribution of perchlorate that occurred when the soil was tilled, rather than to biodegradation. The concentration of perchlorate monitored in the groundwater below did not increase and provided evidence that perchlorate was not mobilized or leached into the groundwater.	
LHAAP, TX—Site 17 – trenches used for burning bulk 2,4,6-trinitrotoluene (TNT), photo, flash powder, and reject material.			
Pilot *In Situ* Bioremediation Surface Application and Mobilization of Nutrient Amendments	Perchlorate ranging from 8.2 mg/kg to 480 mg/kg, 1,3,5-trinitrobenzene (1,400 mg/kg), 2,4,6-trinitrotoluene (10,000), 2,4-dinitrotoluene (9,000 mg/kg), 2,6-dinitrotoluene (3,700 mg/kg), 2-amino-4,6-dinitrotoluene	Period of Performance: May 2003 to March 2004. Active treatment was 10 months with follow-up soil monitoring through	O'Niell and Nzengung, 2003 AWMA, 2003

Scale and Technology	Design and Operation	Performance Summary	Source
(SAMNAS)	(7.5 mg/kg), 4-amino-2,6-dinitrotoluene (130 mg/kg), 2-nitrotoluene (11 mg/kg) and 4-nitrotoluene (9.5 mg/kg) were treated with mushroom compost and cow manure on a 0.4 hectare (ha) (1 acre) site to the water table at 2 m (7 ft) bgs. The shallow groundwater contained perchlorate at 230 mg/L, and other co-contaminants, such as chlorinated organic solvents, including perchloroethene (PCE) and its reductive dechlorination products.

The site was subdivided into three sections: 2/3 of the southern section of the plot was treated with 459 m^3 (600 yd^3) of mushroom compost; 1/3 of the northern portion of the plot separated into a northeast section (1/6 of the total area) was treated with 96 m^3 (125 yd^3) of cow manure compost only, while the remaining 1/6 of the northwest portion of the plot was treated with 96 m^3 (125 yd^3) of cow manure compost and perchlorate-degrading bacteria. The top 1 m (3 ft) of the vadose zone soils (backfilled silty clays) were tilled with the amendments and irrigated with water via an installed irrigation system. Additionally, mushroom compost, mushroom compost tea and ethanol were occasionally added to the irrigation water to achieve faster biostimulation of explosives and perchlorate degradation in the vadose zone soil and groundwater. The wetness of the soil was monitored using clusters of tensiometers installed at different depths. | spring 2005.

After eight months of active treatment, the mass of perchlorate decreased from 78 kg in soil to 16 kg – about 80% (± 9%) of the estimated initial mass of perchlorate had been removed. Stimulation of biodegradation in the site groundwater was confirmed by the decrease in concentrations of perchlorate and chlorinated solvents during the demonstration test. Total explosives decreased by three orders of magnitude in 8 months with TNT decreasing from about 3,400 mg/kg to 62 mg/kg. The concentrations of total explosives and perchlorate in the vadose zone soils were below detection levels when Site 17 was last monitored in spring 2005. The concentration of perchlorate in plant tissues growing on the bioremediated soils was below detection limit in spring 2005. | |
| **LHAAP, TX—Building 43-X; 90 day temporal storage area** | | | |
| Pilot *In Situ* Bioremediation Surface Application and Mobilization of Nutrient Amendments (SAMNAS) | Perchlorate-contaminated soil consisted of 30-38 cm (12–15 in) of sand underlain by clay-rich soils, with perchlorate concentrations of 6.7 mg/kg. The 10.2 m^2 (110 ft^2) of contaminated soil was completely homogenized with cow and composted chicken manure, and water was added as needed intermittently.

This project posed unique challenges | Period of Performance: August 2002 to July 2003

Perchlorate concentrations in soil were bioremediated from 6,700 µg/kg to below 40 µg/kg down to a depth of 76 cm (30 in) bgs in ten months. The | AWMA, 2003

Corrigan, 2004 |

Scale and Technology	Design and Operation	Performance Summary	Source
	because: (1) the perchlorate contamination occurred in the vadose zone soils in a confined space – Building 43-X; (2) the top 30 to 38 cm (12 to 15 in) of soil in this building consisted entirely of sand underlain by clay-rich soils; and (3) the top soil had been exposed to creosote, an antibacterial agent.	site was completely restored and closed out in only 10 months.	
Aerojet General Corp. Superfund Site, Rancho Cordova, CA			
Pilot Composting	Anaerobic composting was used to treat soil from the former perchlorate burn area. Approximately 15.3 m^3 (20 yd^3) of soil was treated with manure initially placed on top of perchlorate hot spots. Compost was later tilled into the soil to enhance perchlorate destruction 5 to 8 cm (2 to 3 in) below the surface.	Period of Performance: June 2001 to October 2002 The maximum initial soil perchlorate concentration of 4,200 mg/kg was treated to an average concentration of 0.1 to 23 mg/kg following seven days of treatment.	Cox et al., 2000
Edwards Air Force Base (AFB), CA			
Pilot Composting	Anaerobic composting of perchlorate-contaminated soil treated with horse stable compost in 55-gallon drums.	Period of Performance: Not available Initial concentration of perchlorate decreased from 57 mg/kg to the remedial goal of 7.8 mg/kg.	ITRC, 2005
United Technologies Corporation (UTC) Site, San Jose, CA			
Pilot Composting	Anaerobic composting of perchlorate-contaminated soil piled 1.5 m (5 ft) high with 2.1 m (7 ft) diameter at the bottom. A plastic liner was placed underneath the pile, and soil berms were constructed around the circumference of the pile to prevent migration of leachate, if any. A plastic sheet was used to cover the top of the compost pile.	Period of Performance: Not available The average initial concentration of 170 mg/kg was treated to less than 0.64 mg/kg in less than 38 days.	Cox and Scott, 2003
Naval Weapons Industrial Reserve Plant, McGregor, TX			
Full Composting	Perchlorate-contaminated soils were excavated and transported to an onsite treatment cell. The engineered treatment cell was lined with a 30-milimeter (mm) (1.2 in) high-density polyethylene (HDPE) liner. The cell was approximately 1.8 m (6 ft) deep with a 152x9 m (500x30 ft) bottom.	Period of performance: October 1999 to April 2000. Influent perchlorate concentration in soil was 500 mg/kg. Perchlorate	Roote, 2001

Scale and Technology	Design and Operation	Performance Summary	Source
	Perchlorate contaminated soil was placed approximately 0.8 m (2.5 ft) deep in the cell. Prior to placing soil in the treatment cell, it was mixed with citric acid (carbon source), nitrate- and phosphate-fertilizers (micronutrients) and soda-ash (buffer). Soil was saturated as it was placed in the treatment cell. Approximately 5 cm (2 in) of water was maintained above the soil to foster anaerobic conditions. The cell was covered with a 6-mm (0.24 in) HDPE liner.	concentrations in the treated soil sampled at six different locations were less than 270 mg/kg.	
Former Bermite Site, Santa Clarita, CA			
Full-Scale *Ex Situ* Bioremediation	Perchlorate-contaminated soil was excavated and rocks were removed by screening. Rocks were crushed and added to the screened soil. The soil plus crushed rock was then introduced into a pug mill with water (final concentration about 14 to 17 percent), 450 mg/kg glycerin, and 50 to 100 mg-N/kg of di-ammonium phosphate (DAP) were mixed with the soil. The amended soil was then placed in plastic Ag-Bag® or plastic tarp-covered concrete containment cells. Each Ag-Bag contained about 250 metric tons of soil and each concrete containment cell contained about 350 metric tons of soil. Perchlorate bioremediation was then allowed to proceed in the cells for a period of about two to six weeks. Nitrogen addition was determined to be critical with respect to stabilization of the bioremediation process across a broad range of moisture contents. The process also eliminated the need for bulking agents thus facilitating re-use of the soil at this brownfields site.	Period of performance: January to July 2007. Initial perchlorate concentrations ranged from 490 to 8,400 µg/kg. Final concentrations were non-detectable (< 20 µg/kg) generally in two weeks. Soil treated in this six-month period was 73,000 metric tons. Full-scale operations are on-going.	Evans et al., 2008

REFERENCES

AWMA (Air & Waste Management Association). 2003. In situ bioremediation of perchlorate. Hazard Waste Consult 21:1.10-1.12.

AFCEE (Air Force Center for Environmental Excellence). 2002. Perchlorate Treatment Technology Fact Sheet: Phytoremediation. AFCEE, Brooks City-Base, San Antonio, TX,

USA. http://www.afcee.af.mil/shared/media/document/AFD-071211-068.pdf. Accessed June 14, 2008.

ASTM (American Society for Testing and Materials). 1998. Standard Guide for Remediation of Ground Water by Natural Attenuation at Petroleum Release Sites. Standard E 1943 98. ASTM, West Conshohocken, PA, USA.

Bender KS, O'Connor SM, Chakraborty R, Coates JD, Achenbach LA. 2002. Sequencing and transcriptional analysis of the chlorite dismutase gene of *Dechloromonas agitata* and its use as a metabolic probe. Appl Environ Microbiol 68:4820-4826.

Blount BC, Valentin-Blasini L, Osterloh JD, Mauldin JP, Pirkle JL. 2007. Perchlorate exposure of the U.S. population, 2001-2002. J Expo Sci and Environ Epidemiol 17:400–407.

Böhlke JK, Sturchio NC, Gu B, Horita J, Brown GM, Jackson WA, Batista J, Hatzinger PB. 2005. Perchlorate isotope forensics. Anal Chem 77:7838–7842.

Brady WD, Eick MJ, Grossi PR, Brady PV. 2003. A site-specific approach for the evaluation of natural attenuation at metals-impacted sites. Soil Sediment Contam 12:541-564.

Brennan RA, Cai H, Min B, Evans PJ. 2006. Treatability study for the bioremediation of perchlorate in vadose zone soils using gaseous electron donors. Proceedings, Fifth International Conference on Remediation of Chlorinated and Recalcitrant Compounds, Monterey, CA, USA, May 22-25, 2006, paper L-15.

Chaudhuri SK, O'Connor SM, Gustavson RL, Achenbach LA, Coates JD. 2002. Environmental factors that control microbial perchlorate reduction. Appl Environ Microbiol 68:4425–4430.

Coates JD, Michaelidou U, Bruce RA, O'Connor SM, Crespi JN, Achenbach LA 1999. Ubiquity and diversity of dissimilatory (per)chlorate-reducing bacteria. Appl Environ Microbiol 65:5234–5241.

Cooley A, Ferrey M, Harkness M, Dupont RR, Stroo H, Spain J. 2005. Monitored natural attenuation forum: A panel discussion. Remediat 15:83–96.

Corrigan WR. 2004. Demonstration of In-Situ Bioremediation of Perchlorate-Contaminated Soils in a 90-Day Container Storage Area, Building 43-X, Longhorn Army Ammunition Plant, Location Site 18/24. Project completion report. Submitted to U.S. Army SMALL-CR, Louisiana AAP. Contract No. DAAA09-02-P-0045.

Cox EE, Scott N. 2003. In Situ Bioremediation of Perchlorate: Comparison of Results from Multiple Field Demonstrations. Presented at In Situ and On-Site Bioremediation – The Seventh International Symposium. Orlando, FL, USA. June 2–5.

Cox EE, Edwards E, Neville S, Girard M. 2000. Aerojet bioremediation of soil from former burn area by anaerobic composting. http://www.perchlorateinfo.com/perchlorate-case-01.html. Accessed June 15, 2008.

Dibley VR, Krauter PW. 2005. Containerized Wetland Bioreactor Evaluated for Perchlorate and Nitrate Degradation. Technology News and Trends. EPA 542-N-05-001. Solid Waste and Emergency Response, U.S. Environmental Protection Agency, Cincinnati, OH, USA, pp 5-6.

Dondero A. 2001. Phytoremediation of Perchlorate under Greenhouse and Natural Conditions. Masters Thesis, The University of Georgia, Athens, GA, USA.

Eberts SM, Schalk CW, Vose J, Harvey GJ. 1999. Hydrologic effects of cottonwood trees on a shallow aquifer containing trichloroethene. Hydrol Sci Technol 15:115-121.

Ellington JJ, Wolfe NL, Garrison AW, Evans JJ, Avants JK, Teng Q. 2001. Determination of perchlorate in tobacco plants and tobacco products. Environ Sci Technol 35:3213–3218.

EWG (Environmental Working Group). 2005. Lettuce grown during the winter months may contain higher levels of toxic rocket fuel than is considered safe by the EPA. HTTP://WWW.EWG.ORG/REPORTS/SUSPECTSALADS/. Accessed June 14, 2008.

ESTCP. 2008a. In Situ Bioremediation of Perchlorate in Vadose Zone Soils. Project ER-0435 Fact Sheet. ESTCP, Arlington, VA, USA. http://www.estcp.org/Technology/ER-0435-FS.cfm. Accessed June 14, 2008.

ESTCP. 2008b. In Situ Bioremediation of Perchlorate in Vadose Zone Soil Using Gaseous Electron Donors. Project 0511 Fact Sheet. http://www.estcp.org/Technology/ER-0511-FS.cfm. Accessed June 14, 2008.

Evans PJ. 2007. Process for in situ bioremediation of subsurface contaminants. U.S. Patent Number 7,282,149. October 16.

Evans PJ, Trute M. 2006. In situ bioremediation of nitrate and perchlorate in vadose zone soil for groundwater protection using gaseous electron donor injection technology. Water Environ Res 78:2436-2446.

Evans PJ, Weaver WJ. 2005. In situ and ex situ treatment of perchlorate in source areas. Proceedings National Ground Water Association Conference on MTBE and Perchlorate, San Francisco, CA, USA, May 26-27, 2005. http://info.ngwa.org/GWOL/pdf/062681352.pdf. Accessed June 14, 2008.

Evans PJ, Lo I, Moore AE, Weaver WJ, Grove WF, Amini H. 2008. Rapid full-scale bioremediation of perchlorate in soil at a large brownfields site. Remediation (Spring):9-25.

Gibbs JT, Alleman BC, Gillespie RD, Foote EA, McCall SE, Snyder FA, Hicks JE, Crowe RK, Ginn J. 1999. Bioventing nonpetroleum hydrocarbons. In Engineered Approaches for In Situ Bioremediation of Chlorinated Solvent Contamination. Battelle Press, Columbus, OH, USA. pp. 7-14.

van Ginkel CGV, Rikken GB, Kroon AGM, Kengen SWM. 1996. Purification and characterization of chlorite dismutase: a novel oxygen-generating enzyme. Arch Microbiol 166:321–326.

Hatzinger PB, Whittier CM, Arkins MD, Bryan CW, Guarini WJ. 2002. In-situ and ex-situ bioremediation options for treating perchlorate in groundwater. Remediat 12:69–86.

Heaton THE. 1986. Isotopic studies of nitrogen pollution in the hydrosphere and atmosphere: A review. Chem Geol 59:87–102.

Herman DC, Frankenberger WTJ. 1999. Bacterial reduction of perchlorate and nitrate in water. J Environ Qual 28:1018–1024.

Hogue C. 2003. Environment of lettuce and rocket fuel. Chem Eng News 81:11.

ITRC (Interstate Technology & Regulatory Council). 2005. Perchlorate: Overview of Issues, Status, and Remedial Options. http://www.itrcweb.org/gd_Perch.asp. Accessed June 14, 2008.

Kastner JR, Das KC, Nzengung VA, Dowd J, Fields J. 2001. In-situ bioremediation of perchlorate-contaminated soils. In Leeson AJ, Peyton BM, Means JL, Magar VS, eds, Bioremediation of Inorganic Compounds. Battelle Press, Columbus, OH, USA, pp 289-295.

Kirk AB, Martinelango PK, Tian K, Dutta A, Smith E, Dasgupta PK. 2005. Perchlorate and iodide in dairy and breast milk. Environ Sci Technol 39:2011–2017.

Kramer U. 2005. Phytoremediation: Novel approaches to cleaning up polluted soils. Curr Opin Biotechnol 16:133–141.

Krauter PW. 2001. Using a wetland bioreactor to remediate ground water contaminated with nitrate (mg/L) and perchlorate (µg/L). Int J Phytoremediation 3:415–433.

Krauter PW, Daily B, Dibley V, Pinkart H, Legler T. 2005. Perchlorate and nitrate remediation efficiency and microbial diversity in a containerized wetland bioreactor. Int J Phytoremediation 7:113–128.

Leeson A, Hinchee RE. 1995. Principles and Practices of Bioventing. Volume 1: Bioventing Principles. Battelle Memorial Institute, Columbus, OH, USA. p. 39.

Lieberman MT. 2004. Evaluation of potential for monitored natural attenuation of perchlorate in groundwater. Project ER-0428 Fact Sheet. ESTCP, Arlington, VA, USA. http://www.estcp.org/projects/cleanup/cu-0428.cfm. Accessed June 14, 2008.

Lieberman MT, Knox SL, Beckwith WJ, Borden RC. 2006a. Evidence for effective remediation of perchlorate using monitored natural attenuation. Proceedings, Fifth International Conference on Remediation of Chlorinated and Recalcitrant Compounds, Monterey, CA, USA, May 22–25, 2006, paper G-72.

Lieberman MT, Borden RC, Hatzinger PB, Sturchio NC, Böhlke JK, Gu B. 2006b. Isotopic fractionation of perchlorate and nitrate during biodegradation in an EOS® biobarrier. Poster presented at Partners in Environmental Technology Technical Symposium & Workshop, Washington, DC, USA, November 28–30, 2006.

Lieberman MT, Knox SL, Borden RC. 2007. Field and Laboratory Evaluation of the Potential for Monitored Natural Attenuation of Perchlorate in Groundwater. Final Technical Report. ER-0428. July. http://www.estcp.org/Technology/upload/ER-0428-Treat-Study.pdf. Accessed June 14, 2008.

Logan BE. 2001. Assessing the outlook for perchlorate remediation. Environ Sci Technol 35:483A–487A.

Logan BE, Zhang HZ, Mulvaney P, Milner MG, Head IM, Unz RF. 2001a. Kinetics of perchlorate- and chlorate-respiring bacteria. Appl Environ Microbiol 67:2499–2506.

Logan BE, Wu J, Unz RF. 2001b. Biological perchlorate reduction in high-salinity solutions. Water Res 35:3034–3038.

Losi ME, Giblin T, Hosangadi V, Frankenberger WT Jr. 2002. Bioremediation of perchlorate-contaminated groundwater using a packed bed biological reactor. Bioremediation J 6:97–103.

Mariotti A, Germon JC, Hubert P, Kaiser P, Letolle R, Tardieux A, Tardieux P. 1981. Experimental determination of nitrogen kinetic isotope fractionation: Some principles; illustration for the denitrification and nitrification processes. Plant and Soil 62:413–430.

Mbuya OS, Nzengung VA. 2006. Phytoremediation of perchlorate and N-nitrosodimethylamine as single and co-contaminants. EPA STAR Project Report # RD831090, Washington, DC, USA.

McCutcheon SC, Schnoor JL. 2003. Overview of Phytotransformation and Control of Wastes. In McCutcheon SC, Schnoor JL, eds, Phytoremediation: Transformation and Control of Contaminants. Wiley-Interscience Publishers, Hoboken, NJ, USA, pp 3–58.

Milopoulos PG, Suidan MT, Sayles GD, Kaskassian S. 2002. Numerical modeling of oxygen exclusion experiments of anaerobic bioventing. J Contam Hydrol 58:209–220.

Newman B, Birdsell K, Longmire P, Counce D, Gard M, Heikoop J, Katzman D, Kluk EC, Larson T. 2005. Vadose zone transport of perchlorate: A case study from a semiarid canyon in New Mexico. Proceedings, The Geological Society of America Salt Lake City Annual Meeting, October 16–19, 2005, Paper No. 141-11. http://gsa.confex.com/gsa/2005AM/finalprogram/abstract_95347.htm. Accessed June 14, 2008.

Nzengung VA, McCutcheon SC. 2003. Phytoremediation of perchlorate. In McCutcheon SC, Schnoor JL, eds, Phytoremediation: Transformation and Control of Contaminants. Wiley-Interscience Publishers, Hoboken, NJ, USA, pp 863–885.

Nzengung VA, Wang C. 2000. Influences on phytoremediation of perchlorate-contaminated water. In Urbansky E, ed, Perchlorate in the Environment. Kluwer Academic/Plenum Publishers, New York, NY, USA, pp 219-229.

Nzengung VA, Wang C, Harvey G. 1999. Plant-mediated transformation of perchlorate into chloride. Environ Sci Technol 33:1470–1478.

Nzengung VA, Das KC, Kastner JR. 2001. Final report: Pilot scale in-situ bioremediation of perchlorate-contaminated soils at the Longhorn Army Ammunition Plant. U.S. Army Operations Support Command, Rock Island, IL, USA.

Nzengung VA, Penning H, O'Niell W. 2004. Mechanistic changes during phytoremediation of perchlorate under different root zone conditions. Int J Phytoremediation 6:63–83.

O'Connor SM, Coates JD. 2002. Universal immunoprobe for (per)chlorate-reducing bacteria. Appl Environ Microbiol 68:3108–3113.

Okeke BC, Giblin T, Frankenberger WT. 2002. Reduction of perchlorate and nitrate by salt tolerant bacteria. Environ Pollut 118:357–363.

O'Niell WL, Nzengung VA. 2003. Field demonstration of in-situ bioremediation of perchlorate-contaminated soils and groundwater. Proceedings, Air Waste Management Association 96th Annual Conference, San Diego, CA, USA, June 22 - 26, 2003. http://secure.awma.org/OnlineLibrary/ProductDetails.aspx?productID=2889.

Pennington JC, Bowen R, Brannon JM, Zakikhani M, Harrelson DW, Gunnison D, Mahannah J, Clarke J, Jenkins TF, Gnewuch S. 1999. Draft protocol for evaluating selecting and implementing monitored natural attenuation at explosives-contaminated sites. Technical Report EL-99-10, U.S. Army Corps of Engineers, Vicksburg, MS, USA.

Pilon-Smits E. 2005. Phytoremediation. Ann Rev Plant Biol 56:15–39.

Rikken GB, Kroon AGM, van Ginkel CG. 1996. Transformation of perchlorate into chloride by a newly isolated bacterium: Reduction and dismutation. Appl Microbiol Biotechnol 45:420–426.

Romanenko VI, Korenkov VN, Kuznetsov SI. 1996. Bacterial decomposition of ammonium perchlorate. Microbiologiya 45:204–209.

Roote DS. 2001. Technology Status Report: Perchlorate Treatment Technologies, First Edition. Ground-Water Remediation Technologies Analysis Center (GWRTAC). Pittsburgh, PA, USA.

Schnoor JL, Licht L, McCutcheon S, Wolfe N, Carreira L. 1995. Phytoremediation of organic and nutrient contaminants. Environ Sci Technol 29:318A–323A.

Schnoor JL, Parkin GF, Just CL, van Aken B, Strout JD. 2002. Phytoremediation and Bioremediation of Perchlorate at the Longhorn Army Ammunition Plant. Final Report. Submitted to U.S. Army Operations Support Command, Rock Island, IL, USA. http://clu-in.org/download/contaminantfocus/perchlorate/LHAAPfinalSchnoor.pdf Accessed June 14, 2008.

Schnoor JL, Parkin GF, Just CL, van Aken B, Shrout JD. 2004. Demonstration Project of Phytoremediation and Rhizodegradation of Perchlorate in Groundwater at the Longhorn Army Ammunition Plant, University of Iowa, Department of Civil and Environmental Engineering. U.S. Army Operations Support Command, Rock Island, IL, USA.

Shah JK, Sayles GD, Suidan MT, Mihopoulos P, Kaskassian S. 2001. Anaerobic bioventing of unsaturated zone contaminated with DDT and DNT. Water Sci Technol 43:35–42.

Shrout JD, Struckhoff GC, Parkin GF, Schnoor JL. 2006. Stimulation and molecular characterization of bacterial perchlorate degradation by plant-produced electron donors. Environ Sci Technol 40:310–317.

Smith PN, Yu L, McMurry ST, Anderson TA. 2004. Perchlorate in water, soil, vegetation and rodents collected from the Las Vegas Wash, Nevada, USA. Environ Pollut 132:121–127.

Stroo HF, Cosentini CC, Ronning T, Larsen M. 1997. Natural biodegradation of wood preservatives. Remediat 7:77–93.

Sturchio NC, Hatzinger PB, Arkins MD, Christy S, Heraty LJ. 2003. Chlorine isotope fractionation during microbial reduction of perchlorate. Environ Sci Technol 37:3859–3863.

Sundberg SE, Ellington JJ, Evans JJ, Keys DA, Fisher JW. 2003. Accumulation of perchlorate in tobacco plants: development of a plant kinetic model. J Environ Monitor 5:505–512.

Susarla S, Bacchus S, Wolfe NL, McCutcheon S. 1999. Phytotransformation of perchlorate and identification of metabolic products in *Myriophyllum aquaticum*. Int J Phytoremediation 1:96–107.

Tan K, Anderson TA, Jackson WA. 2004. Degradation kinetics of perchlorate in sediments and soils. Water Air Soil Pollution 151:245–259.

Tan K, Anderson TA, Jackson WA. 2006. Uptake and exudation behavior of perchlorate in smartweed. Int J Phytoremediation 8:13–24.

USEPA (U.S. Environmental Protection Agency). 1998. Technical Protocol for Evaluating Natural Attenuation of Chlorinated Solvents in Ground Water. EPA/600/R-98/128. Office of Research and Development, USEPA, Washington, DC, USA.

USEPA. 1999. Final Directive: Use of monitored natural attenuation at Superfund, RCRA corrective action, and underground storage tank sites. OSWER Directive 9200.4-17P. USEPA, Washington, DC, USA. http://www.epa.gov/swerust1/directiv/d9200417.htm. Accessed June 14, 2008.

USEPA. 2005a. EPA Superfund Record of Decision Amendment: Apache Powder Company, St. David, AZ. EPA/AMD/R09-05/049.

USEPA. 2005b. Perchlorate treatment technology update. Federal Facilities Forum Issue Paper, EPA 542-R-05-015. USEPA, Washington DC, USA. http://www.epa.gov/tio/download/remed/542-r-05-015.pdf. Accessed June 14, 2008.

USEPA. 2006. In situ and ex situ biodegradation technologies for remediation of contaminated sites. EPA/625/R-06/015. National Risk Management Research Laboratory, Cincinnati, OH, USA. http://www.epa.gov/nrmrl/pubs/625r06015/625r06015.pdf. Accessed June 14, 2008.

Urbansky ET. 1998. Perchlorate chemistry: Implications for analysis and remediation. Bioremediation J 2:81-95.

Urbansky ET, Magnuson ML, Kelty CA, Brown SK. 2000. Perchlorate uptake by salt cedar (*Tamarix ramosissima*) in the Las Vegas wash riparian ecosystem. Sci Total Environ 256:227–232.

Waller AS, Cox EE, Edwards EA. 2004. Perchlorate-reducing microorganisms isolated from contaminated sites. Environ Microbiol 6:517–527.

Wilson JT, Kaiser PM, Adair C. 2005. Monitored natural attenuation of MTBE as a risk management option at leaking underground storage tank sites. EPA/600/R-04/1790. National Risk Management Research Laboratory, USEPA, Cincinnatti, OH, USA. http://www.epa.gov/ada/download/reports/600R04179/600R04179.pdf. Accessed September 21, 2008.

Yifru DD. 2006. Phytoremediation and enhanced natural attenuation of perchlorate and N-nitrosodimethylamine (NDMA) as single and co-contaminants. Ph.D. Dissertation. The University of Georgia, Athens, GA, USA.

Yifru DD, Nzengung VA. 2006. Uptake of N-nitrosodimethylamine (NDMA) from water by phreatophytes in the absence and presence of perchlorate as a co-contaminant. Environ Sci Technol 40:7374–7380.

Yifru DD, Nzengung VA. 2008a. Organic carbon biostimulates rapid rhizodegradation of perchlorate (ClO_4^-). Environ Toxicol Chem (In Press).

Yifru DD, Nzengung VA. 2008b. Organic carbon biostimulates rapid rhizodegradation of perchlorate in soil (Submitted).

Zhang H, Logan BE, Regan JM, Achenbach LA, Bruns MA. 2005. Molecular assessment of inoculated and indigenous bacteria in biofilms from a pilot-scale perchlorate-reducing bioreactor. Microb Ecol 49:388-398.

INDEX

A
Abiotic process, ix, 29, 34, 47
Activated carbon, xxi, 7–8, 12, 29–32, 229
Active treatment, xxvii, 82–84, 86–88, 91–93, 97, 101, 108–112, 118, 129, 155, 235
Aerobic, xxvii, 1, 9, 94, 112, 122, 162–163, 183, 220, 231, 233
Aerojet, 16–18, 20–23, 27, 96–98, 102–104, 108, 113–129, 237
Anaerobic bioventing, xxvii, 232
Analytical model, xxvii, xxxii, 103
Attenuation
  rate, xxviii, 195

B
Biobarrier
  active, 93, 96, 113, 200–201, 206–207, 209–210, 216
  passive, 20–21, 32, 80, 83, 86, 136–137, 144, 148, 200–201, 206–210, 212–213, 217
Biofouling, xxviii, 32, 39–42, 83–85, 87, 91–93, 97–98, 108–113, 121–122, 128–129, 136–138, 144, 152–153, 170, 206–207, 210, 216
Biogeochemistry (biogeochemical), 158, 159, 166, 187, 188
Biological perchlorate remediation, 24, 79–89
Bioremediation, xxvii, xxviii, xxx, 1–10, 15–25, 35–36, 38–39, 42–45, 47–48
Biowall, ix, xiii, xiv, xv, xviii, 21, 80–83, 85–88, 155, 177–197, 200–202, 204–205, 208–210, 212–215, 217

C
Capital costs, ix, 83–84, 138, 164, 190, 199, 201, 206–217
Chemical reduction, 29, 31–33, 111, 178
Chlorate, 3,
Chlorite, 16, 34, 35, 38, 44, 45, 79, 140, 165, 187, 219, 221, 229
Chlorite dismutase, xxi, xxix, 34, 44–45, 221

Co-contaminants, 9, 32, 74, 91, 100, 110, 165, 181, 222, 228
Compost (ing), 22–23, 80, 85, 177–178, 183–184, 189, 193, 226–227, 229–231, 234, 236–237
Constructed wetland, 229
Construction costs, 189–191
Cost analysis, 199–217

D
Demolition and restoration costs, ix, 86, 217, 222
Depth to water, 87, 155–156, 161, 171–172, 202–203, 205–206, 214–215
Discount rate, 202
Dissolved oxygen, xxi, xxvii, 88, 98, 110, 119, 122–123, 140–142, 156, 202–203, 205–206, 212–213, 221, 226
Diversity of perchlorate reducing bacteria, 36–38

E
Edible Vegetable Oil (EVO), xxi, 42, 209
Electron donor
  selection, 39–42, 91, 107–108, 113, 136
  slow release, xxxii, 21, 25, 85–87, 91, 111
  soluble, xxxiii, 19–20, 42, 47, 64, 80, 84–85, 91–93, 95–96, 107–108, 129, 136, 206
Emulsified oil, 156, 158, 160–173
Enhanced *in situ* bioremediation (EISB), xiv, xxi, 17, 47–48, 112–113, 135, 138–139, 144, 153, 159, 171
Enrichment, 37
Environmental Security Technology Certification Program (ESTCP), viii, xvi, xxi, 1, 20–21, 24–25, 64, 98, 112–113, 138, 158–160, 162–163, 166–171, 232–234

F
Field demonstration, ix, xvii, 18–21, 24, 89, 103, 106, 129, 138, 167, 228–229, 232

F

Full-scale, xv, xvii, 2, 8, 17, 23–25, 38, 82, 88–89, 93, 96, 107–108, 129, 192, 195, 227, 230–231, 234, 238

G

Gaseous Electron Donor Injection Technology (GEDIT), xxi, 232–234

Geochemistry (geochemical), 91–93, 100, 110–112, 129, 136, 141, 146, 155, 156, 158, 165, 166, 179, 181

Gradient, xxix, xxx, 8, 18–19, 38, 40–41, 80, 83, 85, 87, 92–93

Granular activated carbon (GAC), 8, 31, 229

Groundwater
   extraction, 7–8, 80, 82–83, 91, 93–94, 96–97, 99–102, 113–129, 136–140, 142–147, 153, 157, 183, 206, 208–210
   geochemistry, xv, 91–92, 100–101, 111–112, 141, 146, 155–156, 179, 188–189, 222

Groundwater model, 142, 147

H

Horizontal Flow Treatment Wells (HFTW), xxi, 93–98, 102–105, 109–110, 112–113, 129

Hydraulic conductivity, xvii, xxx, 39, 41, 94, 101, 137, 139, 156, 164, 169–170, 180, 201–203, 210

Hydrogeology, 80, 107, 114, 139, 141, 142, 159, 163, 164, 179, 181, 182, 187, 195, 222

Hydroponics, 225

I

Immunoprobes, 45

In situ abiotic perchlorate treatment, 29–38

Injection point, 84, 85, 88, 92, 137, 157, 160, 161, 164, 171–173, 214, 215

Ion chromatography, 5, 60, 68, 69, 71, 72, 74, 120

Ion exchange, 7–8, 23–24, 29–33, 64, 73, 96

Isolation, 16, 46–47

Isotopes, xxx, 45–46, 62–67, 74, 221–222

L

Laboratory test, ix, 9, 105–107, 112–113, 167, 221, 233–234

Life cycle costs, 7, 86–87, 190–191, 199, 202, 206

Long term monitoring
   costs, 87, 201, 206–210

Longhorn Army Ammunitions Plant (LHAAP), xxii, 20–21, 138, 144–153, 226–229, 231, 234–236

M

Mass balance, xxxi, 170

Mass flux, xxxi, 156, 164

Maximum contaminant level (MCL), xxii, 124, 156, 199–200, 205

Methane, 24, 41, 82, 91, 92, 110, 111, 119, 126, 137, 158, 163, 166, 169, 170, 196

Microcosm, xxxi, 102, 105–107, 109, 158, 167, 221, 223

Model, 81, 101–105, 109, 116, 120, 121, 142, 146, 147, 158

Modeling, 98, 101–105, 107, 109, 116, 142, 147, 160, 205, 222

Monitored natural attenuation (MNA), xiv, xvi, xvii, xxii, xxxi, 21, 155, 199–200, 219–223

Monitoring
   cost, ix, 189–190, 200–202, 205–212, 214, 216–217
   parameters, 110
   performance, 108–111, 116, 145, 178, 187–190

Mulch, ix, xv, xviii, 21, 80, 85, 137, 155, 177–178, 182–189, 209, 214, 217

N

Net present value (NPV), xxii, 202, 204, 206–207, 209–215, 217

Nitrate, 3, 8, 9, 25, 35, 37, 38, 40, 60, 61, 79, 85, 87, 88, 103, 106, 107, 112, 126, 140, 143–146, 156, 187, 189, 196, 213, 221, 226, 229, 239
   inhibition, 37–38, 88, 103

Numerical model, xxxii, 101, 120

Nutrient, xxiii, 84, 106, 195, 231, 233–236

NWIRP McGregor, 180, 183, 191–196

O

Oil
  consumption, 159
  retention, 158–160, 171
Operations and maintenance (O&M), 20, 199
Organic carbon, 92, 111, 156, 158, 165, 166, 170, 178, 180, 181, 183, 188, 195, 223, 229, 231
Oxidation reduction potential (ORP), 91, 122–123, 141, 188, 216
Oxygen
  dissolved, 88, 98, 122, 123, 140, 141, 156, 202, 205, 213, 221, 226

P

Passive Biobarrier, 21, 80, 137, 209–210
Passive treatment, xxxii, 85, 87, 92, 155
Passive trench, 98, 208–210, 213–215, 217
Perchlorate
  analysis, 6, 69–70
  biodegradation, xv, 15, 16, 18, 20, 61, 73, 93, 99, 103, 105, 107, 116, 120, 123, 128, 136, 140, 156, 167, 168, 220–222, 234
  degradation, 97, 107, 124, 144, 152, 188, 196, 221, 226, 231
  production, 3
  reducing bacteria, 17, 34–36, 42, 44–47, 73, 88, 105, 230
  sources, 25, 46, 55–74, 219
  toxicology, xv, xvii
  uptake, 2, 6, 225–227
Perchlorate reducing bacteria, 17, 34–36, 42, 44–47, 105, 230
Perchlorate reduction
  monitoring, 42–46
  predicting, 42–46
Perforated pipes, 178–179
Performance monitoring, 108–111, 116, 145, 178, 187–190
Permeability, xxvii, 22, 32, 46, 81, 84, 87–88, 111, 114, 121, 141, 157–158, 160–161, 164, 166, 170, 172, 178, 181–183, 200
Permeable reactive barrier (PRB), 190
Phytoaccumulation, xxxii, 224–227
Phytoextraction, 228
Phytoremediation, xvii, xxxii, 219, 224–229
Pilot test, 8, 18–19, 94, 98, 100, 102, 113–114, 116, 119, 121–124, 126–129, 166–167, 169–170
Pilot-scale, xiv, 113, 121, 228–229, 233–234
Plume
  length, 95–96, 100–101, 203, 220–221
  longevity, 21, 80–81, 84–86, 164–166, 172, 192, 195, 197, 228
Pore flushing, 204
Pore volume, 160, 204–205, 211, 214
Porosity, 41, 160, 192, 202, 229–230
Pump and treat, ix, vii, 24, 32, 81, 192, 201, 206, 210, 229

R

Radius of influence, 84, 102, 109, 160, 164
Recirculation, 18–19, 83–84, 87, 91, 93, 97–99, 103, 112, 116, 135–143, 145–148, 152–153, 160, 183, 189, 206–207, 210, 214
Regeneration, 8, 30, 31, 33
Remediation, 25, 37, 79, 178–219, 228, 231
Residence time, 158
Rhizodegradation, 224–225

S

Saturated thickness, 215
Secondary water quality, 162–163
Seepage velocity, 214
Semi-passive, 84, 88, 135–136, 139–140, 153
Site assessment, 100–102, 106, 141–142
Site characterization, 73, 161, 163–165, 179, 205
Sodium lactate, 85, 121, 136, 142, 146–147, 152–153
Soil
  gas, xvii, 162, 163
  remediation, 22–24
Soluble electron donor, 85, 92–93, 95, 129, 206
Sorption, xxix, xxviii, xxxii, xxxiii, 2, 29–31, 87, 184, 188, 220

Source area treatment, 93, 98–100, 111, 156–157, 172
Source identification, 3, 46, 55–74
Stable isotope fractionation, 62–63
Strategic Environmental Research & Development Program (SERDP), 17–18

T

Template site, 200, 202–205
Terminal electron accepting process (TEAP), 40
Tracer test, 103, 109, 119–121, 123, 142, 146–147, 169
Travel time, 109, 116–117, 120, 123, 142, 146–147, 164, 203
Trench, 208
Trichloroethene (TCE), 9, 94, 125, 181

U

Unsaturated, xxxiii, 23–24, 61, 98, 231–232

V

Vadose zone, 231, 234
Volatile fatty acid (VFA), 110, 119, 143, 165, 188
Volatile organic compound (VOC), 109, 119, 169, 192